I0868788

Studies on Brachyura:
a homage to Danièle Guinot

CRUSTACEANA MONOGRAPHS constitutes a series of books on carcinology in its widest sense. Contributions are handled by the Series Editor(s) and may be submitted through the office of KONINKLIJKE BRILL Academic Publishers N.V., P.O. Box 9000, NL-2300 PA Leiden, The Netherlands.

Series Editors for the present volume: J.C. VON VAUPEL KLEIN, P.O. Box 30, NL-3720 AA Bilthoven, The Netherlands; e-mail: jcvvk@xs4all.nl and C.H.J.M. FRANSEN, c/o National Museum of Natural History, P.O. Box 9517, NL-2300 RA Leiden, The Netherlands; e-mail: Fransen@naturalis.nnm.nl

Editorial Committee: N.L. BRUCE, Wellington, New Zealand; Mrs. M. CHARMANTIER-DAURES, Montpellier, France; D. DEFAYE, Paris, France; H. DIRCKSEN, Stockholm, Sweden; J. FOREST, Paris, France; R.C. GUIAŞU, Toronto, Ontario, Canada; R.G. HARTNOLL, Port Erin, Isle of Man; E. MACPHERSON, Blanes, Spain; P.K.L. NG, Singapore, Rep. of Singapore; H.-K. SCHMINKE, Oldenburg, Germany; F.R. SCHRAM, Langley, WA, U.S.A.; C.D. SCHUBART, Regensburg, Germany; G. VAN DER VELDE, Nijmegen, Netherlands; W. VERVOORT, Leiden, Netherlands; H.P. WAGNER, Leiden, Netherlands; D.I. WILLIAMSON, Port Erin, Isle of Man.

Published in this series:

CRM 001 - *Stephan G. Bullard*	Larvae of anomuran and brachyuran crabs of North Carolina
CRM 002 - *Spyros Sfenthourakis et al.*	The biology of terrestrial isopods, V
CRM 003 - *Tomislav Karanovic*	Subterranean Copepoda from arid Western Australia
CRM 004 - *Katsushi Sakai*	Callianassoidea of the world (Decapoda, Thalassinidea)
CRM 005 - *Kim Larsen*	Deep-sea Tanaidacea from the Gulf of Mexico
CRM 006 - *Katsushi Sakai*	Upogebiidae of the world (Decapoda, Thalassinidea)
CRM 007 - *Ivana Karanovic*	Candoninae (Ostracoda) from the Pilbara region in Western Australia
CRM 008 - *Frank D. Ferrari & Hans-Uwe Dahms*	Post-embryonic development of the Copepoda
CRM 009 - *Tomislav Karanovic*	Marine interstitial Poecilostomatoida and Cyclopoida (Copepoda) of Australia
CRM 010 - *Carrie E. Schweitzer et al.*	Systematic list of fossil decapod crustacean species
CRM 011 - *Peter Castro et al.*	Studies on Brachyura: a homage to Danièle Guinot
CRM 012 - *Patricio De los Ríos*	Crustacean zooplankton communities in Chilean inland waters
CRM 013 - *Katsushi Sakai*	Axioidea of the world and a reconsideration of the Callianassoidea (Decapoda, Thalassinidea, Callianassida)

In preparation (provisional titles):

CRM 01x - *Danielle Defaye et al.*	Studies on freshwater Copepoda — A volume in honour of Bernard Dussart
CRM 01y - *Chang-tai Shih*	Marine Calanoida of the China seas
CRM 01z - *Charles H.J.M. Fransen, Sammy De Grave & Peter K.L. Ng*	L.B. Holthuis Memorial Volume

Editors' addresses:

Peter Castro, Biological Sciences Department, California State Polytechnic University, Pomona, California 91768, U.S.A.; e-mail: pcastro@csupomona.edu

Peter J.F. Davie, Queensland Museum, P.O. Box 3300, South Brisbane, Queensland, 4101, Australia; e-mail: Peter.Davie@qm.qld.gov.au

Peter K.L. Ng, Tropical Marine Science Institute and Department of Biological Sciences, National University of Singapore, Lower Kent Ridge Road, Singapore 119260, Republic of Singapore; e-mail: dbsngkl@nus.edu.sg

Bertrand Richer de Forges, Correspondant du Muséum national d'Histoire naturelle, Paris, France; e-mail: b.richerdeforges@gmail.com

Cover: Eogarthambrus guinotae n. gen., n. sp., see p. 109.

Studies on Brachyura: a homage to Danièle Guinot

By

Peter Castro, Peter J. F. Davie, Peter K. L. Ng
and Bertrand Richer de Forges
(Editors)

CRUSTACEANA MONOGRAPHS, 11

BRILL

LEIDEN • BOSTON

This book is printed on acid-free paper.

Library of Congress Cataloging-in-Publication Data

The Library of Congress Cataloging-in-Publication Data is available from the Publisher.

ISBN: 978-90-04-17086-5

© 2010 by Koninklijke Brill NV, Leiden, The Netherlands.
Koninklijke Brill NV incorporates the imprints BRILL, Hotei Publishing, IDC Publishers, Martinus Nijhoff Publishers and VSP.

All rights reserved. No part of this publication may be reproduced, translated, stored in a retrieval system, or transmitted in any form or by any means, electronic, mechanical, photocopying, recording or otherwise, without prior written permission from the publisher.

Authorization to photocopy items for internal or personal use is granted by Koninklijke Brill NV provided that the appropriate fees are paid directly to The Copyright Clearance Center, 222 Rosewood Drive, Suite 910, Danvers, MA 01923, USA.
Fees are subject to change.

PRINTED IN THE NETHERLANDS

CONTENTS

© Koninklijke Brill NV, Leiden, 2010

PREFACE

In every generation a few special scientists emerge whose ground-breaking works have a major impact on the development of their disciplines. Their interesting ideas and novel approaches become the standard for the researchers that follow. Such a scientist is Danièle Guinot. Working in taxonomy, a discipline incorrectly considered by many as "old fashioned", she has helped reawaken interest on the systematics of brachyuran decapods crustaceans, the true crabs. There has certainly been nothing "old fashioned" about her work. She has been a leader in the discovery and use of new morphological characters for recent taxa, including the use of reproductive structures and sperm, and this in turn has led her to develop new ideas on how they should be classified. In fossils, she has pioneered the use of sternal morphology, and has contributed substantially to how we study them. In addition, she has been one of the leaders in linking fossil and recent fauna in her systematic analyses. In these and many other ways, she has significantly helped to redefine the study of the complete evolutionary process in crabs.

We therefore decided that it was timely that we honour Danièle, our esteemed colleague and friend. For decades, she has gracefully and unselfishly volunteered to give us valuable advice, provided us with references, drawings and photographs, corrected our French abstracts and our long manuscripts, and patiently listened to our hypotheses, helping us to develop and improve many of them. This volume was born out of our admiration for Danièle's immense contributions to her discipline. We wrote our contributions thinking of her, working those long hours in her small office and worrying about the MNHN collections. A total of 35 of her colleagues have contributed to this volume, submitting papers on those aspects of the Brachyura to which Danièle, herself, has significantly contributed — taxonomy, evolution, morphology, palaeontology and general biology of crabs. This volume is but a small tribute to a highly respected colleague and friend from the active band of researchers that she has helped so much over the years. *Nos très sincères remerciements Danièle* !

February 2010, *Guest Editors for the present volume*,
PETER CASTRO
PETER DAVIE
PETER NG
BERTRAND RICHER DE FORGES

LE PROFESSEUR DANIÈLE GUINOT

Certains personnages scientifiques ont, au cours de leur carrière, et parfois bien au-delà, une profonde influence sur le développement de leur discipline. Par leur quantité de travail, l'apport de nouvelles idées et de nouvelles voies d'exploration des connaissances, ils marquent leur époque. Tel est le cas du Professeur Danièle Guinot. Travaillant pourtant en taxonomie, une discipline considérée par beaucoup comme désuète, elle a su rénover la connaissance systématique des crustacés décapodes brachyoures, les crabes, en apportant de nouveaux caractères morphologiques à la classification, en utilisant la morphologie spermatique, les caractères de l'appareil reproducteur chez les crabes actuels et la morphologie du plastron sternal chez les fossiles. C'est l'étude de toute l'évolution de ce groupe qui a été revisitée.

JEUNESSE ET ÉTUDES

Originaire de l'est de la France, Danièle a durement subi dans sa jeunesse les années troublées de la seconde guerre mondiale. Il lui en est resté un goût de l'effort et une ténacité peu commune.

Après des études de zoologie à l'Université de Montpellier, elle termine sa formation à La Sorbonne à Paris en 1955 et commence immédiatement à travailler sur son groupe favori, les crustacés. Elle rentre au Muséum national d'Histoire naturelle (MNHN) à Paris comme assistante de recherches et y réalisera toute sa carrière. Ses premiers travaux sur les crustacés décapodes paraîtront dès 1956, en collaboration avec Jacques Forest.

Elle soutiendra sa thèse de doctorat à l'Université Pierre et Marie Curie, Paris en 1977.

Au cours de sa longue carrière, Danièle a travaillé sur plusieurs aspects intéressants de la taxonomie des crabes mettant ainsi en évidence ses remarquables talents. Quelques uns des thèmes qu'elle a abordé sont exposés ici.

© Koninklijke Brill NV, Leiden, 2010

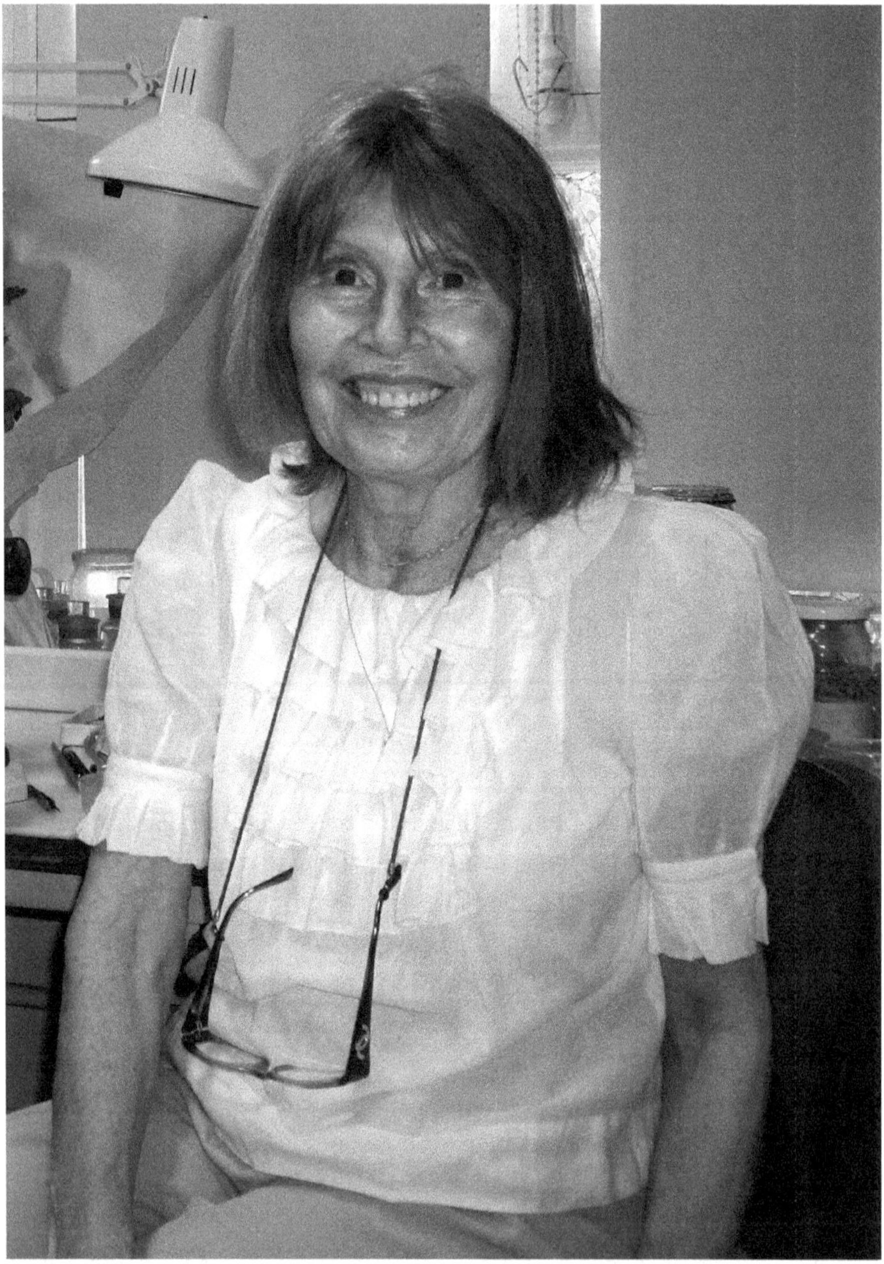

Fig. 1. Danièle Guinot dans son 'Laboratoire de Zoologie (Arthropodes)', rue de Buffon, à Paris, septembre 2009. [Danièle Guinot in her 'Laboratoire de Zoologie (Arthropodes)' at the rue de Buffon in Paris, September 2009.] [Photo: T. Naruse]

LES CRABES DES MILIEUX EXTRÊMES

Les crabes de profondeur

Découvertes au milieu du 19ème siècle, les faunes de profondeurs n'étaient connues que par les collections des « Grandes expéditions » qui eurent lieu entre 1873 et 1952. L'étude des crabes de ces expéditions était fragmentaire et souvent limitée par le petit nombre de spécimens connus. A partir des années 1970, de nouvelles récoltes eurent lieu dans la zone bathyale de l'Indo-ouest Pacifique et en particulier par les campagnes MUSORSTOM, qui dans bien des domaines, rénovèrent les connaissances sur l'environnement profond. Danièle entreprit tout naturellement l'étude de ces nouvelles collections de crabes qui contenaient de nombreux nouveaux taxons à décrire. Elle publia ainsi une série d'articles sur les Dromiidae, Homolidae, Raninidae, Atelecyclidae, Majidae et Goneplacidae. L'un des plus remarquables décrit une nouvelle famille étrange, les Poupiniidae, proche des Homolidae.

Faune des sources hydrothermales

Très tôt, après la découverte fortuite des sources hydrothermales en 1976 par le submersible « Alvin », Danièle s'intéressa aux crabes abondants et étranges qui peuplaient ces sources hydrothermales. Leur silhouette rappelait celle des Potamonoidea, crabes d'eau douce, mais ils vivaient par plus de 2000 m de profondeur, dans une totale obscurité dans un écosystème à base de chimiosynthèse bactérienne. Par ailleurs, ces milieux extrêmes sont petits et discontinus, le long des dorsales océaniques, ce qui pose la question de la dispersion des espèces. Pour des crabes aussi particuliers une nouvelle famille, les Bythograeidae a du être décrite. Danièle a enrichi ce groupe de deux nouveaux genres et cinq nouvelles espèces.

Faune cavernicole

En collaboration avec des biospéléologues français et américains, Danièle a apporté une importante contribution à la connaissance des brachyoures très modifiés de ces habitats obscurs. Ainsi la description de *Trogloplax joliveti* des cavernes de Papouasie-Nouvelle-Guinée a nécessité la création d'une nouvelle sous-famille, les Trogloplacinae dans les Chasmocarcinidae, au sein des Goneplacoidea.

Ces deux milieux extrêmes fournirent à Danièle de beaux exemples de mécanismes de l'évolution.

L'ÉTHOLOGIE

La relation entre le comportement des crabes et leurs adaptations morphologiques a toujours intéressée Danièle. Ainsi, l'étude des Goneplacidae a été l'occasion de s'intéresser à la stridulation ; l'étude des Homolidae à permis de décrire l'usage des cinquièmes paires de pattes préhensiles pour le camouflage. Dans ces deux cas, les travaux de Danièle sont des références essentielles (Guinot-Dumortier & Dumortier, 1960 ; Guinot et al., 1995).

L'HISTOIRE DES SCIENCES

A plusieurs reprises au cours de sa carrière, Danièle s'est intéressée à l'utilisation humaine des crabes. Les techniques de pêche ont été décrites ainsi que l'usage médical de certaines espèces ou encore leur représentation artistique (e.g., Guinot, 1967).

LA PALÉONTOLOGIE

Passionnée par l'évolution des crabes, Danièle a réalisé de nombreux travaux sur les espèces fossiles. Ces études marquèrent profondément sa conception des crabes archaïques du groupe des Podotremata. Ces travaux sur les crabes fossiles se sont intensifiés depuis quelques années, en collaboration avec des paléontologistes du monde entier. En 2001, elle décrivit avec Marcos Tavares, une nouvelle famille fossile de Podotremata du crétacé, les Etyidae, en proposant une nouvelle phylogénie du groupe.

L'APPAREIL REPRODUCTEUR DES CRABES ET L'ÉVOLUTION

En 1977, utilisant les observations morphologiques de sa thèse, Danièle proposa une nouvelle classification des brachyoures basée essentiellement sur la position relative des orifices génitaux mâles. Trois groupes sont distingués : les Podotrèmes, Heterotrèmes et Thoracotrèmes. Cette classification, très novatrice par rapport à celle de Balss (1957), a été confirmée par la morphologie des spermatozoïdes. Elle est maintenant largement acceptée par les carcinologues. A elle seule, cette partie de l'œuvre de Danièle rend sa contribution à la carcinologie exceptionnelle. Ces nouvelles idées ont suscité de nouvelles hypothèses à tester sur l'évolution et la phylogénie des crabes et dynamisé les recherches.

SES CRABES FAVORIS

Bien que Danièle ait exprimé sa passion des crabes sur une large palette de groupes actuels et fossiles, certaines familles lui tiennent particulièrement à cœur, les Xanthidae et les Homolidae.

Xanthidae. — Très tôt dans sa carrière Danièle s'était lancée dans l'étude de cette famille d'une grande richesse spécifique, le plus souvent inféodée aux milieux coralliens. Après les études des collections des Xanthidae des Comores, ce fut celle de la Mer Rouge, puis de Polynésie française. Danièle généralisa l'utilisation de la diversité morphologique du premier pléopode mâle comme critère spécifique. L'observation et la description des pléopodes et de la position des orifices femelles sur le plastron sternal la conduisirent à proposer une nouvelle classification des Brachyoures avec ses gros travaux publiés sur les « groupements naturels des brachyoures » en 1966-1976.

Homolidae. — Grace aux superbes collections réunies au MNHN à la suite des campagnes MUSORSTOM, un travail de révision fut entrepris par Danièle, en collaboration avec Bertrand Richer de Forges. Pratiquement tous les spécimens récoltés auparavant par les Grandes expéditions du 19 et 20èmes siècles furent réunis au Muséum et comparés avec le nouveau matériel. Après cinq années de travail et de nombreuses descriptions de taxons nouveaux cette révision fût publiée dans un volume des Résultats des Campagnes MUSORSTOM, édité par Alain Crosnier. Ce volumineux travail demeure, 15 ans après sa parution, un outil incontournable dans l'étude des crabes de profondeur. A ce groupe archaïque Danièle a ajouté la superbe famille les Poupiniidae qui habitent en profondeur des pentes difficilement accessibles.

SES PRINCIPAUX COLLABORATEURS

Parmi les nombreux collaborateurs de Danièle au cours de sa carrière on retrouve la plupart des carcinologistes qui ont activement fait progresser la connaissance des brachyoures ces dernières années, pour ne citer que les principaux avec lesquels elle a cosigné des articles :

Un des premiers collaborateurs fut Jacques Forest avec lequel Danièle a étudié les crabes Xanthoidea de Polynésie française rapportés par la Mission Singer-Polignac. Pierre Drach, plus connu pour sa description minutieuse des phénomènes de mues, publia avec Danièle de nouvelles observations sur la structure squelettique des brachyoures. Ce thème fut également abordé en collaboration avec une paléontologiste du MNHN, Sylvie Secrétan. Un autre collaborateur fut Tune Sakai, le « monstre sacré » de la carcinologie japonaise, célèbre pour ses superbes ouvrages sur les crabes du Japon illustrés d'aquarelles. Danièle a aussi collaboré avec Alain Crosnier, véritable pionnier de l'étude de la biodiversité marine qui, après avoir récolté par lui-même des crustacés, à Madagascar et en Afrique occidentale, a été l'instigateur et le moteur des campagnes MUSORSTOM pendant de longues années.

Bertrand Richer de Forges, infatigable récolteur sur les fonds marins du Pacifique avec qui une étroite collaboration dura de 1979 à 1998, aussi bien sur la description taxonomique des crabes de profondeur que sur la morphologie des spermatozoïdes. Peter Ng, avec qui elle collabore toujours très activement sur plusieurs groupes de crabes et sur des questions de nomenclature. Thomas Iliffe qui a procuré à Danièle un accès privilégié aux crabes cavernicoles du Pacifique. Barrie Jamieson, spécialiste mondial de la morphologie et de l'évolution des spermatozoïdes a été pendant plusieurs années un collaborateur efficace et productif. Enrique Macpherson, beaucoup plus connu pour ses nombreux travaux sur les galathéides, a publié plusieurs articles sur les crabes avec Danièle. Marcos Tavares qui a tout d'abord été l'élève de Danièle, collabore avec elle dans une étude sur la morphologie et évolution des structures génitales externes des crabes. Peter Davie, fidèle collaborateur de Danièle qui s'intéresse également beaucoup aux Xanthidae. Colin McLay a travaillé avec Danièle sur les brachyoures archaïques du groupe des Dromiacea. Christopher Tudge, élève de Barrie Jamieson, qui a été impliqué avec Danièle dans l'étude évolutive des spermatozoïdes. Michel Segonzac, grand collecteur de matériel zoologique des sources hydrothermales, qui a cosigné avec Danièle la description de nouvelles espèces de Bythograeidae. Jean-Marie Bouchard a réalisé sous la direction de Danièle une thèse de doctorat sur la coaptation morphologique de l'abdomen des crabes. Pedro (Peter) Castro, inconditionnel collaborateur des campagnes MUSORSTOM, francophone et francophile, qui a révisé plusieurs familles importantes de crabes, en partie avec Danièle, et collabore actuellement avec Danièle et Marcos Tavares pour une étude sur la morphologie et évolution de structures génitales externes des crabes.

UNE CARRIÈRE DÉJÀ LONGUE ET ENCORE TRÈS PRODUCTIVE

S'appuyant sur sa longue expérience des crabes actuels, Danièle a développé ces dernières années des travaux très originaux sur les crabes fossiles, avec des collaborateurs comme Frederick Schram, Barry van Bakel, Pedro Artal et beaucoup d'autres, qui rénovent totalement la paléontologie du groupe et permettent de dater les grandes étapes évolutives.

En 2008, Danièle cosigne avec Peter Ng et Peter Davie une œuvre monumentale, le Systema Brachyurorum, qui est une mise au point de la nomenclature pour près de 6800 espèces de brachyoures. Un document devenu immédiatement incontournable pour quiconque essaye de dresser la synonymie et comprendre le statut d'une espèce ou d'une famille. En 2008 la Crustacean Society a attribué à Danièle Guinot le « Crustacean Society Award for Research Excellence » en la citant de la façon suivante : « pour ses efforts à promouvoir des idées révolutionaires dans la taxonomie et l'évolution des Brachyoures ». Cette récompense remise par ses pairs est une consécration pour l'ampleur de ses contributions à la carcinologie.

Si Danièle a toujours été remarquable par son élégance hors du commun, surtout au sein d'un organisme de recherche, elle l'a été également par sa persévérance à faire le travail de base, sans lequel les plus belles théories évolutives manqueraient de fondations. Sa contribution à la connaissance des crabes est également remarquable quantitativement : pour les brachyoures actuels et fossiles, elle a nommé 20 taxa supragénériques, 61 nouveaux genres et 164 nouvelles espèces. En remerciement, la communauté carcinologique, lui a dédié sept genres et 24 espèces (sans compter celles du présent volume !).

C'est avec un immense plaisir que nous dédions à Danièle ce volume de résultats scientifiques produits par ses nombreux collègues et amis.

PROFESSOR DANIÈLE GUINOT

Some research scientists have, in the course of their formal careers and sometimes even beyond, had a great influence on the development of their respective disciplines. As a result of their contributions of interesting ideas and their novel approaches in acquiring new knowledge, these scientists have contributed substantially to future generations. Such is the case of Professor Danièle Guinot. Working in taxonomy, a discipline incorrectly considered by many as "old fashioned", she has helped reawakening interest on the systematics of brachyuran decapod crustaceans, the crabs. For Recent taxa, she has led in the discovery and use of new morphological characters, as well as the utilization of sperm and external reproductive structures. In fossils, she has pioneered the use of sternal morphology and has contributed substantially to how we study them. In addition, she has been one of the leaders in linking fossil and recent fauna in her systematic analyses. She has helped to redefine the study of the complete evolutionary process in crabs.

EARLY LIFE AND STUDIES

Born in eastern France, Danièle was very much affected by the difficult years immediately after World War II. It was this period that forged her uncommon tenacity and love for hard work. After her studies in zoology at the Université de Montpellier, she finished her formal education at the Sorbonne in Paris in 1955, and immediately started studying her favourite group, the decapod crustaceans. She began working at the Muséum national d'Histoire naturelle (MNHN) in Paris, her headquarters for the rest of her career, initially as a research assistant. Her first publication on decapod crustaceans, in collaboration with Jacques Forest, appeared in 1956 and in 1957 she described her first species, *Portunus* (*Xiphonectes*) *mariei* Guinot, 1957. Working tirelessly in the 1960s and 1970s on crabs, and publishing some 40 papers during this period, Danièle eventually defended her doctoral thesis at the Université Pierre et Marie Curie, Paris in 1977.

In her long career, Danièle has worked on many interesting aspects of crab taxonomy, reflecting her diverse interests and remarkable capabilities. A few arc highlighted here.

CRABS FROM EXTREME ENVIRONMENTS

Deep-water crabs

These extreme environments provided Danièle with beautiful examples of the mechanisms of evolution. Such habitats include the deep-sea, hydrothermal vents and caves. Discovered in the middle of the 19th century, deep-sea faunas were known only from the collection of the "great expeditions" that took place between 1873 and 1952. The study of the crabs from these expeditions was incomplete and often limited by the small number of known specimens. Starting in the middle 1970s, new collections were made from the Indo-West Pacific bathyal zone, in particular those made by the MUSORSTOM expeditions, and these expanded our knowledge of this harsh environment. Danièle naturally embarked upon the study of these new collections of crabs, which included many new taxa. As a result, she published a series of papers on the Dromiidae, Homolidae, Raninidae, Atelecyclidae, Majidae and Goneplacidae. One highlight was the discovery of a strange new family, the Poupiniidae (cf. Guinot, 1991).

Fauna of deep-sea hydrothermal vents

Shortly after the discovery of deep-sea hydrothermal vents in 1976 by the "Alvin", Danièle became interested in the many strange crabs found in this bizarre habitat. Bythograeids superficially resemble some freshwater crabs but they live more than 2000 m deep in total darkness in an ecosystem dependent on chemosynthetic bacteria. These extreme environments are also small and discontinuous along mid-oceanic ridges, bringing in the question of their dispersion. Danièle substantially expanded our knowledge of this group, describing two new genera and five new species (Guinot & Segonzac, 1997; Guinot et al., 2002; Guinot & Hurtado, 2003).

Cave fauna

Cave taxa resemble deep-sea species in many ways. In collaboration with French and American biospeleologists, Danièle made important contributions to the knowledge of the crabs highly modified to live in these dark habitats. Noteworthy is the description of the bizarre *Trogloplax joliveti* from caves in Papua New Guinea, which demanded the creation of a new subfamily, the Trogloplacinae within the Chasmocarcinidae (Goneplacoidea) (cf. Guinot, 1986).

CRAB BEHAVIOUR

The relationship between behaviour and the crabs' morphological adaptations is something that always interested Danièle. Her studies of the Goneplacidae led her to become interested in stridulation and her studies on the Homolidae to the use of the fifth pair of pereopods in carrying behaviour and camouflage. One of the first major papers on crab stridulation was by her (Guinot-Dumortier & Dumortier, 1960). Her studies on carrying behaviour and adaptations culminated in an important paper reviewing carrying behaviour in crabs (Guinot et al., 1995).

HISTORY OF SCIENCE

At different times throughout her career, Danièle became interested in the human utilization of crabs. Fishing techniques were described as well as the medical use of some species and the representation of crabs in art (e.g., Guinot, 1967).

PALAEONTOLOGY

Danièle has a strong passion for the study of the evolution of crabs and as a result, she has written many papers on fossil species as well. This work has broadened her views on the primitive crabs belonging to the Podotremata. Collaboration with other workers has been recently expanded as a result of joint studies with palaeontologists from around the world. One highlight was in 2001, when she described, with Marcos Tavares, a new family of Cretaceous Podotremata, the Etyidae, and at the same time proposed a new phylogeny of the group.

THE REPRODUCTIVE SYSTEM OF CRABS AND THEIR EVOLUTION

In 1977, Danièle, using many of the morphological observations contained in her thesis, proposed a new classification of brachyurans essentially based on the relative position of the male genital openings. Three groups are recognized: the podotremes, heterotremes, and thoracotremes. This new system is the most "radical" since the classification proposed by Balss (1957) was accepted. Danièle's classification scheme, which was also supported by the morphology of spermatozoa, is today used by most carcinologists. The testing and refinement of this classification system has been driving many crab research efforts around the world, with workers using new taxa, new characters, new techniques, new analyses, etc. The mark of good science is that it generates hypotheses that arouse interest and can be tested. From the amount of new research efforts Danièle's proposals have generated, her contributions to carcinology have been exceptional!

HER FAVOURITE CRABS

Danièle has constantly expressed a love for all crabs, fossil and extant. But of the many groups she has worked on, perhaps (in the editors' opinions at least) two best epitomize her love for these crustaceans.

The Xanthidae sensu lato. — Very early in her career, Danièle launched herself into the study of this family, which contains a great number of species, many of which are associated with coral reefs. The study of collections of Xanthidae from the Comoro Is. was followed by those from the Red Sea and French Polynesia. Danièle substantially expanded the use of the morphological diversity of the first male pleopod as a species-level character. The observation and description of the pleopods and the position of the female openings on the sternal plastron lead her to eventually propose a new classification of the Brachyura; starting with an important series of papers she published on the "groupements naturels des brachyoures" between 1966 and 1976, and culminating in her thesis of 1977 and two major synoptic publications in 1978 and 1979.

The Homolidae. — Thanks to the superb MNHN collections gathered by the MUSORSTOM expeditions, a revision was undertaken by Danièle in collaboration with Bertrand Richer de Forges. Practically all specimens collected during the great expeditions of the 19th and 20th centuries were gathered at the museum and compared with the new material. After five years of work and numerous descriptions of new taxa the revision was published in a volume of the Résultats des Campagnes MUSORSTOM, edited by Alain Crosnier. It is still considered essential in the study of deep-water crabs and a source of pride by its authors 15 years after its publication. To this primitive group, Danièle added the remarkable family Poupiniidae, which inhabits deep trenches.

HER MOST IMPORTANT COLLABORATORS

Danièle's list of collaborators during her career reads like a who's who in modern carcinology. There are too many to cite individually and not all have published papers with her. We can only cite but a few with whom she has co-authored research papers.

An early collaborator was Jacques Forest, who studied with Danièle the xanthoid crabs collected in French Polynesia by the Mission Singer-Polignac. Pierre Drach, better known because of his detailed description of the moulting phenomenon, published with Danièle many new observations on the brachyuran skeleton. This subject was equally embarked upon with Sylvie Secrétan, the MNHN palaeontologist. Another collaborator was Tune Sakai, the highly respected dean of Japanese carcinology, famous for his outstanding work, illustrated by beautiful watercolours, on the crabs of Japan. Danièle also collaborated with Alain Crosnier, a major pioneer of the study of marine biodiversity who, after making invaluable collections of crustaceans in Madagascar and West Africa himself, was the organizer and main drive of the MUSORSTOM expeditions for many years.

Bertrand Richer de Forges, tireless collector of the Pacific seafloor fauna, was a close collaborator from 1979 to 1998, not only in the taxonomy of crabs but in the morphology of sperm as well. Peter Ng has always actively collaborated with Danièle on many groups of crabs as well as in many questions of nomenclature and systematics. Thomas Iliffe has provided Danièle a special access to cavernicolous crabs of the Pacific. Barrie Jamieson, a specialist on the morphology and evolution of spermatozoa, was for many years a productive collaborator. Enrique Macpherson, best known for his work on galatheids, published several articles on brachyuran crabs with Danièle. Marcos Tavares, one of Danièle's graduate students, collaborated with her in a study of the morphology and evolution of the external reproductive structures of brachyurans. Peter Davie, a faithful collaborator, shares his interest on the Xanthidae with her. Colin McLay has explored with Danièle the primitive brachyurans with their studies on the Dromiacea. Christopher Tudge, Barrie Jamieson's student, was involved with Danièle in studies on the evolution of spermatozoa. Michel Segonzac, productive collector of animal specimens from deep-water hydrothermal vents, co-authored with Danièle the description of new species of bythograeid crabs. Pedro (Peter) Castro, francophone and francophile, a frequent collaborator of the MUSORSTOM expeditions, is currently collaborating with Danièle and Marcos Tavares in a large study of the morphology and evolution of the external reproductive structure of brachyuran crabs.

A LONG AND STILL VERY PRODUCTIVE CAREER

Building on her extensive experience on the Recent crabs, Danièle has written in collaboration with Frederick Schram, Barry van Bakel, Pedro Artal, and many others, a series of papers on fossil crabs which bring up to date the palaeontology of the group and helped to date the stages of their evolution. In 2008, Danièle co-authored with Peter Ng and Peter Davie, a monumental work, the Systema Brachyurorum, which lists and updates the nomenclature of nearly 6800 species of extant brachyuran crabs, the first time this has ever been ventured. This work has immediately developed as an essential reference for anyone interested in the synonymy and status of a species, as well as any supra-specific taxon. In 2008, the Crustacean Society awarded Danièle Guinot the Crustacean Society Award for Research Excellence, citing her "for her efforts at promoting revolutionary insights into the taxonomy and evolution of Brachyura". This award, given to her by her peers, is the ultimate recognition of her massive contributions to carcinology!

If Danièle has always been noted for her exceptional elegance, particularly in the setting of a world-famous research institution in the Paris Museum, she must equally be acknowledged for her tenacity to undertake basic research without which the most exciting evolutionary theories would not have any foundation. Her contributions to her discipline have been nothing short of remarkable — for the extant Brachyura, she has named 20 extant and fossil suprageneric taxa, 61 new genera, and 164 new species (see table I). In return, and in no small acknowledgement of her help and contributions, seven genera and 24 species (excluding what is done in this volume; see table II) are named after her! It is, thus, with great pleasure that we dedicate Danièle this volume of the results of scientific endeavours generated by her numerous colleagues and friends.

PETER CASTRO
PETER J. F. DAVIE
PETER K. L. NG
BERTRAND RICHER DE FORGES

TABLE I
Taxa named by Danièle Guinot

Suprageneric taxa: 15
Ethusinae Guinot, 1977 (currently Ethusidae)
Heterotremata Guinot, 1977
Platyxanthidae Guinot, 1977
Podotremata Guinot, 1977
Telmessidae Guinot, 1977 (= Cheiragonidae Ortmann, 1893)
Trogloplacinae Guinot, 1986
Pilumnoidinae Guinot & Macpherson, 1987 (currently Pilumnoididae)
Cyrtorhininae Guinot, 1993
Lyreidinae Guinot, 1993
Poupiniidae Guinot, 1993
Etyidae Guinot & Tavares, 2001 (fossil)
Etyoidea Guinot & Tavares, 2001 (fossil)
Hypoconchinae Guinot & Tavares, 2003
Cenomanocarcinidae Guinot, Vega & Van Bakel, 2008 (fossil)
Ibericancridae Artal, Guinot, Van Bakel & Castillo, 2008 (fossil)
Marocarcinidae Guinot, De Angeli & Garassino, 2008 (fossil)
Sphaerodromiinae Guinot & Tavares, 2003
Acanthodromiinae Guinot, 2008
Metadynomeninae Guinot, 2008
Paradynomeninae Guinot, 2008

New genera: 55
Coralliope Guinot, 1967
Edwardsium Guinot, 1967
Gonopanope Guinot, 1967
Microcassiope Guinot, 1967

TABLE I
(Continued)

Nanocassiope Guinot, 1967
Nanoplax Guinot, 1967
Medaeops Guinot, 1967
Miersiella Guinot, 1967
Monodaeus Guinot, 1967
Paramedaeus Guinot, 1967
Platyactaea Guinot, 1967
Platypodiella Guinot, 1967
Cataleptodius Guinot, 1968
Neoxanthops Guinot, 1968
Paraxanthodes Guinot, 1968
Pseudomedaeus Guinot, 1968
Chacellus Guinot, 1969
Cycloplax Guinot, 1969
Euphrosynoplax Guinot, 1969
Malacoplax Guinot, 1969
Paractaea Guinot, 1969
Platychelonion Crosnier & Guinot, 1969
Robertsella Guinot, 1969
Thalassoplax Guinot, 1969
Trapezioplax Guinot, 1969
Forestia Guinot, 1976
Drachiella Guinot, in Serène & Soh, 1976
Gaillardiellus Guinot, 1976
Novactaea Guinot, 1976
Palaeotrichia Guinot, 1976 (fossil)
Serenius Guinot, 1976
Beuroisia Guinot & Richer de Forges, 1981
Intesius Guinot & Richer de Forges, 1981
Mathildella Guinot & Richer de Forges, 1981
Sotoplax Guinot, 1984
Trogloplax Guinot, 1986
Segonzacia Guinot, 1989
Garthiope Guinot, 1990
Poupinia Guinot, 1993
Umalia Guinot, 1993
Dagnaudus Guinot & Richer de Forges, 1995
Gordonopsis Guinot & Richer de Forges, 1995
Ihlopsis Guinot & Richer de Forges, 1995
Yaldwynopsis Guinot & Richer de Forges, 1995
Feldmannia Guinot & Tavares, 2001 (fossil)
Secretanella Guinot & Tavares, 2001 (fossil)
Allograea Guinot, Hurtado & Vrijenhoek, 2002
Lamarckdromia Guinot & Tavares, 2003
Lewindromia Guinot & Tavares, 2003

TABLE I
(Continued)

Mclaydromia Guinot & Tavares, 2003
Moreiradromia Guinot & Tavares, 2003
Stebbingdromia Guinot & Tavares, 2003
Hexalaughlia Guinot, 2006
Davusia Guinot, 2007
Hebertides Guinot, De Angeli & Garassino, 2007 (fossil)
Guaplax Naruse, Ng & Guinot, 2008
Sulaplax Naruse, Ng & Guinot, 2008
Actiomera Ng, Guinot & Davie, 2008
Heikeopsis Ng, Guinot & Davie, 2008
Ibericancer Artal, Guinot, Van Bakel & Castillo, 2008 (fossil)
Marocarcinus Guinot, De Angeli & Garassino, 2008 (fossil)

New species: 161
Portunus (Xiphonectes) mariei Guinot, 1957
Hypocolpus perfectus Guinot-Dumortier, 1960
Hypocolpus stenocoelus Guinot-Dumortier, 1960
Leptodius leptodon Forest & Guinot, 1961 [= *Leptodius davaoensis* Ward, 1941]
Macropipus australis Guinot, 1961
Neoliomera demani Forest & Guinot, 1961
Pilumnus ransoni Forest & Guinot, 1961
Calappa dumortieri Guinot, 1962
Stilbognathus soikai Guinot, 1962
Tylodiplax derijardi Guinot & Crosnier, 1963 [currently in *Paratylodiplax*]
Menaethiops ninii Guinot, 1964
Phymodius drachi Guinot, 1964 [currently in *Cyclodius*]
Xanthias cherbonnieri Guinot, 1964
Achaeus calypso Forest & Guinot, 1966 [currently in *Calypsachaeus*]
Achaeus trifalcatus Forest & Guinot, 1966
Dromia monodi Forest & Guinot, 1966 [currently in *Sternodromia*]
Eurynome parvirostris Forest & Guinot, 1966
Pisa calva Forest & Guinot, 1966
Medaeops edwardsi Guinot, 1967
Neoxanthops rotundus Guinot, 1968
Chacellus filiformis Guinot, 1969
Cycloplax pinnotheroides Guinot, 1969
Demania intermedia Guinot, 1969
Euphrosynoplax clausa Guinot, 1969
Nanoplax garthi Guinot, 1969 [= *Pseudorhombila xanthiformis* Garth, 1940, as replacement name]
Paractaea excentrica Guinot, 1969
Paractaea monodi Guinot, 1969
Paractaea rebieri Guinot, 1969
Paractaea secundarathbunae Guinot, 1969
Platychelonion planissimum Crosnier & Guinot, 1969
Robertsella mystica Guinot, 1969

TABLE I
(Continued)

Speocarcinus lobatus Guinot, 1969
Speocarcinus spinicarpus Guinot, 1969
Thalassoplax angusta Guinot, 1969
Trachycarcinus elegans Guinot & Sakai, 1970 [currently in *Trichopeltarion*]
Euxanthus ruali Guinot, 1971
Liagore erythematica Guinot, 1971
Actaea catalai Guinot, 1976
Actaea granulata var. *laevis* A. Milne-Edwards, in Guinot, 1976 [=*Actaea calculosa* (H. Milne Edwards, 1834)]
Actaea jacquelinae Guinot, 1976
Actaea semblatae Guinot, 1976
Actaeodes mutatus Guinot, 1976
Banareia balssi Guinot, 1976
Banareia serenei Guinot, 1976
Novactaea bella Guinot, 1976
Paractaea rufopunctata africana Guinot, 1976
Paractaea rufopunctata plumosa Guinot, in Sakai, 1976
Demania japonica Guinot, 1977
Demania squamosa Guinot, 1977 [=*Demania reynaudii* (H. Milne Edwards, 1834)]
Lophozozymus evestigatus Guinot, 1977
Lophozozymus guezei Guinot, 1977
Beuroisia duhameli Guinot & Richer de Forges, 1981
Beuroisia manquenei Guinot & Richer de Forges, 1981
Carcinoplax crosnieri Guinot & Richer de Forges, 1981 [currently in *Thyraplax*]
Carcinoplax eurysternum Guinot & Richer de Forges, 1981 [=*Carcinoplax ischurodous* (Stebbing, 1923)]
Carcinoplax microphthalmus Guinot & Richer de Forges, 1981 [currently in *Rhadinoplax*]
Demania garthi Guinot & Richer de Forges, 1981
Demania serenei Guinot & Richer de Forges, 1981
Domecia africana Guinot, in Manning & Holthuis, 1981
Homologenus broussei Guinot & Richer de Forges, 1981
Homolomannia occlusa Guinot & Richer de Forges, 1981
Hypsophrys inflata Guinot & Richer de Forges, 1981 [currently in *Lamoha*]
Hypsophrys personata Guinot & Richer de Forges, 1981 [currently in *Lamoha*]
Intesius pilosus Guinot & Richer de Forges, 1981
Latreillopsis gracilipes Guinot & Richer de Forges, 1981
Lophozozymus bertonciniae Guinot & Richer de Forges, 1981
Mathildella maxima Guinot & Richer de Forges, 1981
Progeryon mararae Guinot & Richer de Forges, 1981
Progeryon vaubani Guinot & Richer de Forges, 1981
Cyrtomaia ericina Guinot & Richer de Forges, 1982
Cyrtomaia gaillardi Guinot & Richer de Forges, 1982
Cyrtomaia granulosa Guinot & Richer de Forges, 1982
Cyrtomaia ihlei Guinot & Richer de Forges, 1982
Pleistacantha exophthalmus Guinot & Richer de Forges, 1982

TABLE I
(Continued)

Sotoplax robertsi Guinot, 1984
Cyrtomaia guillei Guinot, 1985
Parapanope singaporensis Ng & Guinot, in Guinot, 1985 [= *Parapanope euagora* De Man, 1895]
Parapanope serenei Guinot & Ng, in Guinot, 1985
Parapanope siamensis Guinot, 1985
Platypilumnus inermis Guinot, 1985
Psopheticus vocans Guinot, 1985
Sphenocarcinus difficilis Guinot & Richer de Forges, 1985 [currently in *Nasutocarcinus*]
Sphenocarcinus pinocchio Guinot & Richer de Forges, 1985 [currently in *Nasutocarcinus*]
Platymaia rebierei Guinot & Richer de Forges, 1986
Sphenocarcinus bipartitum Guinot & Richer de Forges, 1986 [currently in *Oxypleurodon*]
Sphenocarcinus mammatum Guinot & Richer de Forges, 1986 [currently in *Oxypleurodon*]
Sphenocarcinus orbiculatum Guinot & Richer de Forges, 1986 [currently in *Oxypleurodon*]
Sphenocarcinus stuckiae Guinot & Richer de Forges, 1986 [currently in *Oxypleurodon*]
Trachycarcinus crosnieri Guinot, 1986 [currently in *Trichopeltarion*]
Trogloplax joliveti Guinot, 1986
Pilumnoides coelhoi Guinot & Macpherson, 1987
Pilumnoides inglei Guinot & Macpherson, 1987
Pilumnoides monodi Guinot & Macpherson, 1987
Pilumnoides rubus Guinot & Macpherson, 1987
Sendleria genuitei Guinot, 1987
Cyrtomaia cornuta Richer de Forges & Guinot, 1988
Cyrtomaia coriolisi Richer de Forges & Guinot, 1988
Cyrtomaia furici Richer de Forges & Guinot, 1988
Monodaeus cristulatus Guinot & Macpherson, 1988
Monodaeus arnaudi Guinot & Macpherson, 1988
Carcinoplax monodi Guinot, 1989
Carcinoplax nana Guinot, 1989
Carcinoplax polita Guinot, 1989 [= *Carcinoplax specularis* Rathbun, 1914]
Trachycarcinus delli Guinot, 1989 [currently in *Trichopeltarion*]
Trachycarcinus foresti Guinot, 1989 [currently in *Trichopeltarion*]
Trachycarcinus moosai Guinot, 1989 [currently in *Trichopeltarion*]
Austinograea alayseae Guinot, 1990
Cyrtomaia griffini Richer de Forges & Guinot, 1990
Garthiope anchialina Guinot & Iliffe, 1991
Homolodromia kai Guinot, 1993
Poupinia hirsuta Guinot, 1993
Psopheticus crosnieri Guinot, 1990
Psopheticus insolitus Guinot, 1990 [= *Psopheticus stridulans* Wood-Mason, 1892]
Psopheticus musicus Guinot, 1990
Psopheticus vocans Guinot, 1985
Sesarmoides ultrapes Ng, Guinot & Iliffe, 1994 [currently in *Karstarma*]
Dicranodromia alphonsei Martin & Guinot, in Guinot, 1995
Dicranodromia chacei Guinot, 1995

TABLE I
(Continued)

Dicranodromia crosnieri Guinot, 1995
Dicranodromia foersteri Guinot, 1993
Dicranodromia karubar Guinot, 1993
Dicranodromia martini Guinot, 1995
Dicranodromia nagaii Guinot, 1995
Dicranodromia pequegnati Guinot, 1995
Dicranodromia simplicia Guinot & Martin, in Guinot, 1995
Dicranodromia spinulata Guinot, 1995
Homola coriolisi Guinot & Richer de Forges, 1995
Homola eldredgei Guinot & Richer de Forges, 1995
Homola minima Guinot & Richer de Forges, 1995
Homola ranunculus Guinot & Richer de Forges, 1995
Homolochunia gadaletae Guinot & Richer de Forges, 1995
Homologenus boucheti Guinot & Richer de Forges, 1995
Homologenus levii Guinot & Richer de Forges, 1995
Homologenus wallis Guinot & Richer de Forges, 1995
Hypsophrys futuna Guinot & Richer de Forges, 1995 [= *Lamoha longirostris* (Chen, 1986)]
Ihlopsis tirardi Guinot & Richer de Forges, 1995
Latreillopsis antennata Guinot & Richer de Forges, 1995
Latreillopsis cornuta Guinot & Richer de Forges, 1995
Latreillopsis daviei Guinot & Richer de Forges, 1995
Moloha alisae Guinot & Richer de Forges, 1995
Moloha grandperrini Guinot & Richer de Forges, 1995
Paromola bathyalis Guinot & Richer de Forges, 1995
Paromola crosnieri Guinot & Richer de Forges, 1995
Australocarcinus kanaka Davie & Guinot, 1996
Australocarcinus palauensis Davie & Guinot, 1996
Neopilumnoplax gervaini Tavares & Guinot, 1996
Orcovita angulata Ng, Guinot & Iliffe, 1996
Orcovita fictilia Ng, Guinot & Iliffe, 1996
Orcovita gracilipes Ng, Guinot & Iliffe, 1996
Orcovita mollitia Ng, Guinot & Iliffe, 1996
Bythograea laubieri Guinot & Segonzac, 1997
Geelvinkia darnei Ng & Guinot, 1997
Progeryon mus Ng & Guinot, 1999
Discoplax gracilipes Ng & Guinot, 2001
Allograea tomentosa Guinot, Hurtado & Vrijenhoek, 2002
Bythograea galapagensis Guinot & Hurtado, 2003
Bythograea vrijenhoeki Guinot & Hurtado, 2003
Mclaydromia colini Guinot & Tavares, 2003
Goneplax clevai Guinot & Castro, 2007
Hebertides jurassica Guinot, De Angeli & Garassino, 2007 (fossil)
Guaplax denticulata Naruse, Ng & Guinot, 2008
Ibericancer sanchoi Artal, Guinot, Van Bakel & Castillo, 2008 (fossil)
Marocarcinus pasinii Guinot, De Angeli & Garassino, 2008 (fossil)
Sulaplax ensifer Naruse, Ng & Guinot, 2008

TABLE II
Taxa named after Danièle Guinot

Suprageneric taxa: 2
Guinotini Pretzmann, 1971 [= Pseudothelphusinae Ortmann, 1893]
Danielini Števčić, 2005 [= Pilumninae Samouelle, 1819]

Genera: 7
Guinotia Pretzmann, 1965
Eudaniela Pretzmann, 1971
Guinotellus Serène, 1971
Danielum Vázquez-Bader & Gracia, 1995
Danielea Ng & Clark, 2003
Guinotosia Beshin, Busulini, De Angeli & Tessier, 2007 (fossil)
Guinotinia Richer de Forges & Ng, 2009
[+ 4 new genera in this volume]

Species: 24
Portunus (*Xiphonectes*) *guinotae* Stephenson & Rees, 1961
Percnon guinotae Crosnier, 1965
Pilumnus guinotae Takeda & Miyake, 1968
Carpoporus guinotae Zarenkov, 1971 [currently *Hepatoporus*]
Guinotellus melvillensis Serène, 1971
Monodaeus guinotae Forest, 1976
Progeryon guinotae Crosnier, 1976
Hypocolpus guinotae Vannini, 1982
Pseudorhombila guinotae Hernandez-Aguilera, 1982
Parapilumnus guinotae Deb, 1987 [currently *Latopilumnus*]
Cyrtomaia danielae Zarenkov, 1990
Drachiella guinotae Morris & Collins, 1991 (fossil)
Actumnus guinotae Karasawa & Inoue, 1992 (fossil)
Afropinnotheres guinotae Manning, 1993
Euryozius danielae Davie, 1993
Ranilia guinotae Melo & Campos, 1994
Alarconia guinotae Coelho, 1996
Arcotheres guinotae Campos, 2001
Nursia guinotae Komatsu & Takeda, 2001
Loxorhynchus guinotae Hendrickx & Cervantes, 2003
Dicranodromia danielae Ng & McLay, 2005
Umalia guinotae De Angeli & Beschin, 2007 (fossil)
Yaldwynopsis guinotae Richer de Forges & Ng, 2007
Xanthosia danielae Collins & Breton, 2009
[+ 20 new species in this volume]

DANIÈLE GUINOT BIBLIOGRAPHY

1956

FOREST, J. & D. GUINOT, 1956. Sur une collection de Crustacés Décapodes et Stomatopodes des mers tunisiennes. Bulletin de la Station océanographique de Salammbô, **53**: 24-43.

1958

FOREST, J. & D. GUINOT, 1958. Sur une collection de Crustacés Décapodes des côtes d'Israël. Bulletin of the Sea Fisheries Research Station, Israel, **15**: 4-16.

GUINOT, D., 1958. Sur une collection de Décapodes Brachyoures (Portunidae et Xanthidae) de l'île Mayotte. I. *Portunus* (*Hellenus*) *mariei* sp. nov. Bulletin du Muséum national d'Histoire naturelle, Paris, (2) **29** (6): 475-484. [1957 volume, published 1958]

GUINOT, D., 1958. Sur une collection de Décapodes Brachyoures (Portunidae et Xanthidae) de l'île Mayotte. II. Xanthidae. Bulletin du Muséum national d'Histoire naturelle, Paris, (2) **30** (1): 84-93.

GUINOT, D., 1958. Sur une collection de Décapodes Brachyoures (Portunidae et Xanthidae) de l'île Mayotte. II. Xanthidae (suite). Bulletin du Muséum national d'Histoire naturelle, Paris, (2) **30** (2): 175-183.

GUINOT, D., 1958. Sur une collection de Décapodes Brachyoures (Portunidae et Xanthidae) de l'île Mayotte. II. Xanthidae (suite). Bulletin du Muséum national d'Histoire naturelle, Paris, (2) **30** (3): 276-284.

1959

GUINOT-DUMORTIER, D., 1959. Sur une collection de Crustacés (Decapoda Reptantia) de Guyane française. I. Brachyura (Oxyrhyncha exclus). Bulletin du Muséum national d'Histoire naturelle, Paris, (2) **31** (5): 423-434.

1960

GUINOT-DUMORTIER, D., 1960. Sur une collection de Crustacés (Decapoda Reptantia) de Guyane française. I. Brachyura (Oxyrhyncha exclus) (suite). Bulletin du Muséum national d'Histoire naturelle, Paris, (2) **31** (6): 510-515. [1959 volume, published 1960]

GUINOT-DUMORTIER, D., 1960. Sur une collection de Crustacés (Decapoda Reptantia) de Guyane française. II. Brachyura Oxyrhyncha et Macrura. Bulletin du Muséum national d'Histoire naturelle, Paris, (2) **32** (2): 177-187.

GUINOT-DUMORTIER, D., 1960. Révision des genres *Euxanthus* Dana et *Hypocolpus* Rathbun (Crust. Decap. Brach.). Remarques sur les cavités sous-hépatiques et les coaptations des *Hypocolpus*. Bulletin du Muséum national d'Histoire naturelle, Paris, (A) **20** (2): 153-218, pls. 1-12.

GUINOT-DUMORTIER, D. & B. DUMORTIER, 1960. La stridulation chez les crabes. Crustaceana, **1** (2): 117-155.

1961

FOREST, J. & D. GUINOT, 1961. Crustacés Décapodes Brachyoures de Tahiti et des Tuamotu. In: Expédition française sur les récifs coralliens de la Nouvelle-Calédonie [Volume préliminaire]: i-xi, 1-195, pls. 1-18, tabs. 1-3, charts 1-7. (Editions de la Fondation Singer-Polignac, Paris).

GUINOT, D., 1961. Crabes, Crevettes et Langoustes ne sont pas toujours silencieux. La Nature, Science, Progrès, **3310**: 82-89.

GUINOT, D., 1961. Caractères et affinités de *Macropipus australis* sp. nov., Crustacé Décapode Brachyoure de la côte sud-ouest africaine. Bulletin Institut royal des Sciences naturelles de Belgique, **37** (26): 1-13, pls. 1, 2.

GUINOT-DUMORTIER, D., 1961. Les espèces indo-pacifiques du genre *Globopilumnus* (Crustacea Brachyura Xanthidae). Mémoires de l'Institut Scientifique de Madagascar, (F) **3**: 97-119. [1959 volume, published 1960]

GUINOT-DUMORTIER, D. & B. DUMORTIER, 1961. Description d'un appareil stridulatoire dans le genre *Cycloes* De Haan (Crustacea, Brachyura, Oxystomata, Calappidae). Bulletin du Muséum national d'Histoire naturelle, Paris, (2) **32** (6): 558-561. [1960 volume, published 1961]

1962

FOREST, J. & D. GUINOT, 1962. Remarques biogéographiques sur les crabes des archipels de la Société et des Tuamotu. Cahiers du Pacifique, **4**: 41-75, tabs. 1, 2.

GUINOT, D., 1962. Sur une collection de Crustacés Décapodes Brachyoures des îles Maldives et de Mer Rouge (Expédition « Xarifa » 1957-1958). Kieler Meeresforschungen, **18** (2): 231-244.

GUINOT, D., 1962. Sur quelques Crustacés Décapodes Brachyoures indo-pacifiques des collections du Musée de Munich. Opuscula Zoologica, **60**: 1-14.

GUINOT, D. & A. RIBEIRO, 1962. Sur une collection de Crustacés Brachyoures des îles du Cap-Vert et de l'Angola. Memórias da Junta de Investigações do Ultramar, (2) **40**: 1-89, pls. 1-4.

1964

GUINOT, D., 1964. Crustacés Décapodes Brachyoures (Xanthidae) des Campagnes de la Calypso en Mer Rouge (1952), dans le Golfe Persique et à l'île Aldabra (1954). Mémoires du Muséum national d'Histoire naturelle, Paris, (A) **32** (1): i-iii, 1-108, pls. 1-12.

GUINOT, D., 1964. Les trois espèces du genre *Domecia* (Decapoda, Brachyura): *D. hispida* Eydoux et Souleyet, *D. glabra* Alcock et *D. acanthophora* (Desbonne et Schramm). Crustaceana, **7** (4): 267-283.

GUINOT, D., 1964. Sur une collection de Crustacés Décapodes Brachyoures de Mer Rouge et de Somalie. Remarques sur les genres *Calappa* Weber, *Menaethiops* Alcock, *Tyche* Bell, *Ophthalmias* Rathbun et *Stilbognathus* Von Martens. Bolletino del Museo Civico di Storia Naturale di Venezia, **15**: 7-63, pls. 1-4. [1962 volume, published 1964]

GUINOT, D. & A. CROSNIER, 1964. Remarques sur les genres *Cleistostoma*, *Paracleistostoma* et *Tylodiplax*, et description de *Tylodiplax derijardi* sp. nov. (Crust. Decap. Brachyura). Bulletin du Muséum national d'Histoire naturelle, Paris, (2) **35** (6): 606-619. [1963 volume, published 1964]

GUINOT, D. & A. CROSNIER, 1964. Caractères et affinités de deux *Sesarma*, *S. longipes* Krauss et *S. kraussi* De Man (Crust. Decap. Brachyura). Bulletin du Muséum national d'Histoire naturelle, Paris, (2) **36** (2): 211-221.

1965

GRMEK, M. D. & D. GUINOT, 1965. Les Crabes chez Ulysse Aldrovandi: un aperçu critique sur la Carcinologie du XVIe siècle. In: Coll. internat. "Histoire de la Biologie Marine". Vie et Milieu, (Supplement) **19**: 45-64.

GRMEK, M. D. & D. GUINOT, 1965. Les Crustacés dans la matière médicale européenne au XVIe siècle. Revue d'Histoire des Sciences et de leur Applications, **18** (1): 55-71.

HUARD, P. & D. GUINOT, 1965. Les Crabes de Chine dans une série d'aquarelles de Dabry de Thiersant. Bulletin de l'Ecole française d'Extrême-Orient, **52** (2): 551-557, pls. 75-79.

HUARD, P. & D. GUINOT, 1965. Les Crabes de Chine dans une série d'aquarelles provenant de Dabry de Thiersant. In: Coll. int. "Histoire de la Biologie Marine". Vie et Milieu, (Supplement) **19**: 35-43.

1966

FOREST, J. & D. GUINOT, 1966. Campagne de la Calypso dans le Golfe de Guinée et aux îles Principe, São Tomé, et Annobon (1956). 16. Crustacés Décapodes: Brachyoures. In: Résultats scientifiques des Campagnes de la "Calypso", **7**. Annales de l'Institut océanographique de Monaco, **44**: 23-124.

GUINOT, D., 1966. Les espèces comestibles de crabes dans l'océan Indien occidental et la mer Rouge. In: Réunion de spécialistes C. S. A. sur les Crustacés, Zanzibar 1964. Mémoires de l'Institut fondamental d'Afrique Noire, **77**: 353-394.

GUINOT, D., 1966. Recherches préliminaires sur les groupements naturels chez les Crustacés Décapodes Brachyoures. I. Les affinités des genres *Aethra*, *Osachila*, *Hepatus*, *Hepatella* et *Actaeomorpha*. Bulletin du Muséum national d'Histoire naturelle, Paris, (2) **38** (5): 744-762.

1967

GUINOT, D., 1967. Les Crabes comestibles de l'Indo-Pacifique. In: Expédition française sur les récifs coralliens de la Nouvelle-Calédonie. Volume préliminaire, **2**: 1-145, pls. 1-10. (Editions de la Fondation Singer-Polignac, Paris). [1966 volume, published 1967]

GUINOT, D., 1967. La faune carcinologique (Crustacea Brachyura) de l'océan Indien occidental et de la mer Rouge. Catalogue, remarques biogéographiques et bibliographie. In: Réunion de Spécialistes C. S. A. sur les Crustacés, Zanzibar 1964. Mémoires de l'Institut fondamental d'Afrique Noire, **77**: 235-352. [1966 volume, published 1967]

GUINOT, D., 1967. Recherches préliminaires sur les groupements naturels chez les Crustacés Décapodes Brachyoures. I. Les affinités des genres *Aethra*, *Osachila*, *Hepatus*, *Hepatella* et *Actaeomorpha* (suite et fin). Bulletin du Muséum national d'Histoire naturelle, Paris, (2) **38** (6): 828-845. [1966 volume, published 1967]

GUINOT, D., 1967. Recherches préliminaires sur les groupements naturels chez les Crustacés Décapodes Brachyoures. II. Les anciens genres *Micropanope* Stimpson et *Medaeus* Dana. Bulletin du Muséum national d'Histoire naturelle, Paris, (2) **39** (2): 345-374.

GUINOT, D., 1967. Recherches préliminaires sur les groupements naturels chez les Crustacés Décapodes Brachyoures. III. A propos des affinités des genres *Dairoides* Stebbing et *Daira* De Haan. Bulletin du Muséum national d'Histoire naturelle, Paris, (2) **39** (3): 540-563.

1968

GUINOT, D., 1968. Recherches préliminaires sur les groupements naturels chez les Crustacés Décapodes Brachyoures. IV. Observations sur quelques genres de Xanthidae. Bulletin du Muséum national d'Histoire naturelle, Paris, (2) **39** (4): 695-727. [1967 volume, published 1968]

GUINOT, D., 1968. Recherches préliminaires sur les groupements naturels chez les Crustacés Décapodes Brachyoures. V. Établissement d'un caractère évolutif: l'articulation ischio-mérale des chélipèdes. Bulletin du Muséum national d'Histoire naturelle, Paris, (2) **40** (1): 149-166, 1 pl.

GUINOT, D., 1968. Recherches préliminaires sur les groupements naturels chez les Crustacés Décapodes Brachyoures. VI. Les Carpilinae. Bulletin du Muséum national d'Histoire naturelle, Paris, (2) **40** (2): 320-334, pls. 1, 2.

GUINOT, D. & M. D. GRMEK, 1968. Observations sur l'action anticoagulante de l'hépatopancréas des Crustacés: une étape méconnue dans l'histoire des anticoagulants. In: Actes XIe Congress on the International History of Science (Varsovie-Cracovie, 24-29 août 1965), Wroclaw, Ossolineum, **5**: 227-232.

1969

CROSNIER, A. & D. GUINOT, 1969. Un nouveau Crabe ouest-africain, *Platychelonion planissimum* gen. nov., sp. nov. Bulletin du Muséum national d'Histoire naturelle, Paris, (2) **41** (3): 725-730.

GUINOT, D., 1969. Recherches préliminaires sur les groupements naturels chez les Crustacés Décapodes Brachyoures. VII. Les Goneplacidae. Bulletin du Muséum national d'Histoire naturelle, Paris, (2) **41** (1): 241-265, pl. 1.

GUINOT, D., 1969. Recherches préliminaires sur les groupements naturels chez les Crustacés Décapodes Brachyoures. VII. Les Goneplacidae (suite). Bulletin du Muséum national d'Histoire naturelle, Paris, (2) **41** (2): 507-528, pl. 2.

GUINOT, D., 1969. Recherches préliminaires sur les groupements naturels chez les Crustacés Décapodes Brachyoures. VII. Les Goneplacidae (suite et fin). Bulletin du Muséum national d'Histoire naturelle, Paris, (2) **41** (3): 688-724, pls. 3-5.

GUINOT, D., 1969. Sur divers Xanthidae notamment sur *Actaea* De Haan et *Paractaea* gen. nov. (Crustacea: Decapoda, Brachyura). Cahiers du Pacifique, **13**: 223-267.

1970

GUINOT, D. & T. SAKAI, 1970. Un nouveau *Trachycarcinus*, *T. elegans* sp. nov. (Crustacea Decapoda Brachyura). Bulletin du Muséum national d'Histoire naturelle, Paris, (2) **42** (1): 201-205.

1971

GUINOT, D., 1971. Recherches préliminaires sur les groupements naturels chez les Crustacés Décapodes Brachyoures. VIII. Synthèse et bibliographie. Bulletin du Muséum national d'Histoire naturelle, Paris, (2) **42** (5): 1063-1090. [1970 volume, published 1971]

GUINOT, D., 1971. Sur l'existence d'une deuxième espèce de *Liagore* De Haan, *L. erythematica* sp. nov. (Crustacea Decapoda Brachyura). Bulletin du Muséum national d'Histoire naturelle, Paris, (2) **42** (5): 1091-1098. [1970 volume, published 1971]

GUINOT, D., 1971. Un nouvel *Euxanthus* de Nouvelle-Calédonie, *E. ruali* sp. nov. Cahiers du Pacifique, **15**: 19-22, pls. 1, 2.

1976

GUINOT, D., 1976. Constitution de quelques groupes naturels chez les Crustacés Décapodes Brachyoures. I. La superfamille des Bellioidea et trois sous-familles de Xanthidae (Polydectinae Dana, Trichiinae De Haan, Actaeinae Alcock). Mémoires du Muséum national d'Histoire naturelle, Paris, (A) **97**: 1-308, pls. 1-19.

1977

GUINOT, D., 1977. Données nouvelles sur la morphologie, la phylogenèse et la taxonomie des Crustacés Décapodes Brachyoures: i-xv, 1-486, xvi-xxiv, pls. 1-31. (Thèse Doctorat d'État ès Sciences soutenue le 21 juin 1977 à l'Université de Pierre et Marie Curie, 2 volumes in folio, ronéotypé). [distributed, considered available publication]

GUINOT, D., 1977. Propositions pour une nouvelle classification des Crustacés Décapodes Brachyoures. Comptes rendus hebdomadaires des Séances de l'Académie de Sciences, Paris, (D) **285** (10): 1049-1052.

1978

GUINOT, D., 1978. Analyse morphogénétique d'une lignée de Crabes: la lignée "parthénoxystomienne" et position systématique du genre *Drachiella* Guinot (Crustacea, Decapoda, Brachyura). Archives de Zoologie expérimentale et génèrale, **119** (1): 7-20, pls. 1-4.

GUINOT, D., 1978. Principes d'une classification évolutive des Crustacés Décapodes Brachyoures. Bulletin biologique de la France et de la Belgique, (n.s.) **112** (3): 211-292.

1979

GUINOT, D., 1979. Problème pratique [sic pour "Problèmes pratiques"] d'une classification cladistique des Crustacés Décapodes Brachyoures. In: Comptes rendus VIe Réunion des Carcinologistes de langue française, Nabeul, Tunisie, 4-9 septembre 1979. Bulletin de l'Office national des Pêches de Tunisie, **3** (1): 33-46.

GUINOT, D., 1979. Examen des théories actuelles de la classification zoologique. History and Philosophy of the Life Sciences, **1** (1): 119-138.

GUINOT, D., 1979. Données nouvelles sur la morphologie, la phylogenèse et la taxonomie des Crustacés Décapodes Brachyoures. Mémoires du Muséum national d'Histoire naturelle, Paris, (A) **112**: 1-354, pls. 1-27.

1981

DELAMARE DEBOUTTEVILLE, C. & D. GUINOT, 1981. Considérations sur les Bythograeoidea Williams, nouvelle superfamille de Crabes de la dorsale Pacifique Est. Conférence présentée à la VIIe Réunion des Carcinologistes de langue française, Banyuls sur Mer, 1-6 juin 1981. [Résumé ronéotypé]

GUINOT, D. & B. RICHER DE FORGES, 1981. Crabes de profondeur, nouveaux ou rares, de l'Indo-Pacifique (Crustacea, Decapoda, Brachyura) (Première partie). Bulletin du Muséum national d'Histoire naturelle, Paris, (4) **2** (A) (4): 1113-1153, pls. 1-7. [1980 volume, published 1981]

GUINOT, D. & B. RICHER DE FORGES, 1981. Crabes de profondeur, nouveaux ou rares, de l'Indo-Pacifique (Crustacea, Decapoda, Brachyura) (Deuxième partie). Bulletin du Muséum national d'Histoire naturelle, Paris, (4) **3** (A) (1): 227-260.

GUINOT, D. & B. RICHER DE FORGES, 1981. Homolidae, rares ou nouveaux, de l'Indo-Pacifique (Crustacea, Decapoda, Brachyura). Bulletin du Muséum national d'Histoire naturelle, Paris, (4) **3** (A) (2): 523-581, pls. 1-8.

1982

DRACH, P. & D. GUINOT, 1982. Connexions morphologiques et fonctionnelles d'un type nouveau dans le squelette des Brachyoures du genre *Paradasygius* Garth (carapace, pleurites, sternites, pléon). Comptes rendus hebdomadaires des Séances de l'Académie de Sciences, Paris, (3, Sciences de la Vie) **295** (12): 715-720, pl. 1.

GUINOT, D. & B. RICHER DE FORGES, 1982. Nouvelles récoltes des genres *Cyrtomaia* Miers et *Pleistacantha* Miers (Crustacea, Decapoda, Brachyura). Bulletin du Muséum national d'Histoire naturelle, Paris, (4) **3** (A) (4): 1087-1125, pls. 1-4. [1981 volume, published 1982]

GUINOT, D. & B. RICHER DE FORGES, 1982. Révision du genre indo-pacifique *Cyrtomaia* Miers, 1886: campagnes océanographiques du "Challenger", de l' "Albatross", du "Siboga" et du "Vauban" (Crustacea, Decapoda, Brachyura). Annales de l'Institut océanographique, Monaco, **58** (1): 5-88.

1983

DRACH, P. & D. GUINOT, 1983. Les Inachoididae Dana, famille de Majoidea caractérisée par des connexions morphologiques d'un type nouveau entre carapace, pleurites, sternites et pléon (Crustacea Decapoda). Comptes rendus hebdomadaires des Séances de l'Académie de Sciences, Paris, (3, Sciences de la Vie) **297**: 37-42, pl. 1.

1984

DRACH, P. & D. GUINOT, 1984. Connexions morphologiques d'un type nouveau entre carapace, pleurites et pléon chez des Brachyoures apparentés aux Majidés: conséquences physiologiques. In: VIIIe Réunion des Carcinologistes de langue française, Liège, août-septembre 1983. Annales de la Société royale zoologique de Belgique, **114** (Supplement 1): 199-200.

GUINOT, D., 1984. Le genre *Leurocyclus* Rathbun, 1897 (Crustacea Decapoda Brachyura). Bulletin du Muséum national d'Histoire naturelle, Paris, (4) **6** (A) (2): 377-395, pls. 1, 2.

GUINOT, D., 1984. Découverte d'un nouveau genre de Crabe dans le golfe du Mexique, *Sotoplax robertsi* gen. nov., sp. nov. (Crustacea Decapoda Brachyura). Anales del Instituto de Ciencias del Mar y Limnología, Universidad Nacional Autónoma de México, **11** (1): 91-97, pl. 1.

1985

GUINOT, D., 1985. Crabes bathyaux de l'île de La Réunion; description de *Cyrtomaia guillei* sp. nov., de *Platypilumnus inermis* sp. nov. et de *Psopheticus vocans* sp. nov. (Crustacea Decapoda Brachyura). Comité National Français des Recherches Antarctiques, **55**: 7-35, pls. 1-4. [1984 volume, published 1985]

GUINOT, D., 1985. Crustacea (chapter restricted to brachyuran decapod Crustacea: 446-455). In: G. RICHARD (ed.), Fauna and Flora, a first compendium of French Polynesian sea-dwellers, 5th Intern. Coral Reef Congress, Tahiti, 27 May-1 June 1985, **1**, French Polynesian Coral Reefs: 379-520.

GUINOT, D., 1985. Révision du genre *Parapanope* De Man, 1895 (Crustacea Decapoda Brachyura). Résumé. In: N. KASCHEK & H. O. VON HAGEN (eds.), Papers honouring Rudolf Altevogt on the 22nd of January, 1984, vol. **1** and **2**, Münster, 1984.

GUINOT, D., 1985. Révision du genre *Parapanope* De Man, 1895 (Crustacea Decapoda Brachyura), avec description de trois espèces nouvelles. Bulletin du Muséum national d'Histoire naturelle, Paris, (4) **7** (A) (3): 677-707, pls. 1-4.

GUINOT, D. & B. RICHER DE FORGES, 1985. Révision of the Indo-Pacific *Sphenocarcinus* with a single rostrum and description of two species (Crustacea, Decapoda, Brachyura, Majidae). Marine Research in Indonesia, **24**: 49-71, pls. 1, 2.

1986

GUINOT, D., 1986. Une nouvelle espèce du genre *Trachycarcinus*, *T. crosnieri* sp. nov., de Madagascar (Crustacea Decapoda Brachyura). Bulletin du Muséum national d'Histoire naturelle, Paris, (4) **7** (A) (4): 805-817, pl. 1. [1985 volume, published 1986]

GUINOT, D., 1986. Description d'un crabe cavernicole aveugle de Nouvelle-Bretagne (Papouasie Nouvelle-Guinée), *Trogloplax joliveti* gen. nov. sp. nov., et établissement d'une sous-famille nouvelle, Trogloplacinae subfam. nov. Comptes rendus hebdomadaires des Séances de l'Académie de Sciences, Paris, (3, Sciences de la Vie) **303** (8): 307-312, pls. 1, 2.

GUINOT, D., 1986. Découverte d'un crabe aveugle cavernicole dans deux grottes de Nouvelle-Bretagne (Papouasie Nouvelle-Guinée) (Résumé). In: Abstracts, 51e Convegno dell'Unione Zoologica Italiana, Roma, 6-11 octobre 1986. Bolletino di Zoologia, **53**: 104.

GUINOT, D. & B. RICHER DE FORGES, 1986. Crustacés Décapodes: Majidae (genres *Platymaia*, *Cyrtomaia*, *Pleistacantha*, *Sphenocarcinus* et *Naxioides*). In: Résultats des Campagnes MUSORSTOM, I et II — Philippines (1976, 1980), **2**. Mémoires du Muséum national d'Histoire naturelle, Paris, (A) **113**: 83-178, pls. 1-11. [1985 volume, published 1986]

GUINOT, D. & B. RICHER DE FORGES, 1986. Découverte d'une nouvelle espèce de *Sphenocarcinus* en Nouvelle-Calédonie, *S. mammatus* sp. nov. (Crustacea, Decapoda, Brachyura). Indo-Malayan Zoology, **3**: 27-37.

1987

GUINOT, D., 1987. Nouvelles découvertes dans des grottes de Nouvelle-Bretagne du Crabe aveugle *Trogloplax joliveti* Guinot, 1986 et description d'un Crabe d'eau douce cavernicole, *Sendleria genuitei* sp. nov. Comptes rendus hebdomadaires des Séances de l'Académie de Sciences, Paris, (3, Sciences de la Vie) **305** (2): 25-30, pls. 1, 2.

GUINOT, I., F. A. CHACE, JR., R. B. MANNING, J. S. GARTH, A. CROSNIER, J. FOREST, M. DE SAINT LAURENT, D. GUINOT & M. TÜRKAY, 1987. Memories of Dr. Tune Sakai from abroad. Researches on Crustacea, **16** (3-4): 3-9.

GUINOT, D. & J.-J. GEOFFROY, 1987. Crabes cavernicoles de Nouvelle-Bretagne (Papouasie Nouvelle-Guinée). Spelunca, **25**: 17-24.

GUINOT, D. & E. MACPHERSON, 1987. Révision du genre *Pilumnoides* Lucas, 1844, avec description de quatre espèces nouvelles et création de Pilumnoidinae subfam. nov. (Crustacea Decapoda Brachyura). Bulletin du Muséum national d'Histoire naturelle, Paris, (4) **9** (A) (1): 211-247, pls. 1, 2.

GUINOT, D. & E. MACPHERSON, 1987. Ultime découverte. Une araignée de mer qui n'en est pas une. L'Univers du Vivant, **24**: 111-114, 3 photographs.

GUINOT, D. & P. K. L. NG, 1987. Observations nouvelles sur *Parapanope singaporensis* Ng et Guinot, 1985, synonyme de *P. euagora* De Man, 1895 (Crustacea Decapoda Brachyura). Bulletin du Muséum national d'Histoire naturelle, Paris, (4) **9** (A) (3): 669-675, pl. 1.

1988

GUINOT, D., 1988. Les crabes cavernicoles du Monde. Mémoires de Biospéléologie **15**: 3-40, pls. 1-4.

GUINOT, D., 1988. Les crabes des sources hydrothermales de la dorsale du Pacifique oriental (Campagne BIOCYARISE 1984). In: Actes du Colloque "Les sources Hydrothermales de la ride du Pacifique oriental. Biologie et écologie", 4-7 novembre 1985, Paris. Oceanologica Acta, (Special Issue) **8**: 109-118. [1987 volume, published 1988]

RICHER DE FORGES, B. & D. GUINOT, 1988. Description de trois espèces de *Cyrtomaia* Miers, 1886, de Nouvelle-Calédonie et des îles Chesterfield (Crustacea Decapoda Brachyura). Bulletin du Muséum national d'Histoire naturelle, Paris, (4) **10** (A) (1): 39-55, pls. 1, 2.

SECRETAN, S. & D. GUINOT, 1988. Premières observations sur le squelette axial du crabe dit "des fumeurs" *Cyanagraea praedator* de Saint Laurent. In: Actes du Colloque "Les sources Hydrothermales de la ride du Pacifique oriental. Biologie et écologie", 4-7 novembre 1985, Paris. Oceanologica Acta, (Special Issue) **8**: 119-124. [1987 volume, published 1988]

1989

GUINOT, D., 1989. Le genre *Carcinoplax* H. Milne Edwards, 1852 (Crustacea, Brachyura: Goneplacidae). In: J. FOREST (ed.), Résultats des Campagnes MUSORSTOM, **5**. Mémoires du Muséum national d'Histoire naturelle, Paris, (A) **144**: 266-346, pls. 1-13.

GUINOT, D., 1989. Les genres *Trachycarcinus* Faxon et *Trichopeltarion* A. Milne Edwards (Crustacea, Brachyura: Atelecyclidae). In: J. FOREST (ed.), Résultats des Campagnes MUSORSTOM, **5**. Mémoires du Muséum national d'Histoire naturelle, Paris, (A) **144**: 347-385.

GUINOT, D., 1989. Description de *Segonzacia* gen. nov. et remarques sur *Segonzacia mesatlantica* (Williams): Campagne HYDROSNAKE 1988 sur la dorsale médio-Atlantique (Crustacea Decapoda Brachyura). Bulletin du Muséum national d'Histoire naturelle, Paris, (4) **11** (A) (1): 203-231, pls. 1-3.

GUINOT, D. & E. MACPHERSON, 1989. Remarques sur le genre *Monodaeus* Guinot, 1967, avec la description de deux espèces nouvelles (Crustacea Decapoda Brachyura). Bulletin du Muséum national d'Histoire naturelle, Paris, (4) **10** (A) (4): 731-757, pls. 1-3. [1988 volume, published 1989]

1990

GUINOT, D., 1990. Crustacea Decapoda: Le genre *Psopheticus* Wood-Mason, 1892 (Goneplacidae). In: A. CROSNIER (ed.), Résultats des Campagnes MUSORSTOM, **6**. Mémoires du Muséum national d'Histoire naturelle, Paris, (A) **145**: 331-367.

GUINOT, D., 1990. *Austinograea alayseae* sp. nov., Crabe hydrothermal découvert dans le bassin de Lau, Pacifique sud-occidental (Crustacea Decapoda Brachyura). Bulletin du Muséum national d'Histoire naturelle, Paris, (4) **11** (A) (4): 879-903. [1989 volume, published 1990]

GUINOT, D., 1990. Etablissement du genre *Garthiope* gen. nov., ses relations avec le genre *Coralliope* Guinot, 1967, et leurs affinities avec les Trapeziidae sensu lato (Crustacea Decapoda Brachyura). Bulletin du Muséum national d'Histoire naturelle, Paris, (4) **12** (A) (2): 469-487, pls. 1, 2.

RICHER DE FORGES, B. & D. GUINOT, 1990. A new *Cyrtomaia*, *C. griffini*, from Australia (Crustacea: Decapoda: Brachyura). Memoirs of the Queensland Museum, **28** (2): 523-530.

1991

GUINOT, D., 1991. Etablissement de la famille des Poupiniidae pour *Poupinia hirsuta* n. gen., n. sp. de Polynésie (Crustacea Decapoda Brachyura, Homoloidea). Bulletin du Muséum national d'Histoire naturelle, Paris, (4) **12** (A) (3-4): 577-605, pls. 1-3. [1990 volume, published 1991]

GUINOT, D., 1991. La malédiction du crabe. Les crabes étranges: les crabes toxiques, le monde des crabes géants. Subacmas, **2**: 7-10, 50-53.

GUINOT, D., 1991. Les crabes à la conquête des arbres: 201-203. In: On a marché sur la Terre, le roman de l'évolution. (ICS et Muséum National d'Histoire Naturelle, Paris).

GUINOT, D. & T. M. ILIFFE, 1991. *Garthiope anchialina* sp. nov., espèce anchialine des Galapagos, île Isabella, Cueva de la Cadena, avec remarques sur la faune carcinologique anchialine (Crustacea Decapoda Brachyura). Bulletin du Muséum national d'Histoire naturelle, Paris, (4) **12** (A) (3-4): 607-621, pl. 1. [1990 volume, published 1991]

1993

GUINOT, D., 1993. Données nouvelles sur les crabes primitifs (Crustacea Decapoda Brachyura Podotremata). Comptes rendus hebdomadaires des Séances de l'Académie de Sciences, Paris, (3, Sciences de la Vie) **316** (10): 1225-1232. [1990 volume, published 1993]

GUINOT, D., 1993. Données nouvelles sur les Raninoidea De Haan, 1841 (Crustacea Decapoda Brachyura Podotremata). Comptes rendus hebdomadaires des Séances de l'Académie de Sciences, Paris, (3, Sciences de la Vie) **316** (11): 1324-1331. [1990 volume, published 1993]

JAMIESON, B. G. M., D. GUINOT & B. RICHER DE FORGES, 1993. The ultrastructure of the spermatozoon of *Paradynomene tuberculata* Sakai, 1963 (Crustacea, Brachyura, Dynomenidae). Helgoländer Meeresuntersuchungen, **47** (3): 311-322.

JAMIESON, B. G. M., D. GUINOT & B. RICHER DE FORGES, 1993. Spermatozoal ultrastructure in four genera of Homolidae (Crustacea, Decapoda): Exemplified by *Homologenus* sp., *Latreillopsis* sp., *Homolomannia sibogae* and *Paromolopsis boasi*. Helgoländer Meeresuntersuchungen, **47** (3): 323-334.

JAMIESON, B. G. M., D. GUINOT & B. RICHER DE FORGES, 1993. The spermatozoon of *Calocarcinus africanus* (Heterotremata, Brachyura, Crustacea): Ultrastructural synapomorphies with xanthid sperm. Invertebrate Reproduction and Development, **24** (3): 189-196.

1994

CHARMANTIER, G., D. GUINOT, F. RENAUD & S. EUZET, 1994. Biosystematics and genetic structure of Mediterranean freshwater crabs (Potamidae). Preliminary results. Poster presented at the Journées annuelles de la Société Française de Systématique, Systématique et Biodiversité. Services communs de Biosystématique. Paris, 12-14 septembre 1994.

GUINOT, D., 1994. Decapoda Brachyura: 165-179, pl. 1. In: C. JUBERTHIE & V. DECOU (eds.), Encyclopaedia Biospeologica, **1**. (Société de Biospéologie, Moulis (C.N.R.S.) and Académie Roumaine, Bucarest).

GUINOT, D., 1994. The classification of Majidae in the context of recent advances in the knowledge of systematics and evolutionary trends in brachyuran decapods: 5. In: G. Y. CONAN & P. NOËL (eds.), Report on the first meeting of the International Working Group on Majid Crab Biology (November 10-12 1993, Paris, France). International Council for the Exploration of the Sea, **1994/K19**: 1-15.

GUINOT, D., B. G. M. JAMIESON & B. RICHER DE FORGES, 1994. Relationship of Homolidae and Dromiidae: Evidence from spermatozoal ultrastructure (Crustacea Decapoda). Acta Zoologica, **75** (3): 255-267.

JAMIESON, B. G. M., D. GUINOT & B. RICHER DE FORGES, 1994. Relationships of the Cyclodorippoidea Ortmann: Evidence from spermatozoal ultrastructure in the genera *Xeinostoma*, *Tymolus* and *Cymonomus* (Crustacea, Decapoda). Invertebrate Reproduction and Development, **26** (2): 153-164.

JAMIESON, B. G. M., D. GUINOT & B. RICHER DE FORGES, 1994. Podotreme affinities of *Raninoides* sp. and *Lyreidus brevifrons*: Evidence from spermatozoal ultrastructure (Crustacea, Decapoda, Raninoidea). Marine Biology, Berlin, **120**: 239-249.

NG, P. K. L., D. GUINOT & T. M. ILIFFE, 1994. *Sesarmoides ultrapes* new species, a remarkable sesarmine crab from caves in the Solomon Islands (Decapoda: Brachyura: Grapsidae). Crustacean Research, **23**: 12-22, 117.

1995

GUINOT, D., 1995. Crustacea Decapoda Brachyura: Révision de la famille des Homolodromiidae Alcock, 1900. In: A. CROSNIER (ed.), Résultats des Campagnes MUSORSTOM, **13**. Mémoires du Muséum national d'Histoire naturelle, Paris, **163**: 155-282.

GUINOT, D., D. DOUMENC & C. CHINTIROGLOU, 1995. A review of the carrying behaviour in brachyuran crabs, with additional information on the symbioses with sea anemones. Raffles Bulletin of Zoology, **43** (2): 377-416, pls. 1-3.

GUINOT, D. & B. RICHER DE FORGES, 1995. Crustacea Decapoda Brachyura: Révision de la famille des Homolidae De Haan, 1839. In: A. CROSNIER (ed.), Résultats des Campagnes MUSORSTOM, **13**. Mémoires du Muséum national d'Histoire naturelle, Paris, **163**: 283-517.

JAMIESON, B. G. M., D. GUINOT & B. RICHER DE FORGES, 1995. Phylogeny of the Brachyura: evidence from spermatozoal ultrastructure (Crustacea, Decapoda). In: B. G. M. JAMIESON, J. AUSIO & J.-L. JUSTINE (eds.), Advances in spermatozoal phylogeny and taxonomy. Mémoires du Muséum national d'Histoire naturelle, Paris, **166**: 265-283.

1996

CHINTIROGLOU, C., D. DOUMENC & D. GUINOT, 1996. Anemone-carrying behaviour in a deep-water homolid crab (Brachyura, Podotremata). Crustaceana, **69** (1): 19-25.

DAVIE, P. J. F. & D. GUINOT, 1996. Two new freshwater crabs in *Australocarcinus* Davie, with remarks on Trogloplacinae Guinot and Goneplacidae MacLeay (Crustacea: Decapoda: Brachyura). Memoirs of the Queensland Museum, **39** (2): 277-287.

GUINOT, D., 1996. Le crabe. Le Règne Animal, **95**: 2257-2281.

JAMIESON, B. G. M. & D. GUINOT, 1996. Ultrastructure and phylogeny of crustacean spermatozoa with particular reference to the Brachyura. Oral presentation at the 2nd European Conference, Liège (Belgium), September 2-6, 1996.

JAMIESON, B. G. M. & D. GUINOT, 1996. Ultrastructure of the spermatozoon of *Australocarcinus riparius* (Crustacea: Brachyura: Goneplacidae: Trogloplacinae). Memoirs of the Queensland Museum, **39** (2): 289-296.

JAMIESON, B. G. M., D. GUINOT & B. RICHER DE FORGES, 1996. Contrasting spermatozoal ultrastructure in two thoracotreme crabs, *Cardisoma carnifex* (Gecarcinidae) and *Varuna litterata* (Grapsidae) (Crustacea: Brachyura). Invertebrate Reproduction and Development, **29** (2): 111-126.

NG, P. K. L., D. GUINOT & T. M. ILIFFE, 1996. Revision of the anchialine varunine crabs of the genus *Orcovita* Ng & Tomascik, 1994 (Crustacea: Decapoda: Brachyura: Grapsidae). Raffles Bulletin of Zoology, **44** (1): 109-134.

TAVARES, M. & D. GUINOT, 1996. Description de *Neopilumnoplax gervaini* sp. nov. des Caraibes (Crustacea, Decapoda, Brachyura, Goneplacidae). Bulletin du Muséum national d'Histoire naturelle, Paris, (4) **18** (A) (1-2): 225-232.

1997

GUINOT, D., 1997. Brachyuran crabs. In: D. DESBRUYÈRES & M. SEGONZAC (eds.), Handbook of deep-sea hydrothermal vent fauna: 208-214. (Editions Ifremer, Brest).

GUINOT, D., B. G. M. JAMIESON & C. C. TUDGE, 1997. Ultrastructure and relationships of spermatozoa of the freshwater crabs *Potamon fluviatile* and *Potamon ibericum* (Crustacea, Brachyura, Potamidae). Journal of Zoology, London, **241** (2): 229-244, pls. 1-3.

GUINOT, D. & B. RICHER DE FORGES, 1997. Affinités entre les Hymenosomatidae MacLay, 1838 et les Inachoididae Dana, 1851 (Crustacea, Decapoda, Brachyura). Zoosystema, **19** (2-3): 453-502.

GUINOT, D. & M. SEGONZAC, 1997. Description d'un crabe hydrothermal nouveau du genre *Bythograea* (Crustacea, Decapoda, Brachyura) et remarques sur les Bythograeidae de la dorsale du Pacifique oriental. Zoosystema, **19** (1): 121-149.

JAMIESON, B. G. M., D. GUINOT, C. C. TUDGE & B. RICHER DE FORGES, 1997. Ultrastructure of the spermatozoa of *Corystes cassivelaunus* (Corystidae), *Platepistoma nanum* (Cancridae) and *Cancer pagurus* (Cancridae) supports recognition of the Corystoidea (Crustacea, Brachyura, Hetertremata). Helgoländer Meeresuntersuchungen, **51** (1): 83-93.

MCLAY, C. L. & D. GUINOT, 1997. Ten arms meet ten legs: Decapoda (Mollusca: Cephalopoda: Sepioidea) spawn on Decapoda (Crustacea: Brachyura: Homolidae). Journal of Crustacean Biology, **17** (4): 692-694.

RICHER DE FORGES, B., B. G. M. JAMIESON, D. GUINOT & C. C. TUDGE, 1997. Ultrastructure of the spermatozoa of Hymenosomatidae (Crustacea: Brachyura) and the relationships of the family. Marine Biology, Berlin, **130** (2): 233-242.

1998

GUINOT, D., 1998. Diagnoses de Crustacés Décapodes Brachyoures. The crabs of Japan. ETI World Biodiversity Database, Linnaeus II, Amsterdam. 4th International Crustacean Congress, Amsterdam, July 1998.

GUINOT, D. & J.-M. BOUCHARD, 1998. Evolution of the abdominal holding systems of brachyuran crabs (Crustacea, Decapoda, Brachyura). Zoosystema, **20** (4): 613-694.

GUINOT, D., B. G. M. JAMIESON, B. RICHER DE FORGES & C. C. TUDGE, 1998. Comparative spermatozoal ultrastructure of the three dromiacean families exemplified by *Homolodromia kai* (Homolodromiidae), *Sphaerodromia lamellata* (Dromiidae) and *Dynomene tanensis* (Dynomenidae) (Podotremata: Brachyura). Journal of Crustacean Biology, **18** (1): 78-94.

NG, P. K. L. & D. GUINOT, 1998. *Geelvinkia darnei*, a new species of cavernicolous crab (Crustacea, Decapoda, Brachyura, Parathelphusidae) from Irian Jaya, Indonesia. Mémoires de Biospéologie, **24** (51): 181-182. [1997 volume, published 1998]

TUDGE, C. C., B. G. M. JAMIESON, M. SEGONZAC & D. GUINOT, 1998. Spermatozoal ultrastructure in three species of hydrothermal vent crab, in the genera *Bythograea*, *Austinograea* and *Segonzacia* (Decapoda, Brachyura, Bythograeidae). Invertebrate Reproduction and Development, **34** (1): 13-23.

1999

NG, P. K. L. & D. GUINOT, 1999. On a new species of deep-water crab of the genus *Progeryon* (Decapoda, Brachyura, Geryonidae) from Hawaii. Crustaceana, **72** (7): 685-692.

2000

GUINOT, D. & M. TAVARES, 2000. *Conchoedromia alcocki* Chopra, 1934: megalopa of *Conchoecetes artificiosa* (Fabricius, 1798) (Decapoda, Brachyura, Dromiidae). Journal of Crustacean Biology, **20** (Special Number 2): 301-309.

2001

GUINOT, D. & M. TAVARES, 2001. Une nouvelle famille de Crabes du Crétacé, et la notion de Podotremata Guinot, 1977 (Crustacea, Decapoda, Brachyura). Zoosystema, **23** (3): 507-546.

JUBERTHIE, C. & D. GUINOT, 2001. Iles Salomon (Solomon Islands): 2115-2119; and Guam: 2157-2159. In: C. JUBERTHIE & V. DECOU (eds.), Encyclopaedia Biospeologica, **3**. (Société de Biospéologie, Moulis (CNRS) & Académie Roumaine, Bucharest).

NG, P. K. L. & D. GUINOT, 2001. On the land crabs of the genus *Discoplax* A. Milne Edwards, 1867 (Crustacea: Decapoda: Brachyura: Gecarcinidae), with description of a new cavernicolous species from the Philippines. Raffles Bulletin of Zoology, **49** (2): 311-338.

2002

GUINOT, D. & R. CLEVA, 2002. D'Orbigny et les Crustacés: 35-43. In: P. TACQUET (ed.), Alcide d'Orbigny. Du Nouveau Monde... au passé du monde. (Nathan/VUEF et Muséum national d'histoire naturelle, Paris).

GUINOT, D. & R. CLEVA, 2002. Les Crustacés récoltés par d'Orbigny en Amérique du Sud et déposés au Muséum national d'Histoire naturelle, Paris. Comptes Rendus Palevol, Paris, **1** (7): 499-515.

GUINOT, D., L. A. HURTADO & R. VRIJENHOEK, 2002. New genus and species of brachyuran crab from the southern East Pacific Rise (Crustacea Decapoda Brachyura Bythograeidae). Comptes Rendus Biologies, Paris, **325** (11): 1143-1152.

2003

GUINOT, D. & L. A. HURTADO, 2003. Two new species of hydrothermal vent crabs of the genus *Bythograea* from the southern East Pacific Rise and from the Galapagos Rift (Crustacea Decapoda Brachyura Bythograeidae). Comptes Rendus Biologies, Paris, **326** (4): 423-439.

GUINOT, D. & M. TAVARES, 2003. A new subfamilial arrangement for the Dromiidae De Haan, 1833, with diagnoses and description of new genera and species (Crustacea, Decapoda, Brachyura). Zoosystema, **25** (1): 43-129.

2004

DEUVE, T., D. GUINOT & J.-M. BOUCHARD, 2004. Case 3291. Dromiidae Bonelli, 1810 (Insecta, Coleoptera, Caraboidea): proposed amendment of spelling to Dromiusidae to remove homonymy with Dromiidae De Haan, 1833 (Crustacea, Decapoda, Brachyura, Dromiacea). Bulletin of Zoological Nomenclature, **61** (4): 225-231.

GUINOT, D., 2004. Crabs of Japan and the adjacent seas. ETI World Biodiversity Database, Linnaeus II, Amsterdam. World Biodiversity Database. CD-ROM Series. K. SAKAI, (ed.).

2005

GUINOT, D. & G. QUENETTE, 2005. The spermatheca in podotreme (Crustacea, Decapoda, Brachyura, Podotremata) and its phylogenetic implications. Zoosystema, **27** (2): 267-342.

GUINOT, D., G. D. F. WILSON & F. R. SCHRAM, 2005. Jurassic Isopod (Malacastroca: Peracarida) from Ranville, Normandy, France. Journal of Paleontology, **79** (5): 954-960.

2006

GUINOT, D., 2006. Brachyuran crabs. In: D. DESBRUYÈRES, M. SEGONZAC & M. BRIGHT (eds.), Handbook of deep-sea hydrothermal vent fauna (Second edition): 473, figs. 1-6. (Denisia, **18**).

GUINOT, D., 2006. Rediscovery of the holotype of *Paeduma cylindricum* (Bell, 1859) and description of a new genus of Hexapodidae (Decapoda, Brachyura). Zoosystema, **28** (2): 553-571.

GUINOT, D. & G. BRETON, 2006. *Lithophylax trigeri* A. Milne-Edwards & Brocchi, 1879 from the French Cretaceous (Cenomanian) and placement of the family Lithophylacidae Van Straelen, 1936 (Crustacea, Decapoda, Brachyura). Geodiversitas, **28** (4): 591-633.

GUINOT, D. & L. HURTADO, 2006. Brachyuran crabs. In: D. DESBRUYÈRES, M. SEGONZAC & M. BRIGHT (eds.), Handbook of deep-sea hydrothermal vent fauna (Second edition): 465, figs. 1-3. (Denisia, **18**).

GUINOT, D., L. HURTADO & R. C. VRIJENHOEK, 2006. Brachyuran crabs. In: D. DESBRUYÈRES, M. SEGONZAC & M. BRIGHT (eds.), Handbook of deep-sea hydrothermal vent fauna (Second edition): 459, fig. 1-3. (Denisia, **18**).

GUINOT, D. & M. SEGONZAC, 2006. Brachyuran crabs. In: D. DESBRUYÈRES, M. SEGONZAC & M. BRIGHT (eds.), Handbook of deep-sea hydrothermal vent fauna (Second edition): 455, 460-461, 466-472, un. figs. (Denisia, **18**).

2007

CLEVA, R., D. GUINOT & L. ALBENGA, 2007. Annotated catalogue of brachyuran type specimens (Crustacea, Decapoda, Brachyura) deposited in the Muséum national d'Histoire naturelle, Paris. I. Podotremata. Zoosystema, **29** (2): 229-279.

GUINOT, D., 2007. A new genus of the family Plagusiidae Dana, 1851, close to *Plagusia* Latreille, 1804 (Crustacea, Decapoda, Brachyura). Zootaxa, **1498**: 27-33.

GUINOT, D., A. DE ANGELI & A. GARASSINO, 2007. Discovery of the oldest eubrachyuran crab from the middle Jurassic (Bathonian) of Normandy (France). Memorie della Società Italiana di Scienze Naturali e del Museo Civico di Storia Naturale di Milano, **35** (2): 53-55.

GUINOT, D., A. DE ANGELI & A. GARASSINO, 2007. *Hebertides jurassica* n. gen., n. sp. (Crustacea, Decapoda, Brachyura) from the Middle Jurassic (Bathonian) of Normandy (France). Atti della Società Italiana di Scienze Naturali e del Museo Civico di Storia Naturale di Milano, **148** (2): 241-260.

GUINOT, D. & P. CASTRO, 2007. A new species of *Goneplax* Leach, 1814 (Crustacea, Decapoda, Brachyura, Goneplacidae) from the south Atlantic and the western limits of the Indo-west Pacific region, long confused with *G. rhomboides* (Linnaeus, 1758). Zootaxa, **1577**: 17-31.

NOËL, P. & D. GUINOT, 2007. Non-indigenous freshwater crabs in France: a new occurrence of a potamid near Nice: 77-90. In: F. GHERARDI (ed.), Invading nature, **2**: Biological invaders in inland waters: profiles, distribution, and threats. (Springer, Netherlands).

2008

ARTAL, P., D. GUINOT, B. VAN BAKEL & J. CASTILLO, 2008. Ibericancridae, a new dakoticancroid family (Decapoda, Brachyura, Podotremata) from the upper Campanian (Upper Cretaceous) of Spain. Zootaxa, **1907**: 1-27.

GUINOT, D., 2008. A re-evaluation of the Dynomenidae (Decapoda, Brachyura, Podotremata), with the recognition of four subfamilies. Zootaxa, **1850**: 1-26.

GUINOT, D., A. DE ANGELI & A. GARASSINO, 2008. Marocarcinidae, a new eubrachyuran family, and *Marocarcinus pasinii* n. gen. n. sp. from the Upper Cretaceous Cenomanian-Turonian) of Gara Sbaa, southeastern Morocco (Crustacea Decapoda Brachyura). Atti della Società Italiana di Scienze Naturali e del Museo Civico di Storia Naturale di Milano, **149** (1): 25-36, figs. 1-3.

GUINOT, D., F. VEGA & B. W. M. VAN BAKEL, 2008. Cenomanocarcinidae, a new Cretaceous podotreme family (Crustacea, Decapoda, Brachyura, Raninoida), with comments on related families. Geodiversitas, **30** (4): 681-719, figs. 1-9.

NARUSE, T., P. K. L. NG & D. GUINOT, 2008. Two new genera and two new species of troglobitic false spider crabs (Crustacea: Decapoda: Brachyura: Hymenosomatidae) from Indonesia, with notes on *Cancrocaeca* Ng, 1991. Zootaxa, **1739**: 21-40.

NG, P. K. L., D. GUINOT & P. J. F. DAVIE, 2008. How many living crab species are there? Update on the Brachyuran Crab Project. In: Abstracts, Advances in Decapod Crustacean Phylogenetics: 73. A Symposium during the Society of Integrative and Comparative Biology (SICB) Annual Meeting, San Antonio, Texas, U.S.A., January 2-6, 2008.

NG, P. K. L., D. GUINOT & P. J. F. DAVIE, 2008. Systema Brachyurorum: Part I. An annotated checklist of extant brachyuran crabs of the world. Raffles Bulletin of Zoology, (Supplement) **17**: 1-286.

2009

CASTRO, P., D. GUINOT & P. K. L. NG, 2009. A new family for *Sotoplax robertsi* Guinot, 1984, with a diagnosis and key to the Goneplacoidea MacLeay, 1838 (Crustacea: Decapoda: Brachyura). Zootaxa, **2356**: 36-56.

GUINOT, D. & R. CLEVA, 2009. Publication préliminaire des nouvelles explications des planches de « Crustacés Décapodes » de la *Description de l'Égypte* (planches 1 à 10) dessinées sous la direction de Jules-César Savigny et commentées sommairement à l'origine par Victor Audouin, et explication des espèces inédites de Crustacés figurant dans l'« Atlas d'Histoire naturelle » de Savigny (volume 4, planches 52, 54, 61 à 66), accompagnées de précisions et commentaires scientifiques et historiques. In: E. IINUMA & N. M. SIDHOM (eds.), Collection Nouvelle Description de l'Égypte: i-xi, 1-195. (Institut d'Orient, Paris).

HOLOTHERES DANIELAE, A NEW SPECIES OF PINNOTHERID CRAB FROM THE INDO-WEST PACIFIC (DECAPODA, BRACHYURA), WITH A KEY TO THE GENUS

BY

SHANE T. AHYONG[1])
Marine Biodiversity & Biosecurity, National Institute of Water and Atmosphere, Private Bag 14901, Kilbirnie, Wellington, New Zealand

ABSTRACT

Holotheres danielae sp. nov., a symbiont of holothurians, is described from Roebuck Bay, Western Australia. It is the sixth species of the genus and represents the first record of the genus from Australia. The new species is most similar to *H. halingi* (Hamel, Ng & Mercier, 2003) from the Solomon Islands, but differs in the form of the maxilliped 3 ischiomerus and the dactyli of the walking legs. A key to the species of *Holotheres* is provided.

RÉSUMÉ

Holotheres danielae sp. nov., une espèce symbiotique des holothuries, est décrite de Roebuck Bay, Western Australia. C'est la sixième espèce du genre et représente le première registre de ce genre en Australie. La nouvelle espèce est proche de *H. halingi* (Hamel, Ng & Mercier, 2003) des îles Salomon, mais en diffère par la forme de l'ischiomerus du 3ème maxillipède et celles des dactyles des pattes marcheuses. Une clef des espèces d'*Holotheres* est présentée.

INTRODUCTION

Ng & Manning (2003) established *Holotheres* for five species of pinnotherid crabs that are obligate symbionts of holothurians: *H. semperi* (Bürger, 1895), *H. flavus* (Nauck, 1880), *H. setnai* (Chopra, 1931), *H. halingi* (Hamel, Ng & Mercier, 2003), and *H. villosissimus* (Doflein, 1904) (see also Ng et al.,

[1]) e-mail: s.ahyong@niwa.co.nz

© Koninklijke Brill NV, Leiden, 2010

2008). Most species of *Holotheres* are very similar in overall morphology, although *H. villosissimus* is aberrant in having long, shaggy carapace setae (in contrast to short and fine) and in the structure of maxilliped 3 in which the inner margin of the ischiomerus is distinctly angled in the distal quarter (rather than at, or proximal to the midlength in congeners) and in which the dactylus and propodus are proportionally much shorter. The taxonomic position of *H. villosissimus* requires further scrutiny, and, as remarked by Ng & Manning (2003), this poorly known species requires a redescription.

The Australian pinnotherid crab fauna was most recently enumerated by Ahyong & Brown (2003), who recognized 12 species and six genera of the Pinnotherinae from Australian waters. A new species of *Holotheres* from Roebuck Bay, Western Australia, is described herein, adding an additional genus and species to the Australian pinnotherid inventory.

MATERIAL AND METHODS

Morphological terminology generally follows Ahyong & Ng (2007). Measurements are in millimetres (mm). Carapace length (cl) is measured along the midline. Carapace width (cw) is the greatest width. The specimens used in this study are deposited in the Australian Museum, Sydney (AM), and the Zoological Reference Collection, Raffles Museum of Biodiversity Research, National University of Singapore (ZRC).

TAXONOMY

Family PINNOTHERIDAE De Haan, 1833

Genus *Holotheres* Ng & Manning, 2003

Holotheres danielae sp. nov. (fig. 1A-H)

Type material. — AM P64675, spent female holotype (cw 6.5 mm, cl 6.6 mm), Roebuck Bay, Broome, Western Australia, from holothurians (*Haplodactyla* sp.), dredged.

Description of holotype. — Carapace (fig. 1A) slightly longer than wide; front slightly produced, anterior margin sinuous; lateral margins subtruncate, subparallel; posterior margin projecting, rounded; margins setose, dorsal surface smooth, glabrous or with short, fine, sparse setation. Antennular sinus

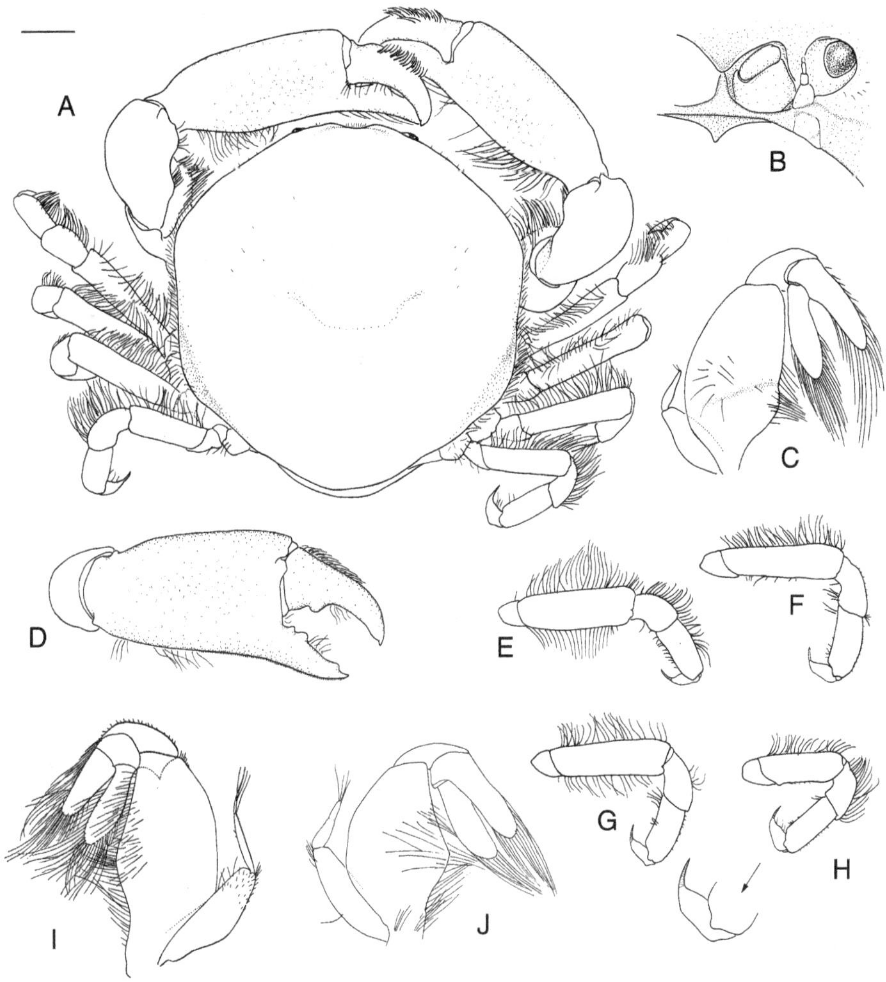

Fig. 1. A-H, *Holotheres danielae* sp. nov., female holotype, cl 6.6 mm, cw 6.5 mm (AM P64675). A, dorsal habitus; B, left cephalothorax, anterior view; C, right maxilliped 3, outer view; D, right chela; E-H, right pereopods 2-5; I, *H. setnai* (Chopra, 1931), right maxilliped 3, oral view [after Chopra, 1931, fig. 5]; J, *H. halingi* (Hamel, Ng & Mercier, 1999), right maxilliped 3, outer view, ovigerous female paratype, cl 10.1 mm, cw 10.7 mm (ZRC 1999.1229). Scale: A, D-H = 1.0 mm, B, C = 0.5 mm, I, J = 0.75 mm.

wider than orbit (fig. 1B); antennules folded slightly obliquely. First two antennal segments fused to epistome. Eyes partially visible in dorsal view, filling orbit, cornea pigmented.

Maxilliped 3 (fig. 1C) ischiomerus length about twice width; mesial margin obtusely rounded proximal to midlength; central mesial margin setose; outer margin strongly convex. Carpus shorter than propodus. Propodus digitiform, distally setose. Dactylus inserting at base of propodus; elongate, subspatulate, widest at midlength, gently tapering to blunt, setose apex; apex fractionally over-reaching propodal apex. Exopod inner and outer margins convex, sparsely setose; distal segment setose.

Chelipeds (fig. 1D) symmetrical. Dactylus and pollex relatively straight, crossing distally when fingers closed. Dactylus dorsally setose; occlusal margin sparsely setose, with blunt proximal tooth. Pollex occlusal margin with low distal and 2 low proximal teeth; pollex with fringe of short setae on inner ventral margin. Propodus dorsal margin about 1.4 times as long as dactylus, about 1.5 times height; ventral margin faintly sinuous, concave at base of pollex; setose on inner proximal margin. Carpus inner margin with tuft or short setae.

Walking legs (pereopods 2-5) (fig. 1E-H) similar in form, symmetrical from left to right, segments variously setose on flexor and extensor margins; relative lengths in decreasing order pereopod 2 > pereopod 3 > pereopod 4 > pereopod 5. Dactyli subequal, broad, thick proximally, tapering abruptly to spiniform corneous unguis. Pereopod 2 dactylus 0.7 propodus length; propodus length 2.3 height, 1.3 carpus length; merus length 3.1 height, 2.5 carpus length. Pereopod 3 sparsely setose; dactylus 0.7 propodus length; propodus length 2.3 height, 1.4 carpus length; merus length 3.3 height, 2.1 carpus length. Pereopod 4 dactylus 0.6 propodus length; propodus length 2.4 height, 1.3 carpus length; merus length 3.5 height, 2.3 carpus length. Pereopod 5 dactylus 0.7 propodus length; propodus length 2.5 height, 1.3 carpus length; merus length 3.0 height, 1.8 carpus length.

Abdomen with all somites freely articulating, widest at somite 5, covering bases of walking legs.

Host. — The host was originally identified as *Haplodactyla* sp. The only Australian holothurian previously known under this generic name is currently known as *Acaudina molpadioides* (Semper, 1868), which occurs off northwestern Australia, including Broome, from 0-54 m (Rowe & Gates, 1995). Thus, *A. molpadoides* is likely to be the host of *H. danielae*. The collection date was not recorded, but the hand writing is similar to that of A. Livingstone, who made collections from Roebuck Bay and environs in 1929.

Etymology. — Named in honour of Danièle Guinot, esteemed colleague and friend, for her monumental contribution to the systematics of the Brachyura.

Remarks. — Of the known species of *Holotheres*, *H. danielae* sp. nov. most closely resembles *H. halingi* from the Solomon Islands in general habitus, carapace shape (subtruncate and subparallel lateral margins; prominently projecting posterior margin), very sparsely setose carapace surface, and setation pattern of the walking legs (setae primarily confined to the flexor and extensor margins). *Holotheres danielae* differs from *H. halingi* in the shapes of the maxilliped 3 ischiomerus and walking leg dactyli. The maxilliped 3 ischiomerus of *H. danielae* (fig. 1C) is also proportionally more slender than in *H. halingi* (fig. 1J), being about twice as long as broad and without a prominently angled mesial margin, contrasting with a length that is distinctly less than twice the breadth, with a prominent, obtuse, mesial angle. The walking leg dactyli of *H. danielae* are abruptly narrowed in the distal half, with styliform apices, rather than evenly tapering as in *H. halingi*. In the form of the maxilliped 3 ischiomerus, *H. danielae* also resembles *H. setnai* (fig. 1I) from the Andaman Sea, but the two species are readily distinguished by the subtruncate rather than evenly rounded lateral carapace margins, and shorter maxilliped 3 dactylus, which only fractionally, rather than distinctly over-reaches the propodal apex. The known species of *Holotheres* can be distinguished by the key below.

Distribution. — At present known only from Roebuck Bay, Broome, Western Australia.

KEY TO SPECIES OF *HOLOTHERES*

1. Dorsal and lateral surfaces of carapace entirely and densely covered by long, shaggy, plumose setae. Maxilliped 3 ischiomerus obtusely angled at distal quarter of inner margin . . . . . *H. villosissimus*
– Dorsal and lateral surfaces of carapace covered at most with short tomentum. Maxilliped 3 ischiomerus obtusely angled at midlength or on proximal half of inner margin 2

2. Walking legs covered with setae. Carapace setose over entire surface 5
– Walking legs setose on extensor and flexor margins. Carapace setose marginally; sparsely setose or glabrous centrally 3

3. Dactyli of walking legs tapering abruptly to slender spiniform unguis. Maxilliped 3 ischiomerus twice as long as wide . . . . . *H. danielae*
– Dactyli of walking legs tapering evenly to spiniform apex. Maxilliped 3 ischiomerus distinctly less than twice as long as wide 4

4. Pereopod 2 longest of walking legs . . . . . *H. halingi*
– Pereopod 3 longest of walking legs . . . . . *H. flavus*

5. Ischiomerus of maxilliped 3 more than twice as long as wide . . . . . *H. setnai*
– Ischiomerus of maxilliped 3 less than twice as long as wide . . . . . *H. semperi*

ACKNOWLEDGEMENTS

Grateful thanks are extended to Stephen Keable (Australian Museum) for the loan of specimens. This study was partially supported by the NIWA

Capability Fund and New Zealand Foundation for Research Science and Technology (C01X0502) and a fellowship from the National University of Singapore.

REFERENCES

AHYONG, S. T. & D. E. BROWN, 2003. Description of *Durckheimia lochi* n. sp., with an annotated checklist of Australian Pinnotheridae (Crustacea: Decapoda: Brachyura). Zootaxa, **254**: 1-20.

AHYONG, S. T. & P. K. L. NG, 2007. The pinnotherid type material of Semper (1880), Nauck (1880) and Bürger (1895) (Crustacea: Decapoda: Brachyura). Raffles Bulletin of Zoology, (Supplement) **16**: 191-226.

BÜRGER, O., 1895. Ein Beitrag zur Kenntniss der Pinnotherinen. Zoologische Jahrbücher (Abtheilung für Systematik, Geographie und Biologie der Thiere), **8**: 361-390.

CHOPRA, B., 1931. Further notes on Crustacea Decapoda in the Indian Museum. II. On some decapod Crustacea found in the cloaca of holothurians. Records of the Indian Museum, **33**: 303-325.

DOFLEIN, F., 1904. Brachyura. In: Wissenschaftliche Ergebnisse der Deutschen Tiefsee-Expedition auf dem Dampfer "Valdivia", 1898-1899, **6**: i-xiv, 1-314, figs. 1-68, pls. 1-57. (Verlag von Gustav Fischer, Jena).

HAAN, W. DE, 1833-1849. Crustacea. In: P. F. VON SIEBOLD (ed.), Fauna Japonica, sive descriptio animalium, quae in itinere per Japoniam, jussu et auspiciis superiorum, qui summum in India Batavia imperium tenent, suscepto, annis 1823-1830 collegit, notis, observationibus a adumbrationibus illustravit, **1-8**: i-xxi, vii-xvii, ix-xvi, 1-243, pls. 1-55, A-Q, circ. pl. 2. (Lugduni Batavorum = Leiden).

HAMEL, J.-F., P. K. L. NG & A. MERCIER, 1999. Life cycle of the pea crab *Pinnotheres halingi* sp. nov., an obligate symbiont of the sea cucumber *Holothuria scabra* Jaeger. Ophelia, **50** (3): 149-175.

NAUCK, E., 1880. Das Kaugerüst der Brachyuren (mit Beschreibung neuer Gattungen und Arten, z. T. von C. Semper). Zeitschrift für Wissenschaftliche Zoologie, Leipzig, **34** (1): 1-69, pl. 1.

NG, P. K. L., D. GUINOT & P. J. F. DAVIE, 2008. Systema Brachyurorum, Part I. An annotated checklist of extant brachyuran crabs of the world. Raffles Bulletin of Zoology, (Supplement) **17**: 1-286.

NG, P. K. L. & R. B. MANNING, 2003. On two new genera of pea crabs parasitic in holothurians (Crustacea: Decapoda: Brachyura: Pinnotheridae) from the Indo-West Pacific, with notes on allied genera. Proceedings of the Biological Society of Washington, **116** (4): 901-919.

ROWE, F. W. E. & J. GATES, 1995. Echinodermata. In: A. WELLS (ed.), Zoological catalogue of Australia, **33**: 1-510. (CSIRO Publishing, Melbourne).

First received 18 June 2009.
Final version accepted 25 June 2009.

LOERENTHOPLUMA DANIELAE, A NEW CRAB (DECAPODA, BRACHYURA, RETROPLUMIDAE) FROM THE LOWER EOCENE OF NORTHWEST BELGIUM

BY

BARRY W. M. VAN BAKEL[1,2,5]), PEDRO ARTAL[3,6]), RENÉ H. B. FRAAIJE[1,7]) and JOHN W. M. JAGT[4,8])

[1]) Oertijdmuseum De Groene Poort, Bosscheweg 80, NL-5283 WB Boxtel, Netherlands
[2]) Nationaal Natuurhistorisch Museum (Naturalis), P.O. Box 9517, NL-2600 RA Leiden, Netherlands
[3]) Museo Geológico del Seminario de Barcelona, Diputación 231, E-08007 Barcelona, Spain
[4]) Natuurhistorisch Museum Maastricht, de Bosquetplein 6-7, NL-6211 KJ Maastricht, Netherlands

ABSTRACT

A new species of retroplumid crab, *Loerenthopluma danielae*, is described from the Egemkapel Clay Member (Tielt Formation; lower Eocene, Ypresian) at the Ampe sand and clay pit near Egem, northwest Belgium. It is characterized by prominent outer orbital spines, wide orbits and a narrow, spatulate rostrum.

RÉSUMÉ

Une nouvelle espèce de crabe Retroplumidae, *Loerenthopluma danielae*, est décrit des argiles de Egemkapel (Formation de Tielt; Eocene inférieur, Yprésien), de la carrière d'argile et de sable d'Ampe près de Egem, au nord-ouest de la Belgique. Ce nouveau crabe est caractérisé par des épines orbitales externes proéminentes, des orbites larges et un rostre étroit et spatulé.

INTRODUCTION

Paleogene and Neogene crustaceans are fairly common in Belgium (see Van Bakel et al., 2006; Jagt et al., 2007). A number of new taxa have been described

[5]) Corresponding author; e-mail: barryvanbakel@gmail.com
[6]) e-mail: partal@optimus.es
[7]) e-mail: info@oertijdmuseum.nl
[8]) e-mail: john.jagt@maastricht.nl

© Koninklijke Brill NV, Leiden, 2010

subsequently (Fraaije et al., 2007; Van Bakel et al., 2009a, b). The Ampe sand and clay pit near Egem (West-Vlaanderen province), where specimens are commonly collected as a "by-product" of screening and wet sieving of sediment for shark teeth, has yielded the richest crustacean fauna. So far, only a single species from this locality, *Upogebia lambrechtsi* Fraaije, Van Bakel, Jagt & Coole, 2007, has been described and formally named (Fraaije et al., 2007). Associated lobsters and crabs are *Hoploparia* sp., *Linuparus scyllariformis* (Bell, 1857), *Silvacarcinus laurae* Collins & Smith, 1993, *Glyphithyreus wetherelli* (Bell, 1858), *Raninoides gottschei* (Böhm, 1928), *Goniochele* n. sp., *Cyclocorystes* cf. *pulchellus* (Bell, 1858), and *Orthakrolophos* n. sp. A new genus and species of stomatopod is also represented. Here we describe a new retroplumid, *Loerenthopluma danielae* n. sp., based on a dozen well-preserved specimens.

Two extant and six fossil genera are currently recognized in the primitive eubrachyuran family Retroplumidae (cf. Beschin et al., 1996; Feldmann & Portell, 2007: 90; Ng et al., 2008: 181). The type genus, *Retropluma* Gill, 1894, first occurs in the early Eocene and continues to the present. Its fossil representatives appear to have preferred soft-bottom shelf settings (Artal et al., 2006: 65).

One of the Egem specimens retains a well-preserved ventral side with sternites, abdomen, and pereiopods; this was prepared using a technique applied and described as "negative preparation" by Collins & Jakobsen (2004: 67). The crab remains, inclusive of the cuticle, were removed with pneumatic tools and finally with a sharp needle. The remaining cavity was filled with silicone rubber, which resulted in a detailed cast. It is, thus, possible to examine fragile details such as the thin pereiopods, including the reduced P5, which normally are not preserved in fossils. Specimens were darkened with black water colour and subsequently coated with ammonium chloride prior to photography.

SYSTEMATIC PART

Abbreviations used to denote the repository of specimens referred to in the text are as follows: MAB, Oertijdmuseum De Groene Poort, Boxtel, Netherlands; MGSB, Museo Geológico del Seminario de Barcelona, Barcelona, Spain.

Infraorder BRACHYURA Linnaeus, 1758

Section EUBRACHYURA de Saint Laurent, 1980

Superfamily RETROPLUMOIDEA Gill, 1894

Family RETROPLUMIDAE Gill, 1894

Genus **Loerenthopluma** Beschin, Busulini, De Angeli & Tessier, 1996

Type species. — *Loerenthopluma lata* Beschin, Busulini, De Angeli & Tessier, 1996, by original designation.

Emended diagnosis. — Small-sized carapace (width ranging between 7 and 32 mm), subrectangular in outline; orbitofrontal margin notably broad, distinctly sinuous; orbits wide; rostrum narrowing posteriorly; 3 transverse ridges cross the entire carapace surface, anterior one slightly sinuous, median ridge defined by 2 markedly straight, oblique crests interrupted by broad urogastric lobe, posterior one discontinuous on account of shallow branchiocardiac grooves, with median portion displaced posteriorly. Sternum broad, sternites 5-7 crossed by well-developed oblique ridges, sternite 8 extremely reduced; abdominal cavity deep, border well defined, male abdomen triangular, narrow, with all pleomeres free, transversely ridged, telson elongate; abdominal holding represented by moderate projection on sternite 5, salient posterior portion of pleomere 6; pereiopods 5 strongly reduced.

Remarks. — The genus is considered to be close to *Retropluma*, which ranges from the lower Eocene to the present day. *Loerenthopluma* is distinguished from *Retropluma* by a very narrow, long rostrum; very broad orbitofrontal margins bounded by prominent outer orbital corners; wide orbits; and a posteriorly displaced median portion of the posterior dorsal ridge.

Loerenthopluma danielae n. sp. (fig. 1A-G)

Types. — The holotype, a well-preserved carapace with remains of the abdomen and pereiopods, is MAB k.2550 (ex E. Wille Collection); carapace width and length (inclusive of rostrum) 9 and 7.5 mm, respectively. There are twelve paratypes (MAB k.2551A-L); all fragmentary carapaces, some retaining venters. MAB k.2551M is a small nodule with a natural mould of a ventral side, of which MAB k.2551N is the corresponding silicone rubber cast. All specimens are from the basal portion of the Egemkapel Clay Member (Tielt Formation), of early Eocene (Ypresian) age, as exposed at the Ampe sand and clay pit, Egem (northwest Belgium). For further details, reference to Fraaije et al. (2007) is made.

Etymology. — The species is named in honour of Danièle Guinot (Muséum national d'Histoire naturelle, Paris), renowned carcinologist and a good friend.

Diagnosis. — As for the genus.

Description. — Carapace subrectangular in outline, small sized, wider than long (length/width ratio ca. 0.7), with lateral, posterior margins gently arched; greatest width between median, posterior dorsal ridges. Carapace steeply convex longitudinally, nearly flat transversely; orbitofrontal margin broad, dorsal border strongly sinuous with blunt supraorbital nodes clearly projected anteriorly, bounded by salient outer orbital spines that are slightly directed outwards. Advanced front, broad at base, narrowing in median portion of spatulate rostrum. Anterolateral margins short, straight, posterolateral ones longer, broadly arched, densely covered with granules. First portion of lateral margins meets ventral regions at acute angle, posterior portion having vertical flanks, inclined at about right angles from dorsal surface. Posterior margin gently convex with horizontal median portion, in size similar to, or somewhat shorter than, orbitofrontal margin. Dorsal regions of carapace weakly defined, dorsal surface transversely ridged. Anterior ridge gently sinuous, slightly arched in central portion, curving downwards at lateral margins; median ridge, broadly interrupted at level of urogastric region, formed by 2 fairly straight, oblique lateral ridges; posterior ridge, subparallel to anterior one, markedly divided by shallow branchiocardiac grooves, with central portion slightly displaced towards posterior margin. Mesogastric region broad, distinctly inflated, rhomboid, separated from protogastric lobes by shallow grooves, posterolaterally delimited by short longitudinal depressions. Gastric pits visible as short, convergent slits. Urogastric region narrow, crescent-shaped, marked only by subtle depressions. Cardiac region large, bounded by weak depressions, entirely crossed by median portion of posterior ridge, with 2 subtle, granular swellings positioned laterally below crest. Intestinal region slightly inflated, suboval in shape, extending towards posterior margin. Ventral regions of carapace (subhepatic, pterygostome) strongly ornamented, marked by deep, oblique groove, followed by longitudinal swelling covered with granules. Dorsal surface of carapace covered by delicate pits not uniformly distributed, scattered over entire surface; small granules or denticles along lateral portions of carapace. Sternum nearly completely preserved, anterior portion (sternites 2-4) subtrapezoidal, posterior (sternites 5-7) subrectangular; phragmal divisions extraordinarily well preserved (see fig. 1C). Sternite 2 weakly ridged, axially depressed; sternites 3-4 fused, large, with small projection in front of large gynglyme for insertion of chelae. Sternites 5-7 bear oblique, weakly granulated ridges. Sternite 8 is very much reduced, tilted in different plane than preceding sternites. Abdominal cavity well defined, deep, rimmed at sternite 3-4, strongly sloping at lateral walls. Male abdomen with all pleomeres free, markedly triangular,

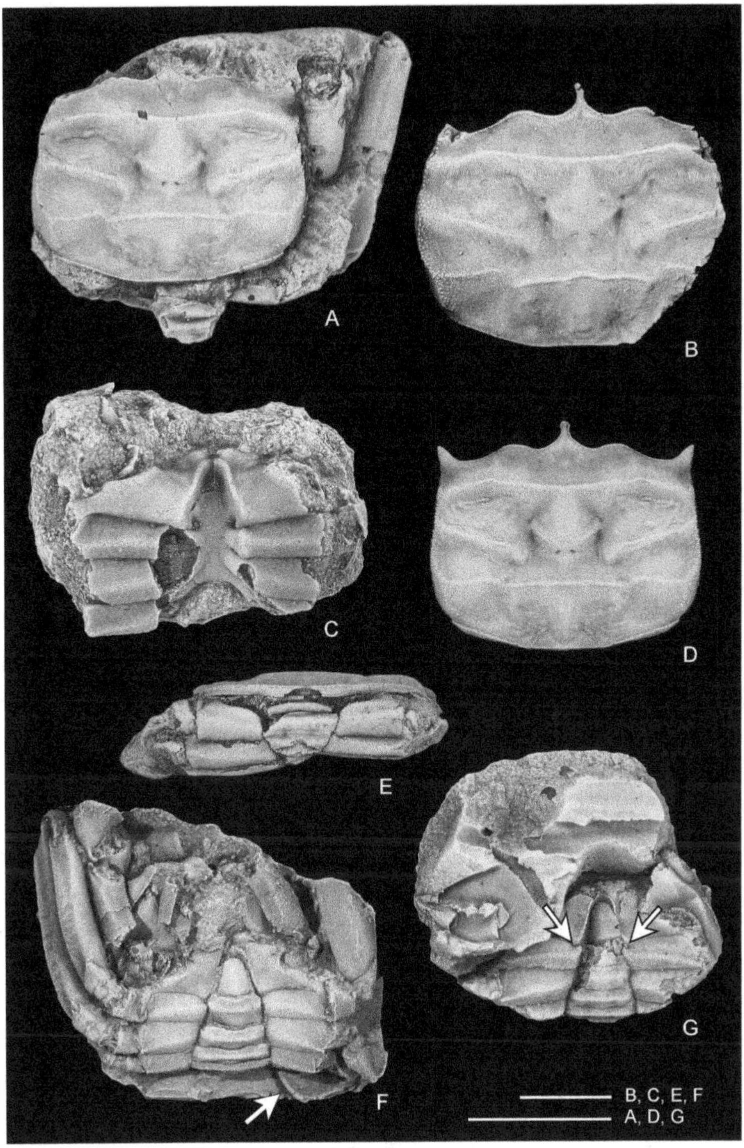

Fig. 1. *Loerenthopluma danielae* n. sp. from the basal portion of the Egemkapel Clay Member (Tielt Formation; lower Eocene, Ypresian) at the Ampe sand and clay pit near Egem, province of West-Vlaanderen (northwest Belgium). Scale bars represent 5 mm. A, D, holotype, MAB k.2550, dorsal view and digital reconstruction of the carapace, respectively; B, paratype, MAB k.2551A, dorsal view of carapace; C, paratype, MAB k.2551C, ventral view showing sternum; E, paratype, MAB k.2551D, posterior view showing abdomen and sternites; F, paratype, MAB k.2551N, silicone rubber cast of natural mould, showing well-preserved ventral side with abdomen, sternites, and pereiopods: arrow indicates merus of reduced P5; G, paratype, MAB k.2551J, ventral view of male specimen: arrows indicate abdominal holding on sternite 5.

transversely ridged. Pleomere 6 longer, broadened posteriorly as mechanism for abdominal holding. Telson long, subtrapezoidal in shape with rounded tip. Chelipeds only partially preserved, left palm higher distally, inner surface with protuberance at articulation with dactylus; fingers long, slender, flattened, with crest. P2, P3 partially preserved; coxae short, basis-ischium fused with suture visible, merus long, with flattened lower face; surface of articles granular. P5 preserved in silicone rubber mould; much reduced in size, with thin, arched merus.

Discussion. — The new species differs from the type, *Loerenthopluma lata*, in having the greatest carapace width more posteriorly, much more prominent outer orbital corners, a wider base of the rostrum, and clearly granular lateral margins. In addition, distinct gastric pits occur (unknown in *L. lata*; however, this needs to be verified when better-preserved material of the type species becomes available).

Loerenthopluma is considered to be closely related to *Retropluma*. The first described extinct representative of *Retropluma*, *R. eocenica*, was erected by Via (1959; see also Via, 1969, 1980), who also recognized that a species recorded by Crema (1895) as *Goneplax craverii* was congeneric. Artal et al. (2006) have recently emended Via's (1959) description of the type species. De Saint Laurent (1989) and McLay (2006a) revised all extant species of the genus.

Five fossil species of *Retropluma* (see below) are so far known; *Loerenthopluma danielae* n. sp. is easily distinguished from all of these by the shape and position of the dorsal ridges, the wide orbitofrontal margin with prominent outer orbital corners, as well as the long and narrow rostrum. The new species is apparently close to *Retropluma gallica* Artal, Van Bakel & Castillo, 2006 from the lower Eocene (Ypresian) of Fontcouverte (Aude, southern France), with which it shares the advanced supraorbital nodes, oblique median carapace ridges, and a discontinuous posterior carina. However, the French species is more subquadratic in outline (length/width ratio approximately 0.8), has a narrower orbitofrontal margin, and the median portion of the posterior carina is less posteriorly displaced.

Retropluma eocenica Via, 1959, from the lower and middle Eocene (Ypresian and Lutetian) of Spain and northern Italy (see Via, 1959; Beschin et al., 1996; Artal et al., 2006), has a narrower orbitofrontal margin, a straighter anterior dorsal ridge, less oblique median dorsal ridges, as well as a continuous posterior dorsal ridge. *Retropluma* cf. *eocenica* was described from the Oligocene (Rupelian) of the Turin area, northwest Italy, by Larghi (2004).

Retropluma laurentae Collins, Lee & Noad, 2003, from the Miocene of Sabah (Indonesia), is more suboval in outline and has a narrower orbitofrontal margin; moreover, the posterior carapace ridge is not discontinuous but interrupted by cardiac tubercles (Collins et al., 2003).

Retropluma borealis Fraaije, Hansen & Hansen, 2005, from the upper Miocene of Denmark (Fraaije et al., 2005), has straighter parallel lateral margins, a narrower orbitofrontal margin, and straighter dorsal ridges, with the posterior ridge entire, continuing from side to side.

Retropluma craverii (Crema, 1895), from the Pliocene of northwest Italy, differs in having the most subquadratic outline. According to Larghi (2004: 59), the length/width ratio is approximately 0.9. This species also differs in having less distinct median ridges as well as a robust node at the intersection of the anterior carina with the lateral margins.

All extant members of the genus, including the recently described *Retropluma solomonensis* McLay, 2006b (= *R. laurentae* McLay, 2006a [non Collins, Lee & Noad, 2003]), differ from *Loerenthopluma danielae* n. sp. in having a more subquadratic carapace outline and a distinctly narrower orbital margin.

ACKNOWLEDGEMENTS

We thank Danièle Guinot for her kind assistance while examining extant members of the genus *Retropluma* contained in the collections of the Muséum national d'Histoire naturelle (Paris). Gérard Breton kindly translated the abstract into French. For donation of fossil material during recent years, we thank Yoeri Christiaens, Yvonne Coole, Jacques Herman, Arne Iserbyt, the Tak family, and Eric Wille. We thank Dr. Peter Castro and an anonymous reviewer for useful comments on the manuscript.

REFERENCES

ARTAL, P., B. W. M. VAN BAKEL & J. CASTILLO, 2006. *Retropluma* Gill, 1894 (Crustacea, Decapoda) from the Eocene of the eastern Pyrenees (Spain, France). Cainozoic Research, **5** (2005): 63-69.

BAKEL, B. W. M. VAN, P. ARTAL, R. H. B. FRAAIJE & J. W. M. JAGT, 2009a. A new Oligocene crab (Decapoda, Brachyura, Dromiacea) from northwest Belgium, with comments on its palaeobiology. Geologica Belgica, **12**: 45-57.

BAKEL, B. W. M. VAN, R. H. B. FRAAIJE & J. W. M. JAGT, 2006. Synopsis of Cenozoic decapod crustaceans from Belgium. In: F. J. VEGA (ed.), Special section. Systematics and paleobiology of fossil Crustacea. Revista Mexicana de Ciencias Geológicas, **23**: 370-374.

BAKEL, B. W. M. VAN, J. W. M. JAGT, P. ARTAL & R. H. B. FRAAIJE, 2009b. *Harenacorystes johanjansseni*, a new Pliocene crab (Crustacea, Decapoda) from the Netherlands, and notes on Miocene-Pliocene corystoid crabs from the North Sea Basin. Bulletin of the Mizunami Fossil Museum, **35**: 79-85.

BELL, T., 1856-1858. A monograph of the fossil malacostracous Crustacea of Great Britain. Part I. Crustacea of the London Clay. Monograph of the Palaeontographical Society of London, **10** (40): i-viii, 9-44.

BESCHIN, C., A. BUSULINI, A. DE ANGELI & G. TESSIER, 1996. Retroplumoidea (Crustacea, Brachyura) nel Terciario del Vicentino (Italia settentrionale). Lavori di Società Veneziana di Scienze Naturali, **21**: 83-102.

BÖHM, J., 1928. *Raninellopsis Gottschei* nov. gen. n. sp. Jahrbuch der Preußischen Geologischen Landesanstalt, **48** [1927]: 563-566.

COLLINS, J. S. H. & S. L. JAKOBSEN, 2004. New crabs (Crustacea, Decapoda) from the Eocene (Ypresian/Lutetian) Lillebælt Clay Formation of Jutland, Denmark. Bulletin of the Mizunami Fossil Museum, **30** (2003): 63-96.

COLLINS, J. S. H., C. LEE & J. NOAD, 2003. Miocene and Pleistocene crabs (Crustacea, Decapoda) from Sabah and Sarawak. Journal of Systematic Palaeontology, **1**: 187-226.

COLLINS, J. & R. SMITH, 1993. Ypresian (Lower Eocene) crabs (Decapoda, Crustacea) from Belgium. Bulletin de l'Institut royal des Sciences naturelles de Belgique (Sciences de la Terre), **63**: 261-270.

CREMA, C., 1895. Sopra alcuni decapodi terziari del Piemonte. Accademia Reale di Scienze di Torino, **30**: 5-22.

FELDMANN, R. M. & R. W. PORTELL, 2007. First report of *Costacopluma* Collins and Morris, 1975 (Decapoda: Brachyura: Retroplumidae) from the Eocene of Alabama, U.S.A. Journal of Crustacean Biology, **27**: 90-96.

FRAAIJE, R. H. B., B. W. M. VAN BAKEL, J. W. M. JAGT & Y. COOLE, 2007. Two new Paleogene species of mud shrimp (Crustacea, Decapoda, Upogebiidae) from Europe and North America. Bulletin of the Mizunami Fossil Museum, **33**: 77-85.

FRAAIJE, R. H. B., J. HANSEN & T. HANSEN, 2005. Late Miocene decapod faunas from Gram, Denmark. Palaeontos, **7**: 51-61.

GILL, T., 1894. A new Bassalian type of crabs. The American Naturalist, **8**: 1043-1045.

JAGT, J. W. M., B. W. M. VAN BAKEL & R. H. B. FRAAIJE, 2007. Post-Paleocene ghost shrimps, lobsters and crabs from Belgium: an update [Geologica Belgica Meeting 2006, 2nd Belgian Geological Congress, Liège, Belgium, 7-8 September 2006]. Geologica Belgica, **10**: 199-200.

LARGHI, C., 2004. First record of Oligocene retroplumid crab (Crustacea: Decapoda: Brachyura) from Italy. Bulletin of the Mizunami Fossil Museum, **30** (2003): 57-60.

LINNAEUS, C., 1758. Systema Naturae, per regna tria naturae, secundum classes, ordines, genera, species, cum characteribus, differentiis, synonymis, locis (Editio decima, reformata), **1** (6): i-iii, 1-824. (Laurentii Salvii, Holmiae).

MCLAY, C. L., 2006a. Retroplumidae (Crustacea, Decapoda) from the Indo-Malayan archipelago (Indonesia, Philippines) and the Melanesian Arc Islands (Solomon Islands, Fiji and New Caledonia), and paleogeographical comments. Tropical Deep-Sea Benthos, **24**: 375-391.

— —, 2006b. *Retropluma solomonensis* nom. nov., a replacement name for *Retropluma laurentae* McLay, 2006 (Crustacea: Decapoda: Brachyura: Retroplumidae). Zootaxa, **1358**: 67-68.

NG, P. K. L., D. GUINOT & P. J. F. DAVIE, 2008. Systema Brachyurorum: Part I. An annotated checklist of extant brachyuran crabs of the world. The Raffles Bulletin of Zoology, (Supplement) **17**: 1-286.

SAINT LAURENT, M. DE, 1980. Sur la classification et la phylogénie des crustacés décapodes brachyoures. I. Podotremata Guinot, 1977 et Eubrachyura sect. nov. Comptes Rendus de l'Académie des Sciences, Paris, (II) **290**: 1265-1268.

— —, 1989. La nouvelle superfamille des Retroplumoidea Gill, 1894 (Decapoda, Brachyura): systématique, affinités et évolution. In: J. FOREST (ed.), Résultats des Campagnes MUSORSTOM, **5**. Mémoires du Muséum national d'Histoire naturelle, Paris, (A) **144**: 103-179.

VIA, L., 1959. Decápodos fósiles del Eoceno español. Boletín del Instituto Geológico y Minero de España, **70**: 331-402.

— —, 1969. Crustáceos decápodos del Eoceno español. Pirineos, **91-94**: 1-469.

— —, 1980. Ocypodoidea (Crustacés Décapodes) du Cenozoïque méditerranéen. Origine et évolution de cette superfamille. Annales de Paléontologie (Invertébrés), **66**: 51-66.

First received 30 May 2009.
Final version accepted 21 July 2009.

A NEW GENUS AND SPECIES OF GONEPLACID CRAB (DECAPODA, BRACHYURA, GONEPLACIDAE) FROM THE WESTERN PACIFIC

BY

PETER CASTRO[1,3]) and PETER K. L. NG[2,4])

[1]) Biological Sciences Department, California State Polytechnic University, Pomona, California 91768, U.S.A.

[2]) Tropical Marine Science Institute and Department of Biological Sciences, National University of Singapore, Lower Kent Ridge Road, Singapore 119260, Republic of Singapore

ABSTRACT

A new genus and species of goneplacid crab are described from the Loyalty Islands and Fiji in the western Pacific. The new genus can be distinguished from the other 17 genera of the Goneplacidae sensu stricto by a carapace that lacks anterolateral teeth together with moderately long orbits and eye peduncles.

RÉSUMÉ

Un nouveau genre et une nouvelle espèce de crabe Goneplacidae sont décrits des îles Loyauté et de Fiji dans le Pacifique ouest. Le nouveau genre est distingué des autres 17 genres de Goneplacidae sensu stricto par l'absence des dents antero-laterales de la carapace et par des orbites et pedoncules oculaires de tailles réduites.

INTRODUCTION

Two unidentified specimens of goneplacid crabs collected by French expeditions in the Loyalty Islands and Fiji in the western Pacific are here referred to as a new genus and species, *Guinoplax apheles*. The new genus differs markedly from the 17 known genera of the Goneplacidae sensu stricto

[3]) Corresponding author; e-mail: pcastro@csupomona.edu

[4]) e-mail: dbsngkl@nus.edu.sg

© Koninklijke Brill NV, Leiden, 2010

(see Castro, 2007) with regard to the form of the carapace, chelae, and male pleopods.

The ambulatory legs are referred to by the abbreviations P2 to P5; the first and second pairs of male pleopods by G1 and G2, respectively. Paired appendages, including the G1 and G2, are referred to in the singular for simplicity. Carapace length (CL) was measured in millimetres along the median portion of the carapace from the middle of the front to the middle of the posterior border of the carapace; carapace width (CW) across the widest breadth of the carapace; width of the front between the outer edges of the frontal margins; length of the orbit dorsally from the inner extremity of the orbit to the tip of the outer orbital angle; length of the eye peduncle dorsally from the proximal edge of the peduncle to the margin of the cornea immediately posterior to the thin extension of the peduncle; width of the cornea dorsally from its proximal edge posterior to the thin extension of the peduncle to its distal margin.

The material examined is deposited in the Muséum national d'Histoire naturelle, Paris (MNHN).

TAXONOMY

Family GONEPLACIDAE MacLeay, 1838

Guinoplax new genus

Type species. — *Guinoplax apheles*, new species, by monotypy. Gender feminine.

Etymology. — The new genus is an arbitrary combination of the name of Danièle Guinot, our teacher, collaborator, and loyal friend for many years, and *plax*, Greek for "plate" or "tablet", a common suffix also used for many related goneplacid genera.

Diagnosis. — Carapace with convex anterolateral borders lacking teeth, front slightly bilobed. Orbits long (slightly longer than front), wide; outer orbital angle sharp, triangular, directed anteriorly. Eye peduncles moderately long (0.6 times frontal width), cornea slightly elongated. Chelipeds relatively short, tips of fingers light in colour; right chela with enlarged "cutting" tooth at base of dactylus. Sternal sutures 4/5, 5/6, 7/8 interrupted medially, 6/7 complete; sternites 7, 8 with median line. Male abdomen wide; somite 2 slightly narrower than somite 1, somite 3 covers space between P5 coxae, such that thoracic sternite 8 is covered by somites 2, 3 of closed abdomen. Penis opens anterior to coxo-sternal condyle of P5 coxa, moderate size; protected by expanded episternite 7. G1 proportionally long, distally slender, slightly bent, numerous spinules distally, acuminate apex; G2 slender, almost as long as G1, flagellum distinct from basal part, as long as basal part, arched, slightly expanded tip with 2 lateral spinules. Female not known.

Remarks. — *Guinoplax* new genus, clearly belongs to the Goneplacidae sensu stricto on account of a suite of characters that distinguishes it from the remaining 10 families of the Goneplacoidea (see Castro, 2007; Ng & Manuel-Santos, 2007; Ng et al., 2008; Castro et al., 2010). Such characters include: (i) a G2 that is approximately as long as the G1 (with the exception of two genera, *Microgoneplax* Castro, 2007, and *Paragoneplax* Castro, 2007, which have a relatively short G2), and has a long flagellum that is distinct from, and as long as, the basal part; (ii) a coxal penis that is protected by an expanded episternite 7 and by wide abdominal somites 2 and 3 of the closed abdomen (somite 3 being wider or slightly wider than somites 1 and 2), and not lying in a concave or hollow posterior portion of thoracic sternites 7 and/or 8; (iii) a male sterno-abdominal cavity that is moderately shallow and reaches the median portion of thoracic sternite 4; (iv) the tubercle of the press button of the male abdominal-locking mechanism is positioned near thoracic sternal suture 4/5; (v) a complete thoracic suture 6/7 (although incomplete in four goneplacid genera); (vi) sternites 7 and 8 with a median line; and (vii) the general shape of the carapace, being transversely rectangular with a straight, moderately long front, and moderately elongated eye peduncles.

No female specimens of the new species are known, so information on the morphology of the vulva, a character used by Castro (2007) and Castro et al. (2010) as part of their diagnoses of the goneplacid genera and families, respectively, is unfortunately lacking.

Guinoplax differs from all the 17 other genera of Goneplacidae by having a carapace that lacks anterolateral teeth, as well as moderately long orbits and eye peduncles (fig. 1). It also appears to be the only genus in which the right chela possesses an enlarged "cutting" tooth at the base of the dactylus of one claw (fig. 2A, B). This specialized tooth is likely to be used for feeding on gastropod snails (see Ng & Tan, 1984, 1985). Anterolateral teeth are absent in all species of *Microgoneplax* and *Singhaplax* Serène & Soh, 1976, but their orbits and eye peduncles are much longer than those in *Guinoplax* and their outer orbital teeth are also conspicuously long and directed outwards (see Castro, 2007, figs. 35A, 37, 38A, 39A, 41A, 42, 43A, 44A, 45A, 46) (relatively short and directed anteriorly in *Guinoplax*;

fig. 1A, B). Furthermore, the G2 of *Microgoneplax* is proportionally much shorter than that of *Guinoplax* (see Castro, 2007, figs. 41D, 43C, 44C, 45C). In *Paragoneplax*, the anterolateral teeth are absent, with the orbits and eye peduncles similar to those of *Guinoplax*, but the G1 of *Paragoneplax* is very different, being distally bent (Zarenkov, 1972, fig. 7-2, as *Geneplax* [sic] *serenei*) (almost straight in *Guinoplax*; fig. 3A, B); the G2 is short (Chen, 1998, fig. 12-8, as *Goneplax serenei*) (long in *Guinoplax*; fig. 3C); and the cornea is distinctively rheniform (spherical and not anterolaterally flattened in *Guinoplax*; fig. 2A-C). The apex of the G2 has two spinules in *Guinoplax* (fig. 3C, D) as in most species of *Carcinoplax* (see Guinot, 1989; Castro, 2007) but the carapace of *Carcinoplax* is quadrate, with two anterolateral teeth on each side of the carapace [only one in *Carcinoplax ischurodous* (Stebbing, 1923)], and the eye peduncles are relatively short. *Guinoplax* also superficially resembles *Psopheticoides* Sakai, 1965, a genus which, until recently, was classified in the Goneplacidae. *Psopheticoides* has, however, been referred to the Euryplacidae, a goneplacoid family (see Castro, 2007; Ng et al., 2008; and Castro & Ng, 2010).

Guinoplax apheles new species (figs. 1-3)

Material examined. — Holotype: male (CL 5.5 mm, CW 8.0 mm), MNHN-B30842, Fiji, BORDAU 1, station DW 1479, off Tuvana-i-Colo I., 20°58.05′S 178°44.94′W, R/V "Alis", 450-460 m, 9.iii.1999. Paratype: 1 male (CL 4.2 mm, CW 4.7 mm), MNHN-B30843, Loyalty Is., MUSORSTOM 6, station CP 464, 21°02.30′S 167°31.60′E, R/V "Alis", 430 m, 21.ii.1989.

Etymology. — The new species is named *apheles*, Greek for "even", "smooth", and "simple", in reference to the smooth appearance of its carapace, which results from the absence of anterolateral teeth and comparatively inconspicuous outer orbital angles.

Description. — Carapace (fig. 1A) transversely rectangular, moderately wider than long (CW 1.4 CL in holotype, 1.1 in paratype); widest at junction of anterolateral, posterolateral borders in holotype (at outer orbital angles in smaller paratype); dorsal surface smooth, moderately convex, without indication of regions. Front (fig. 1B) strongly deflexed ventrally, slightly bilobed, medially marked by short, straight gap, without notch or projection. Hardly noticeable notch between front, inner edge of supraorbital border; orbits (fig. 1B) long (slightly longer than front), wide, further expanded before outer orbital angle; supraorbital borders long, thinly cristate, undular, each with wide, short median convexity; suborbital borders (fig. 1B, C) long, thinly cristate, rounded,

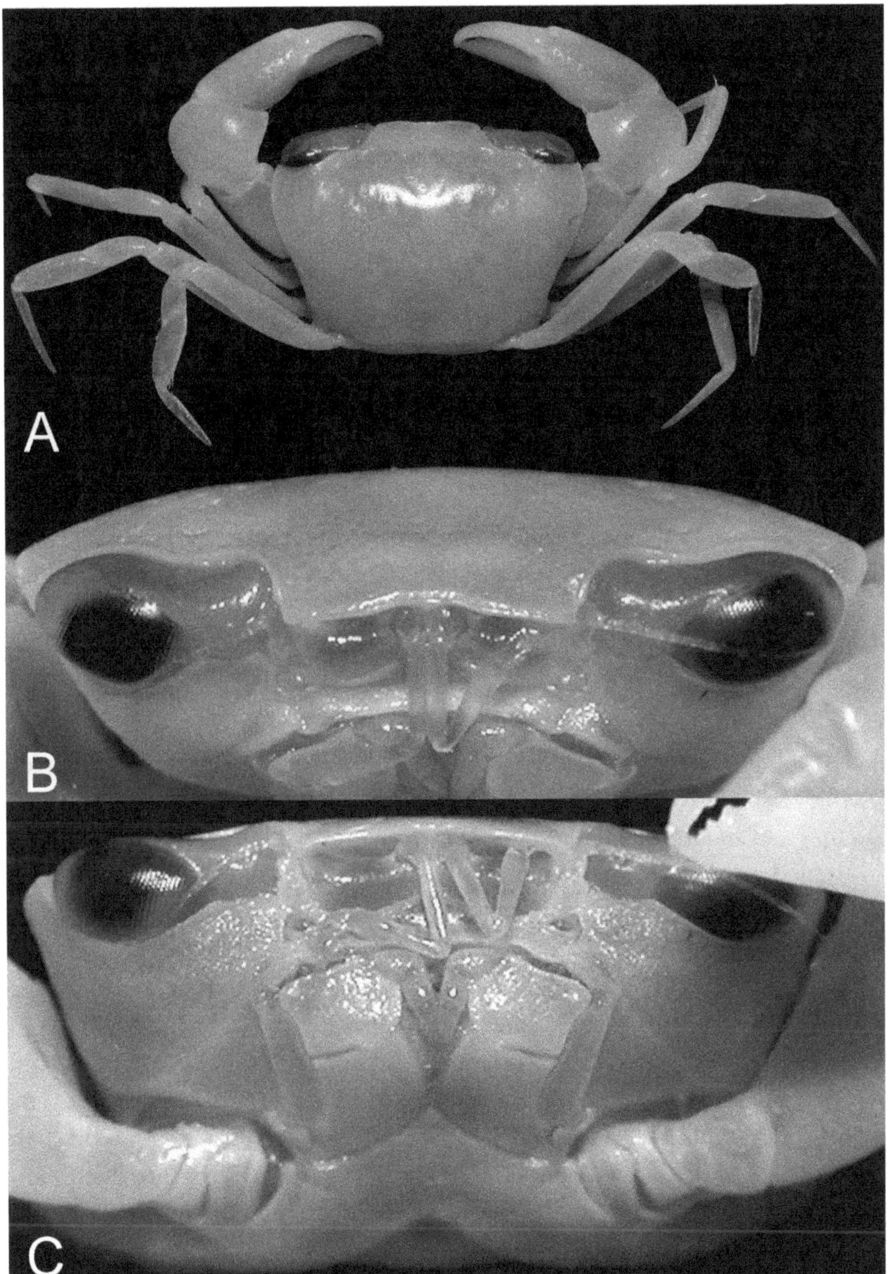

Fig. 1. *Guinoplax apheles* new genus, new species. A-C, holotype, male (CL 5.5 mm, CW 8.0 mm) (MNHN-B 30842). A, dorsal view; B, frontal view of carapace; C, ventral view of anterior portion of carapace.

each with wide, short median convexity, inner tooth absent. Outer orbital angle sharp, triangular, directed anteriorly. Anterolateral borders slightly convex in holotype (straight in paratype), anterolateral teeth absent, boundary between anterolateral, posterolateral borders not pronounced. Posterolateral borders proportionally long, slightly concave.

Eye peduncles (fig. 1B, C) moderately long (0.6 times front width); cornea proportionally large, slightly elongated (0.5 times front width, almost as long as peduncle), expanded distally, rounded outer margin; conspicuous, thin extension of peduncle across dorsal surface.

Basal antennal article short, subquadrate, distalmost (third) article reaches front. Anterior border of endostome well demarcated from buccal cavern, ridges clearly defined. No obvious stridulating mechanism other than possible rubbing of proximal portion of cheliped (P1) merus against pterygostomial ridge.

Third maxillipeds (fig. 1C) rectangular, leaving gap when closed; ischium nearly square, inner margin dentate, lined with setae, mid-length slightly less than 2 times mid-length of merus; merus with distal inner angle excavated, anteroexternal margin angular; exopod wide, distal end reaching beyond external outer angle of merus, flagellum long.

Chelipeds relatively short (figs. 1A, 2A, B). Chelae (fig. 2A, B) markedly hetereochelous, surfaces microscopically granular; right one larger, relatively swollen, subequal, microscopically granular; fingers relatively slender, not compressed, shorter than palm, tips light in colour; cutting margins of fingers with moderately large, triangular, rounded teeth, base of dactylus with prominent large "cutting" tooth; smaller chela similar in form to larger chela but palm relatively more slender, cutting margins of fingers without specialized tooth. Carpus with obtuse tooth on inner margin; anterior margin of merus granular, with short, rounded, granular crest distally. Meri, carpi of ambulatory legs (P2-P5) subcylindrical in cross section; propodi, dactyli dorsoventrally flattened. Anterior, posterior margins of P2-P5 articles unarmed, covered with scattered, simple setae; P2-P4 propodi, dactyli slender, dactyli with slight longitudinal ridge along each side. P5 propodus proportionally wide, short; dactylus slender, smooth, fringed with short, simple setae.

Thoracic sternum (fig. 2B) wide. Thoracic suture 2/3 complete, straight; suture 3/4 visible laterally; sterno-abdominal cavity moderately shallow, with anterior end wide, rounded, reaching median portion of thoracic sternite 4. Sutures 4/5, 5/6, 7/8 interrupted medially, 6/7 complete; sternites 7, 8 with median line.

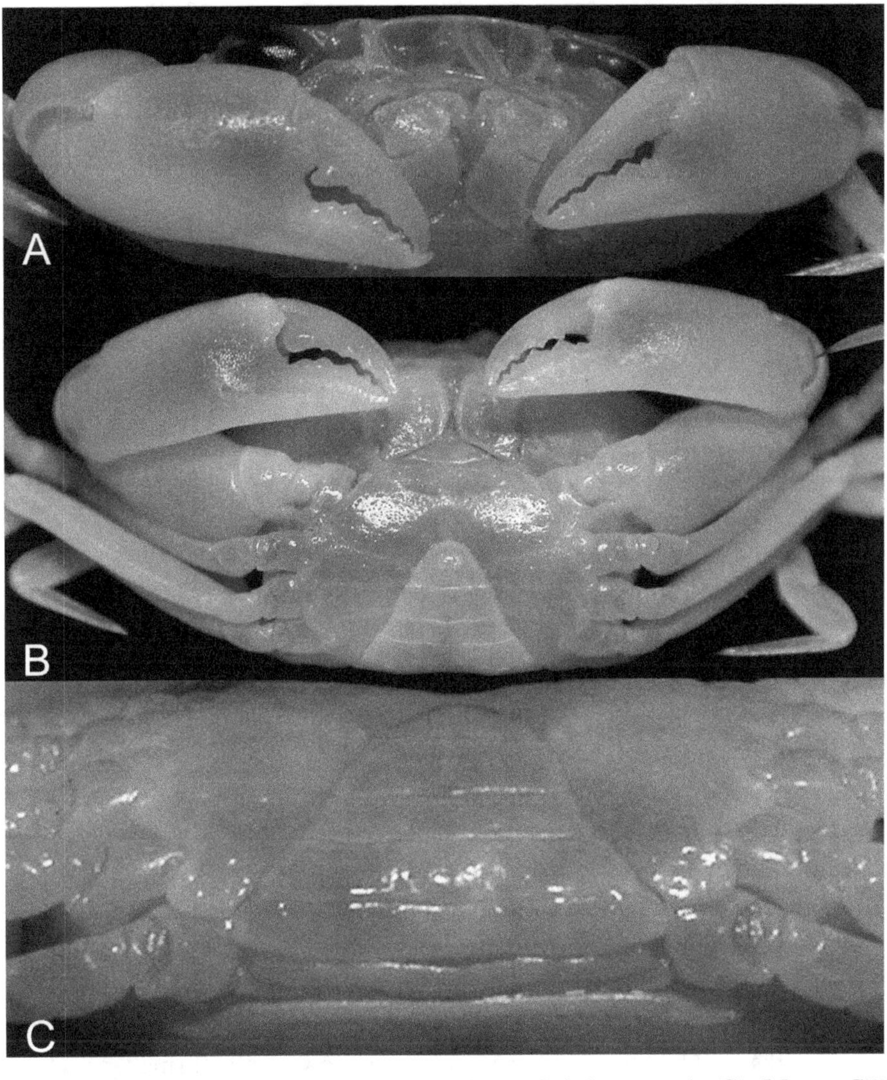

Fig. 2. *Guinoplax apheles* new genus, new species. A-C, holotype, male (CL 5.5 mm, CW 8.0 mm) (MNHN-B 30842). A, chelae; B, ventral view of carapace; C, abdomen.

Male abdomen (fig. 2B, C) wide, with six freely-movable somites plus telson; somite 1 wide, short, partially covered by posterior margin of carapace; somite 2 only slightly narrower than somite 1, short; somites 3-6 longer than 1, 2, somites 4-6 gradually decreasing in width from somite 3 (widest somite). Somite 3 covers space between P5 coxae, so that thoracic sternite 8 is covered by somites 2, 3 of closed abdomen. Telson wider than long, unarmed. Press button of male abdominal-locking mechanism consisting of

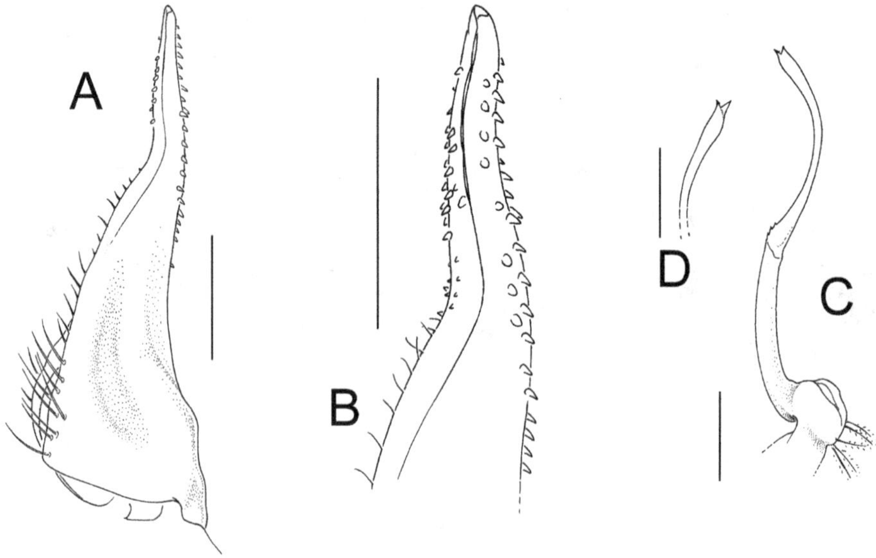

Fig. 3. *Guinoplax apheles* new genus, new species. A-C, holotype, male (CL 5.5 mm, CW 8.0 mm) (MNHN-B 30842). A, B, right G1, dorsal view; C, right G2, dorsal view. Scales: A, C = 1.0 mm; B = 0.5 mm.

tubercle positioned near thoracic sternal suture 4/5. Penis arising from large gonopore that opens just anterior to coxo-sternal condyle of P5 coxa, moderate size; protected by expanded episternite 7.

G1 (fig. 3A, B) proportionally long, distally slender, slightly bent, broadened proximally; numerous spinules distally, on most distal portion of proximal part; acuminate apex. G2 (fig. 3C, D) slender, almost as long as G1; flagellum distinct from basal part (peduncle), as long as basal part, arched, slightly expanded tip with two lateral spinules.

Female not known.

Remarks. — Only two male specimens are known, both of small size. Their male abdomens, however, are fully formed, and the G1 and G2 structures are fully calcified so both specimens are almost certainly mature. The abdomen of the small, paratype specimen is identical to that of the larger holotype. The anterolateral borders are slightly convex in the larger specimen but straight in the small specimen, thus indicating that allometric change occurs correlated with increasing size.

Distribution. — The species is so far known only from two localities in the western Pacific Ocean, the Loyalty Islands and Fiji. The two specimens were collected from depths of 430 to 460 m.

ACKNOWLEDGEMENTS

We thank Régis Cleva and Danièle Guinot for loaning us the material for study; and Jean-François Dejouannet (Institute de Recherche pour le Développement, Paris) for preparing the drawings. The latter's research stint in Singapore was partially supported by the French Embassy (under the MERLION programme) and the National University of Singapore. The first author's research stay in Singapore was partially supported by a research grant from the Raffles Museum of Biodiversity Research, National University of Singapore.

REFERENCES

CASTRO, P., 2007. A reappraisal of the family Goneplacidae MacLeay, 1838 (Crustacea, Decapoda, Brachyura) and revision of the subfamily Goneplacinae, with the description of 10 new genera and 18 new species. Zoosystema, **29** (4): 609-774.

CASTRO, P., D. GUINOT & P. K. L. NG, 2010. A new family for *Sotoplax robertsi* Guinot, 1984, with a diagnosis and key to the Goneplacoidea MacLeay, 1838 (Crustacea: Decapoda: Brachyura). Zootaxa, **2356**: 36-56.

CASTRO, P. & P. K. L. NG, 2010. Revision of the family Euryplacidae Stimpson, 1871 (Crustacea: Decapoda: Brachyura: Goneplacoidea). Zootaxa, **2375**: 1-130.

CHEN, H., 1998. The Goneplacidae (Crustacea: Brachyura) from Nansha Islands and adjacent waters. In: Studies on Marine Fauna and Flora and Biogeography of the Nansha Islands and Neighbouring Waters, **3**: 265-316. (Academia Sinica, Beijing). [In Chinese with English abstract.]

GUINOT, D., 1989. Le genre *Carcinoplax* H. Milne Edwards, 1852 (Crustacea, Brachyura: Goneplacidae). In: J. FOREST (ed.), Résultats des Campagnes MUSORSTOM, **5**. Mémoires du Muséum national d'Histoire naturelle, Paris, (A) **144**: 265-345.

MACLEAY, W. S., 1838. On the brachyurous decapod Crustacea brought from the Cape by Dr. Smith. In: A. SMITH (ed.), Illustrations of the zoology of South Africa; consisting chiefly of figures and descriptions of the objects of natural history collected during an expedition into the interior of South Africa, in the years 1834, 1835, and 1836; fitted out by "The Cape of Good Hope Association for Exploring Central Africa": together with a summary of African Zoology, and an inquiry into the geographical ranges of species in that quarter of the globe. Invertebratae: 53-71, pls. 2, 3. (Smith, Elder & Co., London).

NG, P. K. L., D. GUINOT & P. J. F. DAVIE, 2008. Systema Brachyurorum, Part I. An annotated checklist of extant brachyuran crabs of the world. Raffles Bulletin of Zoology, (Supplement) **17**: 1-286.

NG, P. K. L. & M. R. MANUEL-SANTOS, 2007. Establishment of the Vultocinidae, a new family for an unusual new genus and new species of Indo-West Pacific crab (Crustacea: Decapoda: Brachyura: Goneplacoidea), with comments on the taxonomy of the Goneplacidae. Zootaxa, **1558**: 39-68.

NG, P. K. L. & L. W. H. TAN, 1984. The 'shell peeling' structure of the box crab *Calappa philargius* (L.) and other crabs in relation to mollusc shell architecture. Journal of the Singapore National Academy of Science, **13**: 195-199.

— — & — —, 1985. 'Right handedness' in heterochelous calappoid and xanthoid crabs — suggestion for a functional advantage. Crustaceana, **49**: 98-100.

SAKAI, T., 1969. Two new genera and twenty-two new species of crabs from Japan. Proceedings of the Biological Society of Washington, **82**: 243-280.

STEBBING, T. R. R., 1923. Crustacea of Natal. Fisheries and Marine Biological Survey, **3** [1922]: 1-16.

ZARENKOV, N. A., 1972. [New data on Indo-Pacific crabs (fam. Goneplacidae, Pinnotheridae, Parthenopidae, Dorippidae) and the problem of seasonal reproduction of decapod animals in the Bay of Tonkin]. In: [The complex investigations of ocean nature], **3**: 229-253. (Moscow State University, Moscow). [In Russian.]

First received 4 April 2009.
Final version accepted 19 May 2009.

A REDESCRIPTION OF *POTAMONAUTES LOVENI* (COLOSI, 1924), A MEDICALLY IMPORTANT FRESHWATER CRAB FROM WESTERN KENYA AND EASTERN UGANDA, EAST AFRICA (BRACHYURA, POTAMOIDEA, POTAMONAUTIDAE)

BY

NEIL CUMBERLIDGE[1,3]) and PAUL F. CLARK[2,4])

[1]) Department of Biology, Northern Michigan University, Marquette, Michigan 49855, U.S.A.
[2]) Department of Zoology, The Natural History Museum, Cromwell Road, London, SW7 5BD, U.K.

ABSTRACT

Potamonautes loveni (Colosi, 1924) from western Kenya and eastern Uganda is resurrected from synonymy and redescribed as a valid species. This taxon is compared with similar species and its distribution, ecology, and conservation status are discussed. *Potamonautes loveni* has been implicated in the transmission of onchocerciasis in the highland areas of parts of East Africa.

RÉSUMÉ

Potamonautes loveni (Colosi, 1924) de l'ouest du Kenya et de l'est de l'Ouganda est redécrit comme une espèce valide. Ce taxon est comparé avec les espèces proches et sa repartition, son écologie et son statut de conservation sont discutés. *Potamonautes loveni* a été impliqué dans la transmission de l'onchocercose dans les zones montagneuses de l'Afrique de l'Est.

INTRODUCTION

Potamonautes loveni (Colosi, 1924) is an abundant species of East African freshwater crab that is closely linked with river blindness (onchocerciasis) in the highlands of western Kenya and eastern Uganda. Its distributional range includes Mt. Elgon, a volcanic mountain that straddles the border

[3]) Corresponding author; e-mail: ncumberl@nmu.edu
[4]) e-mail: p.clark@nhm.ac.uk

© Koninklijke Brill NV, Leiden, 2010

between Uganda and Kenya. *Potamonautes loveni*, however, has long been taxonomically problematic, a fact that is reflected in its extensive synonymy list. For example, the literature (Colosi, 1924; Sjöstedt, 1925; Rathbun, 1935; Roux, 1935; Bott, 1955) includes several described taxa from Mt. Elgon. These include *P.* (*Geothelphusa*) *loveni* Colosi, 1924, *P.* (*G.*) *granviki* Colosi, 1924, *P.* (*Potamonautes*) *harvardi* Rathbun, 1935, and *P.* (*G.*) *loveni longimerus* Roux, 1935 (cf. Colosi, 1924; Rathbun, 1935; Roux, 1935). Bott (1955) considered all of these Mt. Elgon taxa to belong to a single species — *Potamonautes* (*Rotundopotamonautes*) *granviki* (Colosi, 1924), and treated the remaining three taxa as junior synonyms of this species. In addition, Bott (1955) also judged two other geographically separated taxa, *Potamon* (*G.*) *jeanneli* Bouvier, 1921, from central Kenya and *Potamon* (*G.*) *idjwiensis* Chace, 1942, from Lake Kivu in the Western Rift Valley in eastern D. R. Congo to be junior synonyms of *P.* (*G.*) *granviki.*

In recognizing that *Potamonautes loveni* (Colosi, 1924) was the valid name of Bott's (1955) "*P.* (*R.*) *granviki*" from Mt. Elgon, Williams (1968, 1991) and Cumberlidge (1997, 1998) treated three taxa (*P. granviki*, *P. harvardi*, and *P.* (*G.*) *loveni longimerus*) as junior synonyms of *P. loveni* (cf. Ng et al., 2008). Williams (1968, 1991) and Cumberlidge (1997, 1998, 2009a) also pointed out that both *P.* (*G.*) *jeanneli* (Mount Kenya) and *P.* (*G.*) *idjwiensis* (Lake Kivu) should properly be treated as valid species rather than as junior synonyms of *P. granviki* as suggested by Bott (1955). As a result of these taxonomic changes, freshwater crabs collected from western Kenya are difficult to identify because the only available keys (Bott, 1955) are outdated, and the only way to make reliable identifications is to refer to the original type series that have been deposited in a number of museums.

Our revision is based on the examination of the type specimens of all of the relevant taxa, as well as on a large series of previously unexamined museum specimens from Mt. Elgon and western Kenya collected by T. R. Williams and his colleagues in the 1960s, and more recently by M. Dobson (Dobson, 2004; Dobson et al., 2007). The specimens of *P. loveni* from Mt. Elgon described here were collected from forests on the western slopes of the mountain in Uganda from streams, rivers, and nearby land during a survey of a focus of onchocerciasis (Barnley & Prentice, 1958; McMahon et al., 1958; Hynes et al., 1961; Williams et al., 1961; Williams, 1964).

Despite recent contributions (Corace et al., 2001; Cumberlidge & Vannini, 2004; Cumberlidge & Dobson, 2008; Cumberlidge, 2009a, b) it is still nevertheless difficult to identify some species of freshwater crabs from western Kenya. This has been exacerbated in the case of specimens from Mt. Elgon by the fact that some taxonomically important characters of species such as *P. loveni* change dramatically as crabs grow to maturity, which means that juvenile, sub-adult, and adult specimens of the same species have morphologically different cheliped and gonopod characters. A similar phenomenon has been reported to occur in other species of freshwater crabs from east and southern Africa (Reed & Cumberlidge, 2004). Other taxonomic difficulties with the identification of freshwater crabs from Mt. Elgon arise from the presence of two other undescribed species that are morphologically similar to *P. loveni* (cf. Cumberlidge & Clark, 2010). These species were most likely confused with *P. loveni* by Williams (1991). In order to clarify the taxonomy of the freshwater crabs of Mt. Elgon, *P. loveni* is redescribed here based on a combination of characters of the adult male (GO1, carapace, thoracic sternum, and chelipeds). This species is closely linked with human river blindness (onchocerciasis). Onchocerciasis is endemic to the highlands of western Kenya and eastern Uganda (Williams, 1968, 1991; Crosskey, 1990).

Specimens are deposited in the following museums: the Natural History Museum, London, U.K. (NHM); Department of Biology, Northern Michigan University, Marquette, MI, U.S.A. (NMU); Museum of Comparative Zoology, Harvard, MA, U.S.A. (MCZ); Musée Royal de l'Afrique Centrale, Tervuren, Belgium (MRAC); Swedish Museum of Natural History (Naturhistoriska Riksmuseet) Stockholm, Sweden (SMNH); and United States National Museum of Natural History, Smithsonian Institution, Washington, D.C., U.S.A. (USNM). Abbreviations used in the text are: cw = distance across the carapace at the widest point; cl = carapace length measured along the median line, from the anterior to the posterior margin; ch = carapace height, the maximum height of the cephalothorax; fw = front width measured along the anterior margin; s = thoracic sternite; e = thoracic episternite; s4/s5, s5/s6, s6/s7, s7/s8 = sternal sutures between adjacent thoracic sternites; s4/e4, s5/e5, s6/e6, s7/e7 = episternal sutures between adjacent thoracic sternites and episternites; p1-p5 = pereiopods 1-5; a1-a6 = abdominal somites 1-6; a7, telson of the abdomen; GO1 = gonopod 1; GO2 = gonopod 2; juv. = juvenile; stn = station; coll. = collected by; asl = above sea level. All measurements are given in mm.

TAXONOMY

Family POTAMONAUTIDAE Bott, 1970

Potamonautes MacLeay, 1838

Potamonautes loveni (Colosi, 1924) (figs. 1-3)

Potamon (*Geothelphusa*) *Lovéni* Colosi, 1924: 13, fig. 9, pl. 1 fig. 4.
Potamon (*Geothelphusa*) *Granviki* Colosi, 1924: 16, fig. 11, pl. 1 fig. 5.
Geothelphusa loveni — Balss, 1929: 351.
Potamon (*Geothelphusa*) *harvardi* Rathbun, 1935: 23-24, pl. 1.
Potamon granviki — Chace, 1942: 211.
Potamon lovéni — Chace, 1942: 215.
Potamon harvardi — Chace, 1942: 212, fig. 9.
Potamonautes (*Rotundopotamonautes*) *granviki* — Bott, 1955: 286-288, figs. 52, 90, pl. XXII fig. 2a-d.
Potamonautes loveni — Williams, 1968: 32; 1991: 181-187 (part); Cumberlidge, 1997: 583; 1998: 201; Ng et al., 2008: 171.

Type material examined. — Holotype, *Potamon* (*Geothelphusa*) *loveni* Colosi, 1924, 1 subadult male (cw 26.3, cl 19.5) (SMNH 1291), Mt. Elgon, Mbale Province, Uganda, coll. Swedish Expedition, 1921; lectotype, *Potamon* (*Geothelphusa*) *harvardi* Rathbun, 1935, 1 adult male (cw 42.6) (MCZ 8241), Mt. Elgon, Sipi Falls, 1829 m asl, coll. 18.xii.1933. Paratypes of *P.* (*G.*) *harvardi*, adult male (MCZ 8242), Mt. Elgon, Sipi Falls, 1829 m asl, coll. 18.xii.1933; 19 males, 6 females (MCZ 8243), Mt. Elgon, Sipi Falls, 1829 m asl, coll. 18.xii.1933; adult male, adult female ovigerous (MCZ 8240), Mt. Elgon, Kaburomi, coll. 28.xii.1933; adult male (cw 32), adult female (cw 29.5) of *P.* (*G.*) *harvardi*, from a large series of 34 males, 28 females, 6 juv. (MCZ 8244, 8245), Kaimosi, Kakamega, Kenya, coll. 10-15.ii.1934.

Additional material examined. — Uganda. 2 adult males (cw 42.2, 38.2), 1 adult female (cw 40.6) (USNM 70907), Mt. Elgon, Sipi, 2000 m asl, 18.xii.1933, coll. A. Loveridge; 1 adult male (cw 38.7) (USNM 70912), Mt. Elgon, Butandiga, 8.i.1935, coll. A. Loveridge; 2 males (cw 30.8, 13.0), 3 females (cw 23.9-17.0) (NHM 2008.3033-3037), stn El.9, Mt. Elgon, Sision River, 2073 m asl, 24.xii.1960, coll. T. R. Williams; 10 males (cw 43.3-10.4), 8 females (cw 16.3-10.5), 2 juv. (NHM 3270-3279), stn El.33, Mt. Elgon, Kamijaro River, Manafwa sector, 1463 m asl, 28.xii.1960, coll. T. R. Williams; 6 males (cw 23.5-11.7), 10 females (cw 28.1-11.7), 10 juv. (NHM 3280-3289), stn El.66, Mt. Elgon, Namafumbulwe River, Manafwa sector, 1402 m asl, 1.i.1961, coll. T. R. Williams; 1 male (cw 12.1), 4 females (cw 23.3-12.7), 13 juv. (NHM 2008.3290-3299), stn El.68, Mt. Elgon, Manafwa River, Manafwa sector, 1829 m asl, 1.i.1961, coll. T. R. Williams; 1 male (cw 47.4) (NHM 2008.3300), stn El.74, Mt. Elgon, Suamme River, Manafwa sector, 1768 m asl, 3.i.1961, coll. T. R. Williams; 2 males (cw 32.4, 28.1), 2 females (cw 27.8, 26.1) (NHM 2008.3303-3306), stn El.78, Mt. Elgon, Nalakwa River, Manafwa sector, 1402 m asl, 3.i.1961, with *Potamonautes niloticus*, coll. T. R. Williams; stns El.93a, El.93b, El.96, Mt. Elgon, Namafumbole River, Namatala sector, 1585 m asl, coll. T. R. Williams; 4 males (cw 36.9-13.3), 4 females (cw 27.7-16.1), 1 juv. (NHM 2008.3024-3032), stn El.95, Mt. Elgon, Namafumbole River, Namatala sector, 1585 m asl, 8.i.1961, coll. T. R. Williams; 1 male (cw 18.0), 4 females (cw 49.0-15.0), 1 juv. (NHM 2008.3308-3331), stn El.99, Mt. Elgon, Sisiwachi River, Namatala sector, 1585 m asl, coll. T. R. Williams; 6 males (cw 27.8-17.2), 6 females (cw 33.5-16.9) (NHM 2008.3038-3052), stn El.113, Mt. Elgon, Simu tributary, Siroko sector, 1341 m asl, 9.i.1961, coll. T. R. Williams; 6 males (cw 29.1-14.20), 3 females (cw 25.1-18.3) (NHM 2008.3340-3348), stn El.135, Mt. Elgon, Tabok

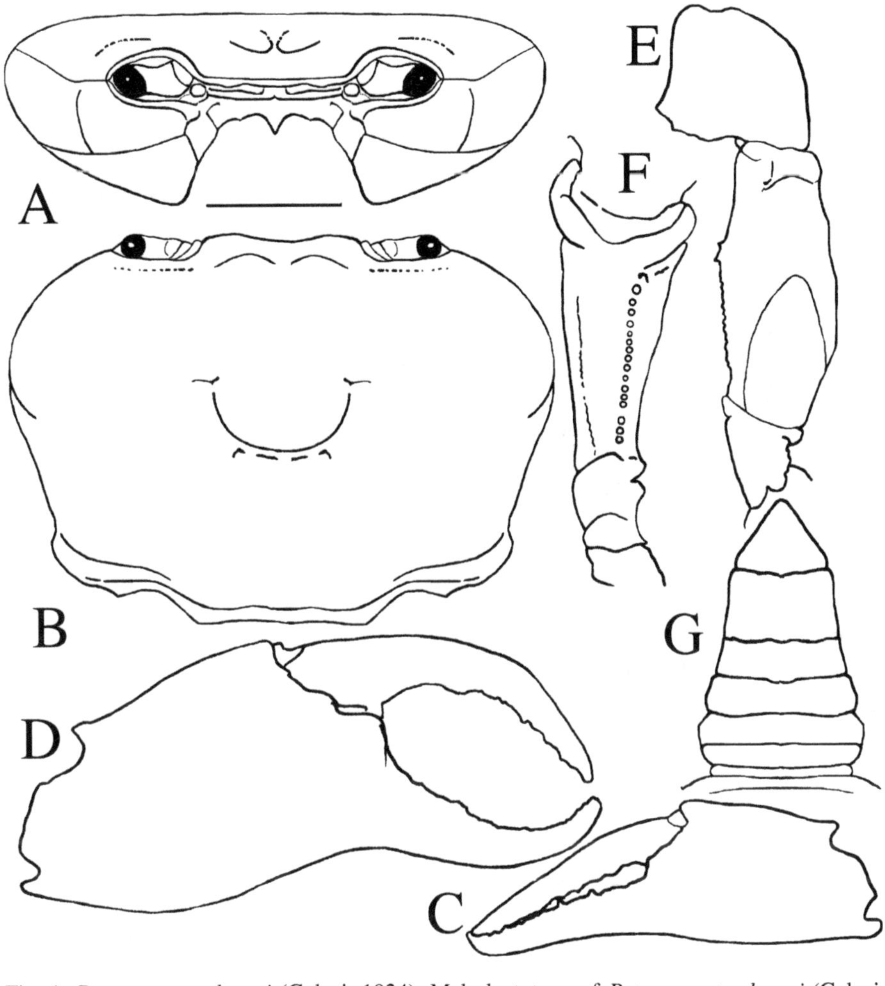

Fig. 1. *Potamonautes loveni* (Colosi, 1924). Male lectotype of *Potamonautes loveni* (Colosi, 1924), male lectotype of *P.* (*G.*) *harvardi* Rathbun, 1935, cw 42.8, from Sipi Falls, Mt. Elgon, Uganda (MCZ 8241). A, cephalothorax, frontal view; B, carapace, dorsal view; C, left (minor) cheliped, frontal view; D, right (major) cheliped, frontal view; E, carpus and merus of right cheliped, dorsal view; F, carpus and merus of right cheliped, inferior view; G, abdomen. Scale bar: 11 mm.

River, Nyenye sector, 2042 m asl, 13.i.1961, coll. T. R. Williams; 6 males (cw 35.8-10.3), 2 females (cw 40.9, 39.8), 2 juv. (NHM 2008.3349-3358), stn El.139, Mt. Elgon, Kondia River, Kiriki sector, 2073 m asl, 14.i.1961, coll. T. R. Williams; 2 males (cw 47, 14.4), 3 females (cw 29.9-18.8), 3 juv. (NHM 2008.1000), stn El.142, Mt. Elgon, Jikwil River, Kiriki sector, 2134 m asl, 14.i.1961, coll. T. R. Williams; 3 males (cw 26.6-20.9), 5 females (cw 28.3-16.9), 7 juv. (NHM 2008.3009-3023), stn El.152, Mt. Elgon, Kimotho River, 3048 m asl, 15.i.1961, coll. T. R. Williams; 1 male (cw 27.9), 3 females (cw 22.0-18.7) (NHM 2008.3359-3362), stn El.159, Mt. Elgon, Suamam River, Suam sector, 2073 m asl, 15.i.1961, coll. T. R. Williams;

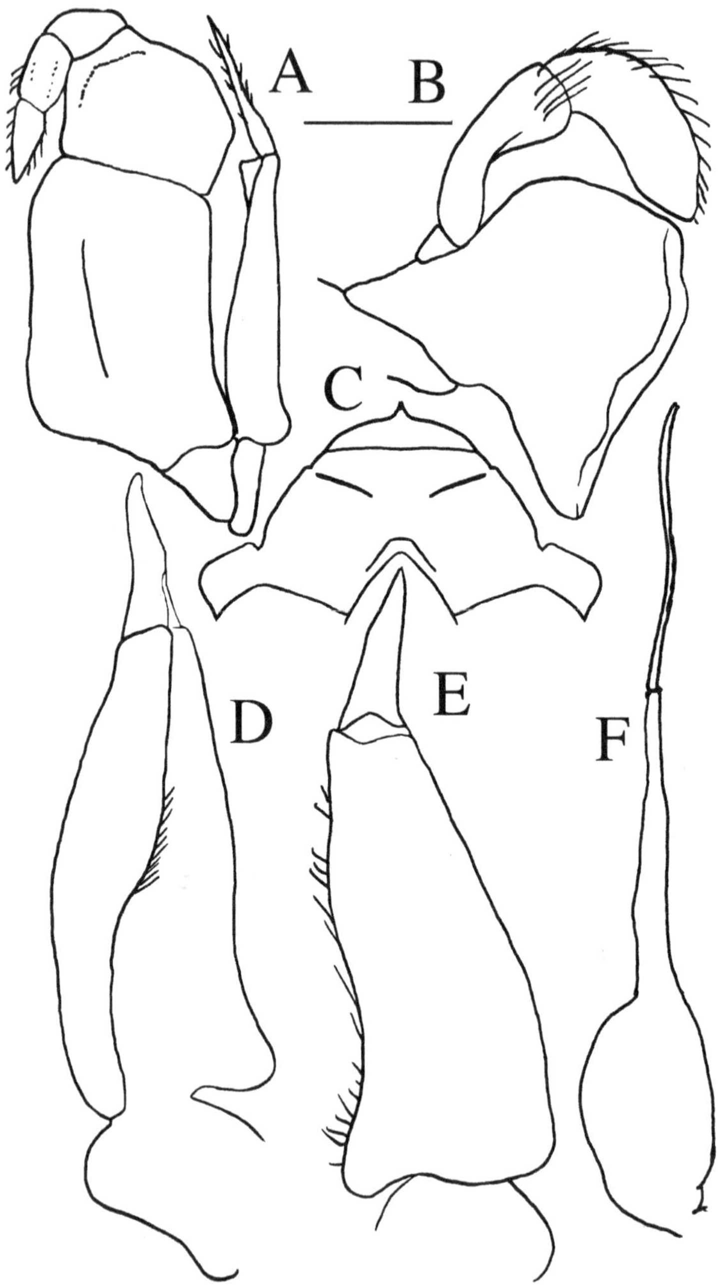

Fig. 2. *Potamonautes loveni* (Colosi, 1924), male lectotype of *Potamon* (*Geothelphusa*) *harvardi* Rathbun, 1935, cw 42.6, from Sipi Falls, Mt. Elgon, Uganda (MCZ 8241). A, left third maxilliped; B, right mandible frontal view; C, anterior sternum; D, left GO1 ventral view; E, left GO1 dorsal view; F, left GO2 ventral view. Scale bar: 11 mm (A, C), 2 mm (B, D-F).

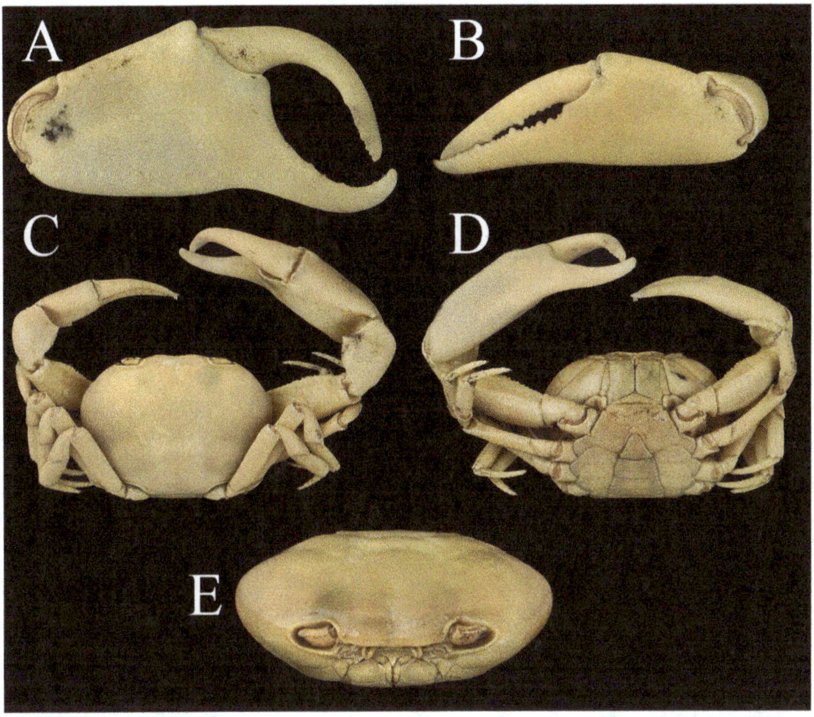

Fig. 3. *Potamonautes loveni* (Colosi, 1924). Adult male, cw 40.6 mm, Rongai, Rift Valley Province, Kenya (NHM 1930.6.24.11-12). A, right (major) cheliped, frontal view; B, left (minor) cheliped, frontal view; C, whole animal, dorsal view; D, whole animal ventral view; E, carapace, frontal view. Photograph Phil Crabb, NHM Photographic Unit. Scale bar: 14 mm (A, B, E), 23.2 mm (C, D).

9 males (cw 26.3-11.0), 2 females (cw 24.3, 18.5), 6 juv. (NHM 2008.3363-3372), stn El.167, Mt. Elgon, Njomunjone River, Nyenye sector, 1890 m asl, 16.i.1961, coll. T. R. Williams; adult male (cw 34 mm) (NMU 20.VII.1996), undisturbed high altitude bamboo forest, caught in a mist net on north facing slopes of Mt. Elgon, 3800 m asl, coll. S. King, 20.vii.1996.

Kenya. 1 adult male (cw 46.9) (USNM 70911), Mt. Elgon, Elgonyi, 2333 m asl, coll. A. Loveridge, 25.i.1934; 1 adult (USNM 82294), coll. P. J. Rainey; adult male (cw 36.5), adult female (cw 41.5) (NMU TRW 09.03.1988.1-2), Kibisi River, Kapsokisio, Coffee factory site 14, Mt. Elgon, coll. 9.iii.1988; adult male (MRAC 41.222), Suam River, Mt. Elgon, 15.xii.1953, coll. J. Bouillon; 3 males (cw 34.4, 28.8, 27.1), 2 females (cw 31.4, 28.7), juvenile male (cw 14.2), juvenile female (cw 10) (USNM 70908), Rift Valley Province, Kaimosa (=Kaimosi) (0°8′0″N 34°56′0″E) near Kakamega, coll. A. Loveridge, 10-15.ii.1935; 1 male (cw 24.9), 4 females (cw 22.0-15.2) (NHM 2008.3373-3377), stn EA62.135, Itsala River (first river to the west of Koiparak), Kakamega (0°8′0″N 34°53′0″E), Western Province, 25.iii.1962, coll. T. R. Williams; 10 males (cw 31.0-25.0), 6 females (cw 32.8-31.1) (NHM 2008.3378-3387), stn EA62.138, local streams near the Kakamega forestry station, Mokhubzia River, Kakamega (0°11′0″N 34°49′0″E), Western Province, 25.iii.1962, coll. T. R. Williams; 1 male (cw 53.4) (NHM 2008.3388), stn EA62.141, Sergoi River at Soy (between Eldoret and Kitale) Kakamega to Kitale region (0°35′0″N 35°1′0″E), Western Province, Kenya, 26.iii.1961, with

Potamonautes niloticus; 2 males (cw 25.4, 20.9), 1 female (cw 17.2) (NHM 2008.3389-3391), stn EA62.141, Sergoi River at Soy (between Eldoret and Kitale), Kakamega to Kitale region (0°35′0″N 35°1′0″E), Western Province, 26.iii.1962, with *Potamonautes niloticus*, coll. T. R. Williams; 10 males (cw 29.9-13.6), 13 females 32.0-13.3) (NHM 2008.3392-3401), stn EA62.143, Little Nzoia River on Eldoret-Kitale road, from Kakamega to Kitale (0°3′48″N 33°57′16″E), Western Province, 26.iii.1962, coll. T. R. Williams; 18 males (cw 35.9-14.6), 16 females (cw 40.2-13.5), 2 females ovig. (cw 43.1, 37.6) (NHM 2008.3402-3411), stn EA62.144, Kaibos Farm, west of Kitale, Kakamega to Kitale region (1°12′0″N 35°8′0″E), Western Province, 26.iii.1962, coll. T. R. Williams; 1 adult female (cw 34.5), 1 sub-adult male (cw 28.5) (NMU TRW 1960.01), river 16 km north of Kisumu, coll. J. P. McMahon (0°6′0″S 34°45′0″E); 1 adult male (cw 36.1), 1 adult female (cw 33), 3 sub-adult males (cw 31.5-32.5) (NMU TRW 1960.02a), Western Province, streams in the Kakamega-Kaimosi region, coll. R. B. Highton (0°12′0″N 34°57′0″E); 4 males (cw 36.5-31.5), 1 sub-adult female (cw 33.4) (NMU 1960.04a), Western Province, from river 16 km north of Kisumu, coll. J. P. McMahon (0°6′0″S 34°45′0″E); 1 sub-adult male (cw 26.5) (NMU TRW EA62.121), Sukute River, 2195 m asl, Koiwa Fisheries Rest House, Kericho, coll. T. R. Williams, 1962 (0°22′0″S 35°17′0″E); 14 specimens (largest adult male cw 33 mm) (NMU TRW EA62.146), Cherangani Mts., Marun River at Kaibuibich, Sigor, coll. T. R. Williams (1°34′0″N 35°31′0″E); (NMU TRW 1963.03), Western Province, Kibos River, near Maragoli, 15 mi from Kisumu, on the Kakamega road, coll. M. J. Clarkson (0°10′0″S 34°46′0″E); 1 adult female (cw 33), 1 sub-adult male (cw 29.5), 1 sub-adult female (cw 32.5) (NMU TRW 5.IV.1972A), Mau Escarpment, Narok Road, coll. 5.iv.1972; 2 sub-adult females (cw 27, 21.5) (NMU TRW 5.iv.1972B), Mau Escarpment, Narok Road, coll. 5.iv.1972; 1 adult male (cw 36.5), 1 adult female (cw 38) (NMU TRW 5.IX.1988), Sosio River, Sosio Bridge, coll. 5.ix.1988; 1 sub-adult male (cw 28.5) (NMU MD 20.XI.2003A), Lukusi River, Lukusi Bridge, coll. M. Dobson, 20.xi.2003; 1 sub-adult male (cw 26) (NMU MD 20.XI.2003B), Lukusi River, Lukusi Bridge, coll. M. Dobson, 20.xi.2003; 1 adult female (cw 31.5) (NMU MD 20.XI.2003C), Lukusi River, Lukusi Bridge, coll. 20.XI.2003; 1 sub-adult male (cw 21.5) carapace damaged (NMU MD 20.XI.2003D), Shitiya River, Kakamega Forest, coll. M. Dobson, 20.xi.2003; 1 adult male (cw 32) (NMU JF 21.II.2001), Lake Naivasha, coll. J. Foster, 21.ii.2001.

Type locality. — *Potamon* (*Geothelphusa*) *loveni*: Uganda, Mt. Elgon, Mbale Province. *Potamon* (*Geothelphusa*) *harvardi*: Uganda, Mt. Elgon, Sipi Falls, 1829 m asl. *Potamon* (*Geothelphusa*) *granviki* Uganda, Mt. Elgon.

Diagnosis. — Postfrontal crest incomplete; epigastric, postorbital crests faint, not joining with each other; postorbital crests faint, short, not extending laterally, not meeting anterolateral margins; exorbital, epibranchial teeth both missing; anterolateral margin completely smooth immediately behind epibranchial tooth; ischium of third maxilliped with faint vertical suture; thoracic sternal suture s3/s4 incomplete, missing on sides deep in middle; dactylus of major cheliped of adult male slender, highly arched, enclosing oval interspace when fingers closed, dentition absent; first, second carpal teeth on carpus of cheliped weak, low, blunt; ventral margins of merus of pereiopod 1 weakly granular; distal meral tooth either low or missing; terminal article of GO1 straight, slim, tapering evenly to pointed tip; terminal article not widened in middle, medial, lateral lobes both low, even; dorsal membrane narrow.

Description (based on sub-adult male holotype of *Potamon* (*Geothelphusa*) *loveni* Mt. Elgon and adult male lectotype of *Potamon* (*Geothelphusa*) *harvardi* from Sipi Falls, Mt. Elgon). — Carapace ovoid, wide (cw/fw 3.15), moderately high (ch/fw 1.2); surface texture completely smooth; semi-circular, urogastric, transverse branchial grooves present. Front straight, about one-third carapace width (fw/cw 0.32), anterior margin sharply deflexed. Postfrontal crest incomplete; epigastric, postorbital crests faint, not joining with each other; postorbital crests faint, short, not extending laterally, not meeting anterolateral margins; exorbital, epibranchial teeth missing; anterolateral margin completely smooth immediately behind epibranchial tooth, continuous with posterolateral margin. Suborbital margin raised, completely smooth. Suborbital, subhepatic, pterygostomial regions of carapace completely smooth; sidewall divided into three parts by longitudinal (epimeral) suture (dividing suborbital, subhepatic regions from pterygostomial region), by vertical (pleural) suture (dividing suborbital from subhepatic regions), dorsal end of vertical suture ending just before meeting epibranchial tooth. First thoracic sternal suture not visible; second suture s2/s3 deep, running horizontally across sternum; thoracic sternal suture s3/s4 incomplete, missing on sides deep in middle; episternal sutures s4/e4 to s7/e7 absent. Third maxillipeds filling entire oral field, except for transversely oval respiratory openings at superior lateral corners; long flagellum on exopod of third maxilliped, ischium of third maxilliped with faint vertical suture. Epistomial tooth prominent, smooth, triangular. Dactylus of major cheliped of adult male slender, highly arched, enclosing oval interspace when fingers closed, dentition absent. First, second carpal teeth of cheliped weak, low, blunt; second carpal tooth followed by several granules; ventral margins of merus of pereiopod 1 weakly granular; distal meral tooth either low or missing; superior surface of merus smooth. Pereiopods p2-p5 slender, p3 longest, p5 shortest, dactyli of p2-p5 tapering to point, each bearing 4 rows of downward-pointing short, sharp spines. Mandibular palp 2-segmented; terminal segment single, undivided, with setae (but no hard flap) at junction between segments. Adult male abdomen triangular, somites a1-a6 of male abdomen four-sided, telson (a7) triangular, apex rounded; somites a5-a6 broadest. Terminal article of GO1 straight, slim, tapering evenly to pointed tip, not widened in middle, medial, lateral lobes both low, even; dorsal membrane narrow. Terminal article of GO2 long, flagellum-like. Medium-size species, adult size ranging from cw 35-49 mm.

Geographical distribution. — *Potamonautes loveni* is found in the highlands of western Kenya including the Mau Range, Nandi Hills, Cherangani Mountains, and Mt. Elgon as well as the Kakamega and Kericho areas of Nyanza and

Western Provinces, Kenya. More recent collections have extended its range beyond the Nile drainage, eastward to Lake Naivasha in the Rift Valley, Kenya (John Foster, pers. comm.). In many parts of its range *P. loveni* is found at high altitude localities, including Kakamega (1465 m asl), Kericho (1770 m asl), the Cherangani Mountains (1830-1950 m asl) and Mt. Elgon (1070-3060 m asl). It is also found on the Ugandan side of Mt. Elgon in all seven radial river drainage sectors recognized by Williams (1968, 1991): Malawa, Manafwa, and Namatala (west), Siroko (north-west), Nyenye (north), Kiriki (north), and Suam (northeast). *Potamonautes loveni* is also found in the eastern and southern (Kenyan) sides of Mt. Elgon in the Suam and Nzoia drainage sectors that drain the southern and southeastern parts of the mountain, joining with rivers from the western side of the Cherangani Mountains that drain into Lake Victoria. Mount Elgon represents the western limit of the range of *P. loveni*.

Habitat. — *Potamonautes loveni* occurs in streams of all sizes, from fast-flowing rivers to forest trickles in cooler waters and is the species that is found at the highest altitudes on the mountain. At higher elevations where there is forest *P. loveni* lives under stones in the larger rivers and also burrows into the banks of small streams, particularly where the soil is loose and waterlogged. In the damper parts of the forest this species is found out of water where it emerges onto the riverbanks. In the cultivated land at lower altitudes *P. loveni* is found in burrows in stream banks and is rarely seen out of the water. *Potamonautes loveni* coexists with *P. niloticus* on Mt. Elgon within the range of altitudes between 1280-1950 m asl where the water is cooler than the lowland streams but warmer than at higher altitudes. The species reaches its lowest altitudes on Mt. Elgon in the Namatala River drainage sector where the forest remains largely intact, below which it is replaced by *P. niloticus* that prefers warmer waters (Williams, 1991; Cumberlidge, 2009b).

Medical importance. — Much of the material reported on here was collected as part of long-term studies of onchocerciasis (river blindness) in the highland areas of East Africa (the Ethiopian highlands, the Ruwenzoris, and Mt. Elgon) by T. R. Williams and his colleagues in the 1960s. Those investigations were aimed at identifying associations between freshwater crabs and the aquatic larval stages of the biting blackflies (*Simulium* Latreille, 1802) that serve as vectors for the parasite (*Onchocerca volvulus* (Leuckart, 1893)). The larval stages of *Simulium* need to develop in fast-flowing rivers and streams, and must attach themselves either to stones or to the carapaces of river-living species of freshwater crabs in order to complete their development into adult flies (McMahon, 1951; Barnley & Prentice, 1958; McMahon et al., 1958;

Williams et al., 1964; Crosskey, 1990; Williams, 1991). Control measures for the spread of river blindness focus on limiting the numbers of blackflies, and one possibility for such control involves understanding the relationship between *Simulium* larval stages and freshwater crabs. In the highland areas of western Kenya and Mt. Elgon, Uganda *P. loveni* has been demonstrated to be associated with onchocerciasis by Williams (1968, 1991) and Crosskey (1990). Other research in the Rwenzori (=Ruwenzori) Mountains, western Uganda indicates that a different species of freshwater crab (*P. aloysiisabaudiae* Nobili, 1906) also supports the larval stages of *Simulium neavei*, the blackfly vector of *O. volvulus*, which attach to the carapace of freshwater crabs living in the cold well-oxygenated waters of fast-running highland streams (Crosskey, 1990; Corace et al., 2001).

Conservation status. — *Potamonautes loveni* is listed as Least Concern (IUCN, 2004) in view of its wide distribution (it is known from more than 50 locations, mostly on the slopes of Mount Elgon in Uganda but the range is wide, covering an area spanning two countries), its tolerance of a degree of habitat modification, and its presumed large population. It is thought that it is unlikely to be declining fast enough to qualify for listing in a more threatened category and there are no known long-term threats. The stable population estimates are based on the fact that there are large numbers of specimens in museum collections, and this species has been collected recently in Kenya, which implies that populations are not necessarily declining (Cumberlidge, 2008; Cumberlidge et al., 2009).

Taxonomic remarks. — *Potamonautes loveni* was first collected by members of the Swedish Zoological Expedition to Kilimanjaro, Meru and the surrounding Masai plains between 1905 and 1906 (Colosi, 1924; Sjöstedt, 1925), and by the American Museum Expedition (Rathbun, 1935). Other specimens from the highland regions of western Kenya and eastern Uganda were collected in the early 1960s by Trefor R. Williams and his colleagues as part of an onchocerciasis study (Williams, 1968, 1991). The type material of *Potamon* (*Geothelphusa*) *loveni longimerus* Roux, 1935, has not been examined here and so we are unable to comment on the suggestions by Bott (1955) and Williams (1991) that this taxon is a junior synonym of *P. loveni.* Until the type specimens can be examined, this taxon remains tentatively listed as a junior synonym of *P. loveni* (cf. Ng et al., 2008: 171).

It is important to be able to identify the freshwater crabs involved in the transmission of river blindness to humans. However, this has proven to be difficult in the past for medical entomologists seeking the identity of crabs with *Simulium* larvae attached to their carapaces. For example, the species from Mt. Elgon originally identified by Williams et al. (1961) and Hynes et al. (1961) as "*P. berardi*" was referred to by Williams et al. (1961) as *P. granviki* (Colosi, 1924). Subsequently, Bott (1955) recognized the latter taxon as *Potamonanutes* (*Rotundopotamonautes*) *granviki*, and he placed several other taxa in synonymy with it. Williams (1968, 1991) pointed out that priority would dictate that this species should properly be referred to as *Potamonautes loveni* (Colosi, 1924) because *Potamon* (*Geothelphusa*) *granviki* Colosi, 1924, is a junior synonym of *Potamon* (*Geothelphusa*) *loveni* Colosi, 1924. This opinion was accepted by Cumberlidge (1997, 1998) following examination of the type specimens of these two taxa, and had implications for the several taxa that Bott (1955) had placed in synonymy with his *P.* (*R.*) *granviki.*

The revision of *P. loveni* presented here was necessary not only because of the above taxonomic confusion, but also because of our recognition (arising from the re-examination of material collected by Williams), that Williams (1968, 1991) included two other (as yet undescribed) species from Mt. Elgon in his treatment of *P. loveni* (see Cumberlidge & Clark, 2010).

ACKNOWLEDGEMENTS

This work is dedicated to Prof. Danièle Guinot in recognition for her outstanding contributions to the field of decapod crustacean biology, and for her expert guidance to both authors over the years. Trefor R. Williams (retired, formerly of the University of Liverpool, U.K.) is thanked for donating to NHM and NMU the specimens described here, and for many thoughtful discussions with the first author. Miranda Lowe (NHM) is thanked for loaning specimens, and Dr. Rafael Lemaitre and Marilyn Schotte (USNM) are thanked for their support during work by the first author on the USNM collection. Furthermore, Ardis Johnson (MCZ) and Karin Sindemark Kronestedt (SMNH) are thanked for their assistance and kind hospitality during visits by the first author to these museums. The Peter White Scholar Award (NMU) is thanked for its support for the first author, and Phil Crabb, of the NHM Photography Unit is thanked for taking the photographs reproduced in figs. 3, 6.

REFERENCES

BALSS, H., 1929. Ueber Ostafrikanische Potamonidae (Decapoda). Mit Anhang; Potamoniden von Madagascar. Zoologische Jahrbücher, (Abteilung für Systematik, Geographie und Biologie der Thiere), **58**: 339-358, figs. 1, 2.

BARNLEY, G. R. & M. A. PRENTICE, 1958. *Simulium neavei* in Uganda. East African Medical Journal, **35**: 475.

BOTT, R., 1955. Die Süsswasserkrabben von Afrika (Crust., Decap.) und ihre Stammesgeschichte. Annales du Musée Royal du Congo Belge, (C, Zoologie) **1** (3) (3): 209-352.

— —, 1970. Betrachtungen über die Entwicklungsgeschichte und Verbreitung der Süßwasser-Krabben nach der Sammlung des Naturhistorischen Museums in Genf/Schweiz. Revue Suisse de Zoologie, **77** (2): 327-344.

BOUVIER, E. L., 1921. Voyage de Ch. Alluaud et R. Jeannel en Afrique orientale (1911-1912). Résultats scientifiques, Crustacés, Decapoda. Librairie des Sciences Naturelles, **3**: 23-62. (L. Lhomme, Paris).

CHACE, F. A., 1942. III. Decapod Crustacea. In: Scientific results of the fourth expedition to forested areas in eastern Africa. Bulletin of the Museum of Comparative Zoology, Harvard College, **91**: 185-233.

COLOSI, G., 1924. Potamonides africains du Museum de Stockholm. Arkiv für Zoologie, **16**: 1-24.

CORACE, R. G., N. CUMBERLIDGE & R. GARMS, 2001. A new species of freshwater crab from Rukwanzi, East Africa. Proceedings of the Biological Society of Washington, **114**: 178-187.

CROSSKEY, R. W., 1990. The natural history of blackflies: 1-711. (Wiley, London).

CUMBERLIDGE, N., 1997. The African and Madagascan freshwater crabs in the Museum of Natural History, Vienna (Crustacea: Decapoda: Brachyura: Potamoidea). Annalen des Naturhistorischen Museums in Wien, **99B**: 571-589.

— —, 1998. The African and Madagascan freshwater crabs in the Zoologische Staatssammlung, Munich (Crustacea: Decapoda: Brachyura: Potamoidea). Spixiana, **21** (3): 193-214.

— —, 2008. *Potamonautes loveni*. In: IUCN 2009. IUCN Red List of Threatened Species. Version 2009.1. <www.iucnredlist.org>.

— —, 2009a. A revision of the freshwater crabs of Mt. Kenya and the Aberdare Mountains, Kenya, East Africa (Brachyura: Potamoidea: Potamonautidae). Zootaxa, **1981**: 29-42.

— —, 2009b. Chapter 27, Freshwater Crabs and Shrimps (Crustacea: Decapoda) of the Nile Basin. In: H. J. DUMONT (ed.), The Nile. Origin, environments, limnology and human use. Monographiae Biologicae, **89**: 547-561. (Springer, New York).

CUMBERLIDGE, N. & P. F. CLARK, 2010. Two new species of freshwater crabs from Mt. Elgon, East Africa (Brachyura: Potamoidea: Potamonautidae) with a redescription of *Potamonautes niloticus* (A. Milne-Edwards, 1887) with notes on their natural history and onchocerciasis. Journal of Natural History, in press.

CUMBERLIDGE, N. & M. DOBSON, 2008. A new species of freshwater crab (Crustacea: Decapoda: Brachyura: Potamonautidae) from western Kenya. Proceedings of the Biological Society of Washington, **121** (4): 468-474.

CUMBERLIDGE, N., P. K. L. NG, D. C. J. YEO, C. MAGALHAES, M. R. CAMPOS, F. ALVAREZ, T. NARUSE, S. R. DANIELS, L. J. ESSER, F. Y. K. ATTIPOE, F.-L. CLOTILDE-BA, W. DARWALL, A. MCIVOR, M. RAM & B. COLLEN, 2009. Freshwater crabs and the biodiversity crisis: importance, threats, status, and conservation challenges. Biological Conservation, **142**: 1665-1673.

CUMBERLIDGE, N. & M. VANNINI, 2004. Ecology and taxonomy of a tree living freshwater crab (Brachyura: Potamoidea: Potamonautidae) from Kenya and Tanzania, East Africa. Journal of Natural History, **38**: 681-693.

DOBSON, M., 2004. Freshwater crabs in Africa. Freshwater Forum, **21**: 3-26.

DOBSON, M., A. M. MAGANA, J. LANCASTER & J. M. MATHOOKO, 2007. Aseasonality in the abundance and life history of an ecologically dominant freshwater crab in the Rift Valley, Kenya. Freshwater Biology, **52**: 215-225.

HYNES, H. B. N., T. R. WILLIAMS & W. E. KERSHAW, 1961. Freshwater crabs and *Simulium neavei* in East Africa. I. — Preliminary observations made on the slopes of Mt. Elgon in

December, 1960, and January, 1961. Annals of Tropical Medicine and Parasitology, **55** (2): 197.

IUCN, 2004. IUCN Red List of threatened species. <www.iucnredlist.org>

MacLeay, W. S., 1838. Brachyurous decapod Crustacea. Illustrations of the Zoology of South Africa 5; being a portion of the objects of natural history chiefly collected during an expedition into the interior of South Africa, under the direction of Dr. Andrew Smith, in the years 1834, 1835, and 1836; fitted out by "The Cape of Good Hope Association for Exploring Central Africa." In: A. Smith, Illustrations of the Zoology of South Africa; Consisting Chiefly of Figures and Descriptions of the Objects of Natural History Collected During an Expedition into the Interior of South Africa, in the Years 1834, 1835, and 1836; Fitted Out by "The Cape of Good Hope Association for Exploring Central Africa." (Invertebrates), (1849): 1-75. (Smith, Elder and Co. London). [For dates of publication see Waterhouse 1880: 489-491.]

McMahon, J. P., 1951. The discovery of the early stages of *Simulium neavei* in phoretic association with crabs and a description of the pupa and the male. Bulletin of Entomological Research, **42**: 419-426.

McMahon, J. P., R. B. Highton & H. Goiny, 1958. The eradication of *Simulium neavei* from Kenya. Bulletin of the World Health Organisation, **19**: 75.

Ng, P. K. L., D. Guinot & P. Davie, 2008. Systema Brachyuorum: Part I. An annotated checklist of extant brachyuran crabs of the world. Raffles Bulletin of Zoology, (Supplement) **17**: 1-286.

Nobili, G., 1906. Spedizione al Ruwenzori di S. A. R. Luigi Amedeo di Savoia Duca degli Abruzzi. 9. Crostacei. Bollettino dei Musei di Zoologia e Anatomia Comparata, Torino, **544**: 1-2.

Rathbun, M. J., 1935. II. Crustacea. No. 2. — Scientific results of an expedition to rain forest regions in eastern Africa. Bulletin of the Museum of Comparative Zoology at Harvard College, **79**: 23-28.

Reed, S. K. & N. Cumberlidge, 2004. Notes on the taxonomy of *Potamonautes obesus* (A. Milne-Edwards, 1868) and *Potamonautes calcaratus* (Gordon, 1929) (Brachyura: Potamoidea: Potamonautidae) from eastern and southern Africa. Zootaxa, **1262**: 1-139.

Roux, R., 1935. Crustacés Décapodes d'eau douce. In: Voyage de Ch. Alluaud et P. A. Chappuls en Afrique occidentale française (déc. 1930-mars 1931). Archives Hydrobiologiques, **28**: 21-31.

Sjöstedt, Y., 1925. The Swedish Zoological Expedition to Kilimanjaro, Meru and the Surrounding Masai Plains, 1905-1906. Geografiska Annaler, **7**: 1-8.

Williams, T. R., 1964. The freshwater crabs of East Africa and their relation to *Simulium neavei* and onchocerciasis: 1-60. (M.Sc. Thesis, University of Liverpool).

— —, 1968. The taxonomy of the East African river crabs and their association with the *Simulium neavei* complex. Transactions of the Royal Society of Tropical Medicine and Hygiene, **62**: 29-34.

— —, 1991. Freshwater crabs and *Simulium neavei* in East Africa. III. Morphological variation in *Potamonautes loveni* (Decapoda: Potamidae). Transactions of the Royal Society of Tropical Medicine and Hygiene, **85**: 181-188.

Williams, T. R., H. B. N. Hynes & W. E. Kershaw, 1961. The distribution of *Potamon berardi*, *P. niloticus*, and associated *Simulium neavei* on Mt. Elgon, Uganda. Transactions of the Royal Society of Tropical Medicine and Hygiene, **55**: 307-308.

First received 29 May 2009.
Final version accepted 21 July 2009.

A NEW SPECIES OF *ILYOPLAX* (DECAPODA, BRACHYURA, DOTILLIDAE) FROM PANGLAO, THE PHILIPPINES

BY

PETER J. F. DAVIE[1,3]) and TOHRU NARUSE[2,4])

[1]) Queensland Museum, P.O. Box 3300, South Brisbane, Queensland, 4101, Australia

[2]) Transdisciplinary Research Organization for Subtropical and Island Studies, University of the Ryukyus, 870 Uehara, Taketomi, Okinawa 907-1541, Japan

ABSTRACT

A new species of *Ilyoplax*, *I. danielae*, is described from the Philippines. It differs from its closest congeners, *I. stapletoni* (De Man, 1908), *I. ningpoensis* Shen, 1940, and *I. formosensis* Rathbun, 1921, by differences in carapace shape and proportions, shape and dentition of the chela and the morphologies of the walking legs and male first gonopods.

RÉSUMÉ

Une nouvelle espèce d'*Ilyoplax*, *I. danielae* est décrite des Philippines. Elle diffère de ses congénères les plus proches, *I. stapletoni* (De Man, 1908), *I. ningpoensis* Shen, 1940, et *I. formosensis* Rathbun, 1921, par des différences portant sur la forme et les proportions de la carapace, la forme et les dents de la pince et la morphologie des pattes marcheuses et des gonopodes mâle.

INTRODUCTION

The genus *Ilyoplax* Stimpson, 1848, has had relatively little taxonomic attention since the 1930s, with only two new species having been described in recent times: *I. strigicarpus* Davie, 1990, from northern Australia, and *I. pacifica* Kitaura & Wada, 2006, from the Philippines and Indonesia. Both species are closely allied to *I. orientalis* (De Man, 1888). There are currently

[3]) Corresponding author; e-mail: Peter.Davie@qm.qld.gov.au

[4]) e-mail: naruse@lab.u-ryukyu.ac.jp

© Koninklijke Brill NV, Leiden, 2010

26 species recognized (see Ng et al., 2008), but an ongoing revision of this group by Davie & Kosuge indicates not only a number of new species, but also that *Ilyoplax* is polyphyletic and as such it will need to be split into several new genera. This concurs with recent behavioural and DNA phylogenetic studies by Kitaura & Wada (1998, 2006b) that also recognized three separate lineages within *Ilyoplax*. An earlier review of the genus by Serène & Lundoer (1974) also recognized three primary groups based on the morphology of the male first gonopod, although *I. integer* (Tesch, 1918) was considered to stand alone. In addition, their "Group II" was also found not to be homogeneous. Serène & Lundoer (1974) provided the most recent published key to *Ilyoplax* sensu lato.

Ilyoplax species, though typically small in size, can numerically dominate the shores on which they occur, and thus have some ecological significance (see Snowden et al., 1991). Like other shore Ocypodoidea they also have evolved some complex and noteworthy behaviours, and these have been the subject of numerous studies (e.g., Wada, 1981, 1983a-c, 1984, 1987a, b; Kosuge et al., 1994; Kitaura & Wada, 1998, 2006a).

The new species reported here was collected as part of the PANGLAO Marine Biodiversity Project of 2004 undertaken by the Raffles Museum of Biodiversity Research, Singapore, the Museum national d'Histoire naturelle, Paris, and the Philippine National Museum, Manila. Extensive collections were made from Panglao Island area in Bohol, the Philippines, by sampling of mudflats, seagrass beds and mangroves, and by scuba diving, trawling, dredging, trapping, and tangle netting. Together with collections from the PANGLAO Deep-Sea Cruise in 2005, it has been estimated that between 150 and 250 new crustacean species have been discovered from a total of more than 1200 identified decapods (see Bouchet et al., 2009).

The material examined is deposited in the Crustacean Collection of the National Museum of the Philippines (NMCR), Manila; Zoological Reference Collection (ZRC) of the Raffles Museum of Biodiversity Research, National University of Singapore; and the Queensland Museum (QM), Brisbane. The abbreviations P2-P5 refer to the pereopods 2-5, respectively; and G1 refers to the male first gonopod. Measurements provided are of the carapace width (CW) and length (CL), respectively.

TAXONOMY

DOTILLIDAE Stimpson, 1858

Ilyoplax Stimpson, 1858

Ilyoplax Stimpson, 1858: 98. (Type species: *Ilyoplax tenella* Stimpson, 1858 by monotypy.)
Dioxippe De Man, 1888: 137 (junior homonym of *Dioxippe* Thomson, 1860 [Insecta, Coleoptera]; gender feminine). (Type species: *Dioxippe orientalis* De Man, 1888 by subsequent designation of Davie, 2002.)
Tympanomerus Rathbun, 1897: 164 (new name for *Dioxippe* De Man, 1888; gender masculine).

Ilyoplax danielae sp. nov. (figs. 1, 2)

Material examined. — Holotype: NMCR, male (5.4 × 3.9 mm), Stn B31, Dauis, Panglao I., Bohol Province, Philippines, rubble on sandy bottom and seagrass, 1-2 m, 9°38.7′N 123°50.4′E, coll. PANGLAO 2004 Expedition, 26 June 2004. Paratypes: QM-W28924, male (4.6 × 3.3 mm), data as for holotype. ZRC, female (5.2 × 3.7 mm), Stn M47, Panglao, Philippines, fringe of mangrove, 0-0.5 m, middle part of Danao embayment, 9°33.6′N 123°45.1′E, coll. PANGLAO 2004 Expedition, 29 June 2004.

Etymology. — Named in honour of our friend and mentor Danièle Guinot. The species epithet is used as a noun in the genitive case.

Description. — Carapace (figs. 1, 2A) sub-rectangular, 1.4 times as wide as long, widest between anterolateral angles, lateral margins slightly converging posteriorly; regions poorly defined; smooth or microscopically granular, with short transverse ridges edged with very short setae. Front relatively broad, about one-quarter distance between exorbital angles, noticeably deflexed, lateral borders converging, rounded; distinctly concave dorsally. Gastrocardiac groove defined; cardiac region with low swelling on either side of mid-line, each bearing transverse ridge; branchial regions with several low oblique setiferous crests posterolaterally. Posterior border straight, broadly rounded laterally, with broad rim. Supraorbital borders moderately sinuous; sloping obliquely backward; smooth. Exorbital angles bluntly rounded, separated from rounded anterolateral prominence by broad shallow sulcus; anterolateral margin minutely granular, extending as oblique ridge for short distance onto dorsal carapace surface. Infraorbital borders project beyond supraorbital borders medially; evenly granular; slightly sinuous, separated from exorbital angle by V-shaped notch; secondary minutely granular ridge extending below ocular peduncle from adjacent to basal antennal segment, creating a broad, transverse sulcus in front of orbit. Eyestalks thick, approx. 0.4 times distance between exorbital angles; cornea bulbous.

Fig. 1. *Ilyoplax danielae* sp. nov., holotype male (5.4 × 3.9 mm), Panglao I., Bohol Province, Philippines (NMCR).

Third maxillipeds (fig. 2B) with merus longer than ischium (approx. 1.3×); antero-internal angle of ischium broadly triangular; produced along edge of merus; ischium with oblique distolateral line of microscopic granules equipped with long plumose setae, longest near outer margin. Maxillipeds moderately bulging; completely closing buccal cavity.

Male abdomen (fig. 2G) with telson elongate, apically rounded, slightly shorter and narrower than sixth somite; sixth somite with sub-parallel margins, 1.15 times as long as wide; fifth somite much shorter (approx. 0.66×) than sixth, markedly constricted at about 0.4 height from base; fourth and third somites similar, fourth slightly narrower, with slightly concave lateral margins; second short, similar in width; first with transverse keel subparallel to anterior margin. Female abdomen relatively narrow (fig. 2F).

Chelipeds (fig. 2C) massive, equal; merus trihedral, inner and outer margins granular, posterior border rounded; outer face sparsely granular; inner face with vertical medial band of small, well-spaced granules. Carpus short, rhomboidal, unarmed. Chela of adult male holotype greatly inflated, with broad gape basally between fingers; dorsal margin of palm marked by raised granular

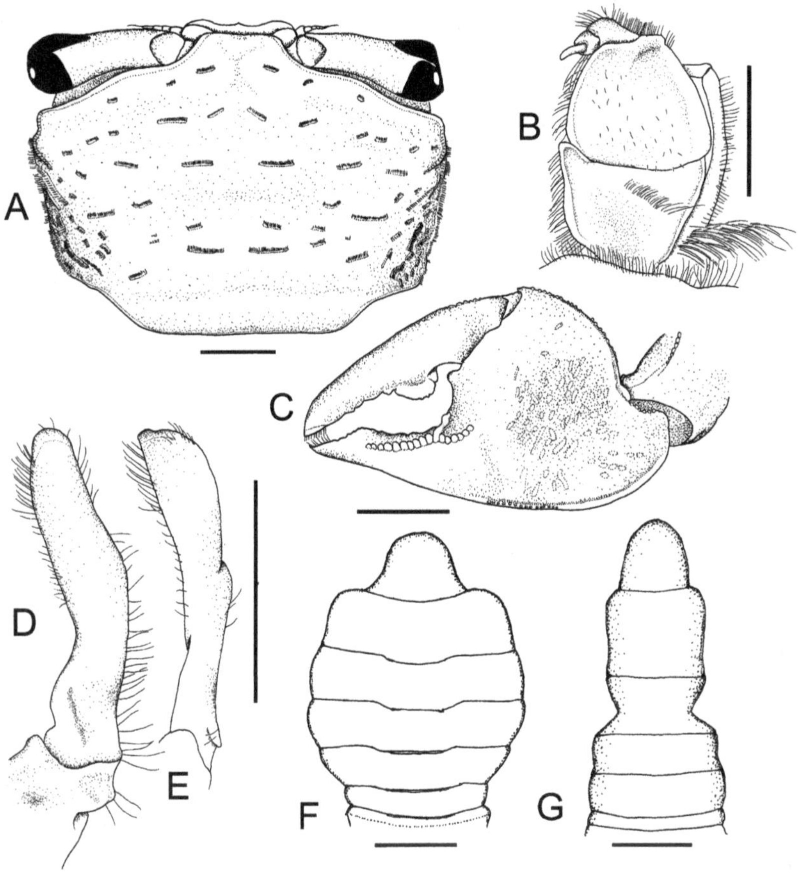

Fig. 2. *Ilyoplax danielae* sp. nov., A-E, G, holotype male, Panglao I., Bohol Province, Philippines (NMCR); F, paratype female (ZRC). A, carapace; B, third maxilliped; C, left chela external aspect; D, G1 abdominal view; E, G1 sternal view; F, female abdomen; G, male abdomen. Scale lines = 1 mm.

rim extending from articulation with carpus to inner edge of dactylus; outside face smooth; lower margin with finely granular low crest extending full length almost to tip of fixed finger. Fixed finger with cutting margin with low teeth becoming granular distally; slightly raised broad subdistal tooth; accessory row of large granules arising subdistally outside cutting margin, extending to base of finger, terminating in short, upwardly directed arc of larger granules. Fingers hollowed, tipped with chitinous rim, brush of setae on inner tips. Upper margin of dactyl finely granular; cutting margin with low teeth over distal half; large, broad, rounded tooth differentiated in basal half.

P2-P4 all of similar length; relatively broad; P4 merus approx. 2.3 times as long as wide, about same length as carpus and propodus combined; dactylus slender, smooth, tapering to acute point, noticeably shorter than propodus. Tympani not present on either face of meri. Dense fur present on distal two-thirds of carpus and proximal half of propodus of P2 and P3.

G1 (fig. 2D, E) short, thick, not tapering distally; slightly twisted medially; apex bluntly rounded, without obvious prominences, lobes or thick spines.

Habitat. — The species was collected from the edge of mangroves and seagrass beds in the intertidal zone.

Geographical distribution. — Only known from Panglao Island, Bohol Sea, central Philippines.

Remarks. — Using the key of Serène & Lundoer (1974), the present specimens of *Ilyoplax danielae* sp. nov. arrive at the last couplet that includes *I. stapletoni* (De Man, 1908) and *I. ningpoensis* Shen, 1940. Neither of these species are very similar. The carapace of *I. stapletoni* has more convex lateral margins and is widest medially, whereas in *I. danielae* the lateral margins are moderately convergent posteriorly, and the widest point is between the anterolateral angles just posterior to the orbit. The walking legs of *I. stapletoni* are also relatively more slender and lack the fur of setae on the P2-P4 carpi and propodi, and the male chela also lacks the large differentiated tooth near the base of the cutting margin of the dactylus.

Ilyoplax ningpoensis is a much larger species (adults to approximately 10 mm CW); the carapace is proportionately wider (approx. 1.5 times as wide as long) than *I. danielae* (1.4 times); the meri of the walking legs are relatively broader, with those of P2-P4 possessing tympani; and the male G1 is of a different form, possessing a long, tongue-like apical extension.

Ilyoplax danielae is most similar in carapace shape to *I. formosensis* Rathbun, 1921. It is also similar to the latter in lacking stridulatory teeth on the lower orbital margin, having short, broad legs with a thick fur of setae on the carpus and propodus of P2 to P4, and similar chela dentition. *Ilyoplax formosensis*, however, has a tympanum on the meri of the walking legs, lacks a marked constriction at the fifth segment of the male abdomen, and has an obviously different male G1 shape (cf. Serène & Lundoer, 1974, fig. 9).

Another species of *Ilyoplax*, *I. philippinensis* (Rathbun, 1914) was described from the shore at Guihulngan, Negros Island, in the central Philippines, relatively close to the north-west of Panglao Island. This species is still poorly known, but is being redescribed as part of the revision of the genus (P. J. F. Davie & T. Kosuge, work in progress). *Ilyoplax philippinensis* differs in the shapes of the carapace and gonopods and, most importantly, it lacks the strongly constricted fifth segment of the male abdomen that is characteristic of a large group of *Ilyoplax* species including *I. danielae* sp. nov.

ACKNOWLEDGEMENTS

We would like to thank Peter K. L. Ng of the National University of Singapore for passing the material to us and for encouraging us to describe this new species. The PANGLAO 2004 Expedition was organized by MNHN (Philippine Bouchet) together with the National University of Singapore (NUS, P. K. L. Ng), University of San Carlos (Danilo Largo) and National Museum of the Philippines (Marivene Manuel), with the financial support of TOTAL Foundation, ASEAN Regional Centre for Biological Conservation, MNHN and NUS; and we thank them all for their help. The first author is also grateful for the support of the Australian Biological Resources Study (Grant nos. 207-50 and 208-72); and a research fellowship to ZRC from NUS. The second author was also supported by a post-doctoral fellowship to NUS.

REFERENCES

BOUCHET, P., P. K. L. NG, D. LARGO & S. H. TAN, 2009. PANGLAO 2004. Investigations of the marine species richness in the Philippines. Raffles Bulletin of Zoology, Supplement No. **20**: 1-19.

DAVIE, P. J. F., 1990. New and rare crabs of the subfamily Dotillinae (Crustacea: Ocypodidae) from northern Australia and New Guinea. Memoirs of the Queensland Museum, **28** (2): 463-473.

— —, 2002. Crustacea: Malacostraca: Eucarida (Part 2: Anomura, Brachyura): 1-630. (CSIRO Publishing, Melbourne).

KITAURA, J. & K. WADA, 2006a. Evolution of waving display in brachyuran crabs of the genus *Ilyoplax*. Journal of Crustacean Biology, **26** (4): 455-462.

— — & — —, 2006b. New species of *Ilyoplax* (Brachyura: Ocypodidae: Dotillinae) from the Philippines and Indonesia: behavioural, molecular, and morphological evidence. Raffles Bulletin of Zoology, **54** (2): 373-379.

KITAURA, J., K. WADA & M. NISHIDA, 1998. Molecular phylogeny and evolution of unique mud-using territorial behavior in ocypodid crabs (Crustacea: Brachyura: Ocypodidae). Molecular Biology and Evolution, **15** (6): 626-637.

KOSUGE, T., S. POOVACHIRANON & M. MURAI, 1994. Male courtship cycles in three species of tropical *Ilyoplax* crabs (Decapoda, Brachyura, Ocypodidae). Hydrobiologia, **285**: 93-100.

MAN, J. G. DE, 1888. Report on the podophthalmous Crustacea of the Mergui Archipelago, collected for the Trustees of the Indian Museum, Calcutta, by Dr John Anderson, F.R.S., Superintendent of the Museum. Part III. Journal of the Linnean Society of London, **22** (138): 129-176, pls. 9-12.

— —, 1908. Decapod Crustacea, with an account of a small collection from Brackish water near Calcutta and in the Decca District, eastern Bengal. In: The fauna of brackish ponds at Port Canning, Lower Bengal, Part X. Records of the Indian Museum, Calcutta, **2**: 211-231, 2 pls.

NG, P. K. L., D. GUINOT & P. J. F. DAVIE, 2008. Systema Brachyurorum: Part I. An annotated checklist of extant brachyuran crabs of the world. Raffles Bulletin of Zoology, (Supplement) **17**: 1-286.

RATHBUN, M. J., 1897. A revision of the nomenclature of the Brachyura. Proceedings of the Biological Society of Washington, **11**: 153-167.

— —, 1914. New species of crabs of the families Grapsidae and Ocypodidae. Scientific Results of the Philippine Cruise of the Fisheries Steamer "Albatross", 1907-1910, **31**. Proceedings of the United States National Museum, **47** (2044): 69-85.

— —, 1921. New species of crabs from Formosa. Proceedings of the Biological Society of Washington, **34**: 155-156.

SERÈNE, R. & S. LUNDOER, 1974. Observations on the male pleopod of the species of *Ilyoplax* Stimpson with a key to the identification of the species. Phuket Marine Biological Center Research Bulletin, **3**: 1-10, figs. 1-11.

SHEN, C. J., 1940. On the collections of crabs of South China. Bulletin of the Fan Memorial Institute of Biology, Peiping, **10** (2): 69-104.

SNOWDEN, R. J., D. A. CLAYTON & E. Y. AL-TAHER, 1991. Population biology of *Ilyoplax stevensi* (Brachyura: Ocypodidae) on a Kuwait mudflat. Marine Ecology Progress Series, **71**: 219-225.

STIMPSON, W., 1858. Prodromus descriptionis animalium evertebratorum, quae in Expeditione ad Oceanum Pacificum Septentrionalem, a Republica Federata missa, Cadwaladaro Ringgold et Johanne Rodgers Ducibus, observavit et descripsit W. Stimpson. Pars. V. Crustacea Ocypodoidea. Proceedings of the Academy of Natural Sciences of Philadelphia, **10**: 93-110.

TESCH, J. J., 1918. The Decapoda Brachyura of the Siboga-Expedition. I. Hymenosomidae, Retroplumidae, Ocypodidae, Grapsidae, and Gecarcinidae. Siboga-Expeditie Monographie, **39c**: 1-148, pls. 1-6.

WADA, K., 1981. Growth, breeding, and recruitment in *Scopimera globosa* and *Ilyoplax pusillus* (Crustacea: Ocypodidae) in the estuary of Waka River, middle Japan. Publications of the Seto Marine Biological Laboratory, **26**: 243-259.

— —, 1983a. Movement of burrow location in *Scopimera globosa* and *Ilyoplax pusillus* (Decapoda: Ocypodidae). Physiology and Ecology Japan, **20**: 1-21.

— —, 1983b. Spatial distributions and population structures in *Scopimera globosa* and *Ilyoplax pusillus* (Decapoda: Ocypodidae). Publications of the Seto Marine Biological Laboratory, **27**: 281-291.

— —, 1983c. Temporal changes of spatial distributions of *Scopimera globosa* and *Ilyoplax pusillus* (Decapoda: Ocypodidae) at co-occurring areas. Japanese Journal of Ecology, **33**: 1-9.

— —, 1984. Barricade building in *Ilyoplax pusillus* (De Haan) (Crustacea: Brachyura). Journal of Experimental Marine Biology and Ecology, **83**: 73-88.

— —, 1987a. Neighbor burrow-plugging in *Ilyoplax pusillus* (Crustacea: Brachyura, Ocypodidae). Marine Biology, Berlin, **95**: 299-303.

— —, 1987b. Use of barricades as foraging sites by *Ilyoplax pusillus* (Crustacea: Brachyura: Ocypodidae). Journal of Ethology, **5**: 161-164.

WADA, K. & C.-H. WANG, 1998. Territorial and sexual behaviours in *Ilyoplax formosensis* and *I. tansuiensis* (Crustacea, Brachyura, Ocypodidae). Journal of Taiwan Museum, **51**: 119-125.

First received 20 July 2009.
Final version accepted 30 July 2009.

A NEW SPECIES OF *MICTYRIS* (DECAPODA, BRACHYURA, MICTYRIDAE) FROM THE RYUKYU ISLANDS, JAPAN

BY

PETER J. F. DAVIE[1,4]), HSI-TE SHIH[2,5]) and BENNY K. K. CHAN[3,6])

[1]) Queensland Museum, P.O. Box 3300, South Brisbane, Queensland, 4101, Australia

[2]) Department of Life Science, National Chung Hsing University, 250 Kuo-Kuang Road, Taichung 402, Taiwan

[3]) Research Centre for Biodiversity, Academia Sinica 128 section 2, Academia Road, Taipei 115, Taiwan

ABSTRACT

The Asian soldier crab *Mictyris brevidactylus* Stimpson, 1858, was re-examined genetically and morphologically from specimens collected across its range. It was found to consist of two cryptic sister species, with crabs from the Ryukyu Islands representing a new species, *M. guinotae*. *Mictyris guinotae* is smaller at maturity than *M. brevidactylus*, has plain blue colouring without red banding on the legs, and has a differently shaped apex of its male first gonopod. The two species also differ genetically, with 16S rRNA and COI showing a mean divergence (K2P distance) of 1.08% and 4.39%, respectively, representing 2.0 and 3.5 times more divergence than present intraspecifically. Results provide further evidence that some elements of the marine fauna of the Ryukyu Islands have had a separate evolutionary history from continental Asia as well as Taiwan.

RÉSUMÉ

Le crabe soldat asiatique *Mictyris brevidactylus* Stimpson, 1858, a été réexaminé à la fois génétiquement et morphologiquement à travers son aire de répartition. Il s'avère qu'il comprend deux espèces sœurs dont les crabes des îles Ryükyü représentent une nouvelle espèce, *M. guinotae*. *M. guinotae* est plus petit à la maturité que *M. brevidactylus*, il a une couleur bleue uniforme sans bandes rouges sur les pattes et la forme de l'extrémité du premier gonopode mâle est différente. Les deux espèces diffèrent aussi génétiquement avec le 16S rRNA et le COI montrant une divergence moyenne respective (distance K2P) de 1,08% et de 4,39%, ce qui représente 2 et 3,5 fois plus que la divergence intra spécifique. Les résultats fournissent de

[4]) e-mail: Peter.Davie@qm.qld.gov.au

[5]) e-mail: htshih@dragon.nchu.edu.tw

[6]) Corresponding author; e-mail: chankk@gate.sinica.edu.tw

© Koninklijke Brill NV, Leiden, 2010

nouvelles preuves que quelques éléments de la faune marine des îles Ryükyü ont eu une histoire évolutive séparée de celle du continent asiatique, Taïwan compris.

INTRODUCTION

Soldier crabs (species of *Mictyris* Latreille, 1806) are a conspicuous intertidal crab on many shores throughout tropical and temperate Australia, Southeast Asia and East Asia. They play an important ecological role in maintaining the proper functioning of the intertidal flats where they occur (Quinn, 1986; Dittman, 1994; Webb & Eyre, 2004) and experimental removal of one species, *M. longicarpus* Latreille, 1806, has resulted in the overgrowth of dense microbial mats (Webb & Eyre, 2004). *Mictyris* armies are also an important food source for shore birds (Zharikov & Skilleter, 2002, 2003, 2004; Rohweder & Lewis, 2004; Webb & Eyre, 2004).

The family Mictyridae contains a single genus *Mictyris*, with, at present, five recognized species: *M. longicarpus*; *M. platycheles* H. Milne Edwards, 1852; *M. brevidactylus* Stimpson, 1858; *M. livingstonei* McNeill, 1926; and *M. occidentalis* Unno, 2008. *Mictyris* species are remarkably similar in their appearance, and this has caused confusion about their specific status. Despite McNeill's (1926) excellent revision of the genus, there has been some longstanding confusion about the status of *M. longicarpus*. *Mictyris longicarpus* has been considered the only widespread species, with many literature records from throughout the tropical central Indo-West Pacific region (Takeda, 1978: 31). The first author is currently undertaking a modern revision of the genus, and after examining samples of *Mictyris* from throughout its known range, has concluded that *M. longicarpus* sensu stricto is confined to the eastern coast of Australia from where it was originally described. A number of new cryptic species, all with restricted, relatively narrow, geographical distributions will be described (P. J. F. Davie, in progress). Davie (2002) also pointed out that there were two undescribed species in northern and western Australia, with one of these, *M. occidentalis*, having been recently described by Unno (2008).

Takeda (1978) firmly established the separate species status of *Mictyris brevidactylus*, and this paper should be referred to for a thorough analysis of the differences between that species and *M. longicarpus*. However, as we shall demonstrate in the present paper, the Ryukyu Islands material that Takeda assumed was characteristic of *Mictyris brevidactylus* elsewhere in Taiwan, China, and the Philippines, is in fact a separate species, which is here described as new.

MATERIAL AND METHODS

The male first gonopods are referred to as G1. Measurements are given in millimetres. Carapace length (CL) was measured in the mid-line from the middle of the rostrum to the posterior margin; carapace width (CW) was measured across the widest point of the carapace. The material examined is deposited in the Queensland Museum (QM), the Muséum national d'Histoire naturelle, Paris (MNHN), the Zoological Reference Collection (ZRC) of the Raffles Museum of Biodiversity Research, National University of Singapore; Ryukyus University Museum, Okinawa (RUMF); National Museum of Natural Science (NMNS), Taichung, Taiwan; Zoological Collections of the Department of Life Science (NCHUZOOL), National Chung Hsing University, Taichung, Taiwan; Coastal Ecology Laboratory, Academia Sinica, Taipei (CEL); Senckenberg Museum, Frankfurt am Main (SMF); Western Australian Museum, Perth (WAM).

Specimens of *Mictyris* for genetic analyses were collected from the Ryukyus, Taiwan, Hong Kong, and Hainan Island, China. Samples were preserved in 75-95% ethanol after collection, and deposited in several museums (as above) (table I).

Genomic DNA was isolated from the muscle tissue of legs using the DNA genomic extraction kit. A region of ~550 base pairs (bp) at the 5′-end of the mitochondrial 16S rRNA gene was selected for amplification by a polymerase chain reaction (PCR) using the primers, 1471 (5′-CCTGTTTANCAAAAACAT-3′) and 1472 (5′-AGATAGAAACCAACCT GG-3′) (Crandall & Fitzpatrick, 1996). A portion of the COI gene was amplified by PCR using the primers, LCO1490 (5′-GGTCAACAAATCATAAAGAT ATTGG-3′) and HCO2198 (5′-TAAACTTCAGGGTGACCAAAAAATCA-3′) (Folmer et al., 1994). PCR conditions for the above primers included denaturation for 50 s at 94°C, annealing for 70 s at 45°C, and extension for 60 s at 72°C, followed by a final extension for 10 min. at 72°C. Sequences were obtained by automated sequencing (ABI PRISM 377 Sequencer and MegaBACE DNA Analysis System 500, Amersham, U.K.) and were aligned with the aid of CLUSTAL W (vers. 1.4, Thompson et al., 1994) and BIOEDIT (vers. 5.09, Hall, 2001), after verification with the complementary strand. Sequences of the different haplotypes were deposited in the DNA Data Bank of Japan (DDBJ), and the accession numbers are given in table I. The Kimura (1980) 2-parameter distance (K2P distance) between haplotypes was calculated by the software MEGA 4 (Tamura et al., 2007). This distance measure is adopted because it has been widely used for genetic comparisons among taxa (see Lefébure et al., 2006; Costa et al., 2007; Hubert et al., 2008).

TABLE I

Seven haplotypes of 16S ribosomal (r)RNA and 12 haplotypes of cytochrome oxidase subunit I (COI) genes of the two species of *Mictyris* from East Asia used in this study. NCHUZOOL, the Zoological Collections of the Department of Life Science, National Chung Hsing Univ.; CEL, Coastal Ecology Laboratory, Academia Sinica; DDBJ, the DNA Data Bank of Japan

Species	Localities	Catalog no.	Haplotypes of 16S	DDBJ access. no.	Haplotypes of COI	DDBJ access. no.
M. brevidactylus	Waziwei, Taipei County, Taiwan	CEL-MIC-TW2	Mb1	AB513630	Mb-C1	AB513638
Stimpson, 1858	Waziwei, Taipei County, Taiwan	CEL-MIC-TW1	Mb2	AB513631	Mb-C2	AB513637
	Siangshan, Hsinchu City, Taiwan	CEL-MIC-TW3	Mb1	AB513630	Mb-C1	AB513638
	Siangshan, Hsinchu City, Taiwan	CEL-MIC-TW4	Mb1	AB513630	Mb-C1	AB513638
	Haomeiliao, Chiayi County, Taiwan	NCHUZOOL 13262	Mb1	AB513630	Mb-C1	AB513638
	Starfish Bay, Hong Kong	CEL-MIC-HK-1	Mb2	AB513631	Mb-C2b	AB513639
	Starfish Bay, Hong Kong	CEL-MIC-HK-2	Mb2	AB513631	Mb-C2c	AB513640
	Wenchang, Hainan, China	CEL-MIC-HN-1	Mb2	AB513631	Mb-C2d	AB513641
	Wenchang, Hainan, China	CEL-MIC-HN-2	Mb2	AB513631	Mb-C2e	AB513642
	Wenchang, Hainan, China	CEL-MIC-HN-3	Mb2	AB513631	Mb-C2f	AB513643
	Sanya, Hainan, China	NCHUZOOL 13263	Mb2	AB513631	Mb-C2f	AB513643
M. guinotae sp. nov.	Awase, Okinawa, Ryukyus, Japan	NCHUZOOL 13264	Mg1	AB513632	Mg-C1	AB513644
	southern Naha, Okinawa, Ryukyus, Japan	CEL-MIC-OK-1	Mg2	AB513633	Mg-C2	AB513645
	southern Naha, Okinawa, Ryukyus, Japan	CEL-MIC-OK-3	Mg2	AB513633	Mg-C2b	AB513639
	southern Naha, Okinawa, Ryukyus, Japan	CEL-MIC-OK-2	Mg3	AB513634	Mg-C3	AB513646
	Funaura Bay, Iriomote, Ryukyus, Japan	NCHUZOOL 13266	Mg4	AB513635	Mg-C4	AB513647
	Funaura Bay, Iriomote, Ryukyus, Japan	NCHUZOOL 13266	Mg5	AB513636	Mg-C2b	AB513648

GENETIC RESULTS

A 553 bp segment (excluding the primer regions) of the 16S rRNA from all 17 specimens was amplified and aligned; 12 positions were variable and 5 parsimoniously informative. Among the total number of sequences, 7 different haplotypes were distinguished (table I). The studied segment of 16S rRNA sequences was AT rich (71.8%) (T, 34.5%; A, 37.3%; G, 17.7%; and C, 10.5%). For the COI gene, a 658 bp segment was compared, resulting in 12 different haplotypes (table I). The studied segment of the COI sequence was also AT rich (63.0%) (T, 39.1%; A, 23.9%; G, 17.8%; and C, 19.2%). In this gene, 43 positions were variable and 28 parsimoniously informative.

The pairwise base-pair (bp) differences and nucleotide divergence with the K2P distance between those haplotypes of 16S rRNA and COI are shown in tables II and III. It is clear that there are two clades among these haplotypes. The percentage base-pair difference and nucleotide divergence between the two clades for 16S rRNA was 0.72-1.27% and 0.73-1.28%, and for COI, 3.5-4.71% and 3.62-4.94% (tables II, III), respectively. Within the first clade (with the haplotypes prefixed as "Mb"), the bp differences and nucleotide divergence of 16S rRNA was ⩽0.18% for both clades; and for COI ⩽1.37% and ⩽1.23%, respectively. For the other clade (with the haplotypes prefixed as "Mg"), the bp differences and nucleotide divergence of 16S rRNA was ⩽0.90% and ⩽0.91% respectively; and for COI ⩽0.61% for both clades (tables II, III). The mean K2P distance of 16S rRNA between the two clades is 1.08%, which is twice the mean, 0.55%, within each clade. The mean K2P distance of COI between clades is 4.39%, which is 3.5 times as high as the mean, 1.24%, within each clade.

TAXONOMY

Family MICTYRIDAE Dana, 1851

Mictyridae Dana, 1851: 247 (as Myctiridae; spelling emended to Mictyridae by Alcock, 1900: 383).

Mictyris Latreille, 1806

Mictyris Latreille, 1806: 21.

Type species. — *Mictyris longicarpus* Latreille, 1806, by monotypy. Gender masculine.

Diagnosis. — Carapace deep, globular, longer than wide; dorsal surface with cervical, cardiobranchial grooves well defined; eyes exposed, orbits obsolescent, small post-ocular spine; posterior border of carapace with short brush

TABLE II

Pairwise nucleotide percentage divergence (K2P distance) matrix (lower-left) and the base-pair (bp) differences (upper right) based on 553 bp of the 16S ribosomal (r)RNA gene between haplotypes of two species of *Mictyris* from East Asia. For abbreviations of haplotypes see table I

	M. brevidactylus Stimpson, 1858		*M. guinotae* sp. nov.				
	Mb1	Mb2	Mg1	Mg2	Mg3	Mg4	Mg5
Mb1		1	7	4	6	6	5
Mb2	0.18		6	5	7	7	6
Mg1	1.28	1.10		3	5	5	4
Mg2	0.73	0.91	0.55		2	2	1
Mg3	1.10	1.28	0.91	0.36		4	3
Mg4	1.10	1.28	0.91	0.36	0.73		3
Mg5	0.91	1.10	0.73	0.18	0.54	0.55	

TABLE III

Pairwise nucleotide percentage divergence (K2P distance) matrix (lower-left) and the base-pair (bp) differences (upper right) based on 658 bp of the cytochrome oxidase subunit I (COI) gene between haplotypes of two species of *Mictyris* from East Asia. For abbreviations of haplotypes see table I

	M. brevidactylus Stimpson, 1858							*M. guinotae* sp. nov.				
	Mb-C1	Mb-C2	Mb-C2b	Mb-C2c	Mb-C2d	Mb-C2e	Mb-C2f	Mg-C1	Mg-C2	Mg-C2b	Mg-C3	Mg-C4
Mb-C1		5	6	3	3	5	5	24	26	24	23	25
Mb-C2	0.77		1	6	4	6	8	27	29	27	26	28
Mb-C2b	0.92	0.15		7	5	7	9	28	30	28	27	29
Mb-C2c	0.46	0.92	1.08		4	6	6	27	29	27	26	28
Mb-C2d	0.46	0.61	0.77	0.61		4	6	27	29	27	26	28
Mb-C2e	0.77	0.92	1.08	0.92	0.61		8	29	31	29	28	30
Mb-C2f	0.77	1.23	1.39	0.92	0.92	1.23		29	31	29	28	30
Mg-C1	3.78	4.27	4.44	4.27	4.27	4.60	4.61		4	2	3	3
Mg-C2	4.11	4.61	4.77	4.61	4.61	4.94	4.94	0.61		2	3	3
Mg-C2b	3.78	4.27	4.44	4.27	4.27	4.61	4.61	0.30	0.30		1	1
Mg-C3	3.62	4.11	4.27	4.11	4.11	4.44	4.44	0.46	0.46	0.15		2
Mg-C4	3.95	4.44	4.61	4.44	4.44	4.77	4.77	0.46	0.46	0.15	0.3	

of bristles, also present on apposed prominent edge of first abdominal somite. Interantennular septum very narrow; antennular flagellum rudimentary, folding almost vertically. Antennae small. Afferent branchial canal opening ventrally into wide cup-shaped dilatation adjacent base of third maxillipeds. Buccal cavern large, oval; completely closed by convex, expanded, foliaceous third maxillipeds. Thoracic sternal somites 1-3 deeply recessed. Third maxilliped with long, flattened, palp articulating at antero-external angle of merus; merus smaller than large hemispherical ischium; exopod slender, largely concealed, without flagellum. Chelipeds elongated, slender, subequal; similar in both sexes. Walking legs slender. Tympana not present on meri of walking legs or on sternum. Male abdomen with all somites distinct, broad, similar to female; base entirely covering sternum. Male and female genital openings sternal.

Mictyris guinotae sp. nov. (figs. 1, 2A, 4A-F, 5)

Myctiris brevidactylus Stimpson, 1858: 99; 1907: 103, pl. 13 fig. 4 [in part, only specimens from Loo Choo Islands (= Ryukyus)].
Myctiris longicarpus — Ortmann, 1894: 748. [Not *M. longicarpus* Latreille, 1806.]
Mictyris longicarpus — Sakai, 1934: 322; 1936: 222, pl. 60 fig. 4; 1939: 645, pl. 73 fig. 5; 1976: 627, pl. 213 figs. 1-3; Takeda, 1976: 158. [Not *M. longicarpus* Latreille, 1806.]
Mictyris brevidactylus — Takeda, 1978: 32-34, fig. 1B; Miyake, 1983: 158, pl. 53 fig. 4; Wada, 1995: 412, pl. 116 fig. 9; Muraoka, 1998: 51.

Material examined. — Holotype: QM-W28919, male (13.7 × 14.8), Funaura Bay, Iriomote I., Ryukyus, Japan, coll. T. Naruse, 24 Apr. 2009. Paratypes: QM-W28920, male (14.2 × 15.6) same data as holotype; QM-W28921, 21 males (6.5 × 8.1 to 13.4 × 14.8), 18 females (5.1 × 5.7 to 11.9 × 13.6), 1 juv. (3.7 × 4.2) same data as holotype. RUMF-ZC-00914, 11 males (10.0 × 11.8 to 14.1 × 15.7), 2 females (10.1 × 11.8, 11.6 × 13.2), Yagaji I., near Okinawa I., Ryukyus, Japan, coll. T. Maenosoho, 1 May 2009. NMNS-6109-001, 13 males (10.4 × 12.0 to 14.2 × 15.6), 3 females (8.9 × 11.5 to 10.4 × 12.1), Yagaji I., near Okinawa I., Ryukyus, Japan, coll. T. Maenosoho, 1 May 2009. NCHUZOOL-13261, 8 males (10.7 × 12.5 to 14.2 × 15.7), 8 females (9.5 × 10.9 to 11.1 × 12.5), Yagaji I., near Okinawa I., Ryukyus, Japan, coll. T. Maenosoho, 1 May 2009. ZRC, 14 females (6.4 × 7.2 to 11.8 × 12.3), 23 males (5.5 × 6.5 to 13.0 × 14.6), Katabaru, Ginoza Village, Okinawa I., Ryukyus, Japan, coll. T. Nagai, 7 May 2009. MNHN, 13 females (5.8 × 6.8 to 11.8 × 12.9), 23 males (6.5 × 7.4 to 13.4 × 15.1), Katabaru, Ginoza Village, Okinawa I., Ryukyus, Japan, coll. T. Nagai, 7 May 2009. SMF-7694, 4 males (5.4 × 6.3 to 13.2 × 14.2), 2 females (4.6 × 5.7, 9.4 × 10.9), Ishigaki Is., Ryukyus, Japan.

Etymology. — Named in honour of Professor Danièle Guinot for her lifetime commitment to carcinology, for always being willing to help, and for her friendship. The species epithet is used as a noun in the genitive case.

Description. — Body subglobular, slightly longer (1.1 times) than broad, breadth about same as distance from hind margin to level of anterolateral spines. Posterior border moderately projecting, square-cut, with rounded lateral margins. Carapace relatively smooth but with sparse clusters of microscopic granules; branchial regions evenly, closely, microscopically granular. Anterolateral spines prominent, directed anteriorly dorsal, not divergent, slightly to strongly granular, apically blunt; clearly defined ridge continuing posteriorly from anterolateral spines to edge of anterior branchials; short minutely granular ridge ventrally from anterolateral spines to form lateral edge of ill-defined lower orbital margin. Carapace regions well-defined; branchial regions moderately swollen, clearly separated from posterolateral margins above walking legs by broad sulcus; subhepatic regions visible from above, separated anteriorly by a groove running posteriorly from about level of anterolateral spine. Eyes globose with 1 or 2 short distal setae. Rostrum deflexed, distally triangular, upper surface broadly sulcate; about as long as broad. External maxillipeds as for genus.

Chelipeds (fig. 1D). Ischium with large spine directed forwards. Merus with lower outer margin with stout spine distally, 2 smaller spines and some granules posteriorly; outer surface minutely granular, with faint transverse furrowing; posterior margin becoming more defined and granular distally. Carpus smooth with faint indication of transverse furrowing; deep longitudinal sulcus running full length parallel with inner edge; inner margin with sparse setae beginning in distal half that get longer and thicker towards articulation with chela; outer margin and anterior margin without setae; brush of long setae on inner face commencing about one-third from distal end, and two-thirds towards upper inner margin. Palm broad; dorsal margin 0.22-0.23 times length of palm plus fixed finger; fingers long, slender, broadly deflexed; dactyl with broad, flat-topped elevated tooth over proximal third of cutting margin; dactyl and immovable finger otherwise unarmed; cutting margin of fixed finger with submarginal groove lined with short setae; dactyl with similar subdorsal setiferous groove.

Walking legs long, slender, somewhat flattened; minutely granular on anterior and ventral surfaces, unarmed except for small spine at anterodistal border of ischium; mostly lacking setae except for fringing setae on carpus and dactylus; dactylus of last leg recurved, apically pointed, trihedral in cross-section, with 3 prominent setal rows.

Male abdomen (fig. 1C) in form of broad flap as typical for *Mictyris* species; fringed with fine setae that become longer distally; all somites freely moveable, length of each somite similar; first somite deeply concave distally, edged with brush of short setae proximally; somites 2-5 laterally diverging, widest point near distal edge of fifth somite; sixth somite tapering, laterally rounded; telson approximately twice as wide as long, articulating slightly below margin of sixth somite, deflexed to close sternal cavity. Female abdomen similar to that of male.

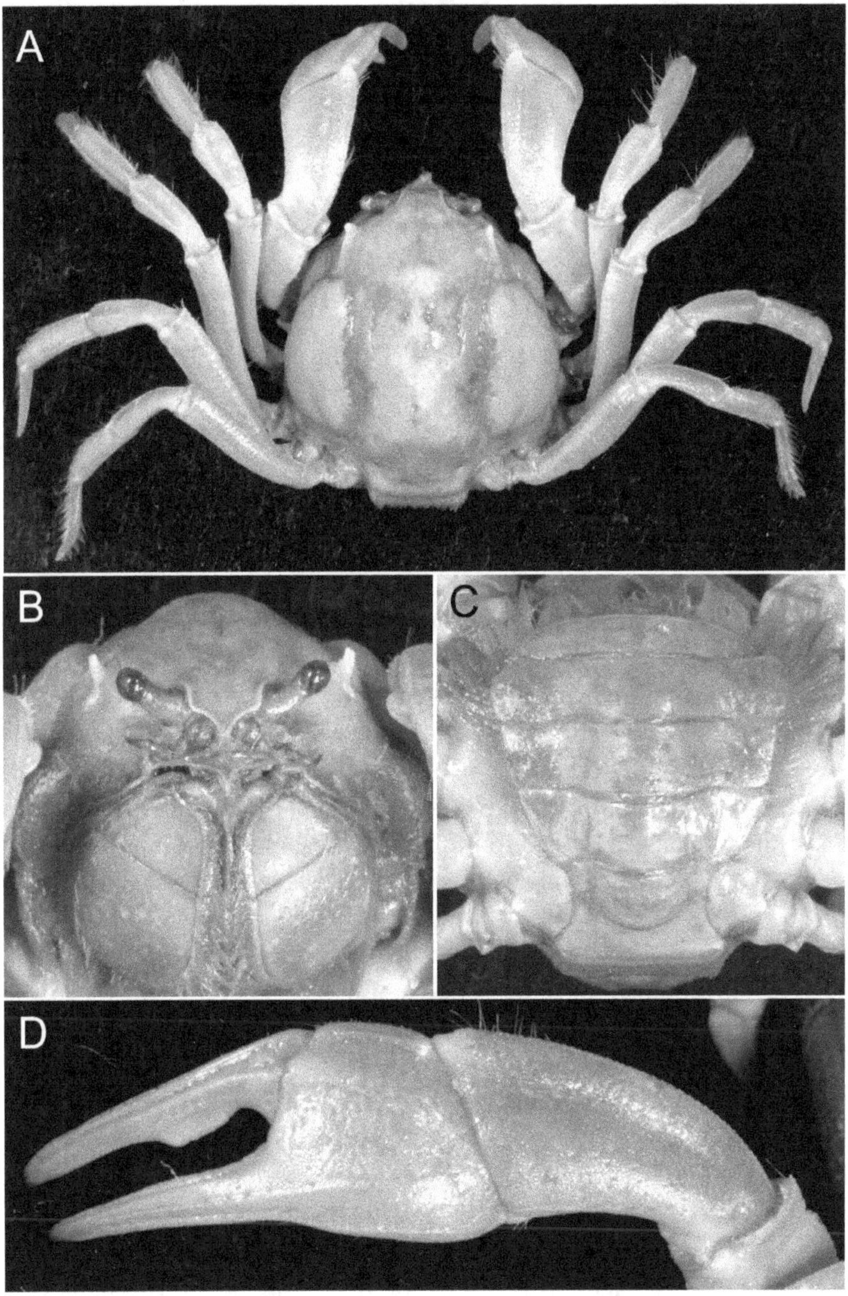

Fig. 1. *Mictyris guinotae* sp. nov. Holotype male from Funaura Bay, Iriomote Island, Ryukyu Islands (QM-W28919, CL 14.8 mm). A, dorsal view; B, mouth frame, ventral view; C, abdomen; D, cheliped.

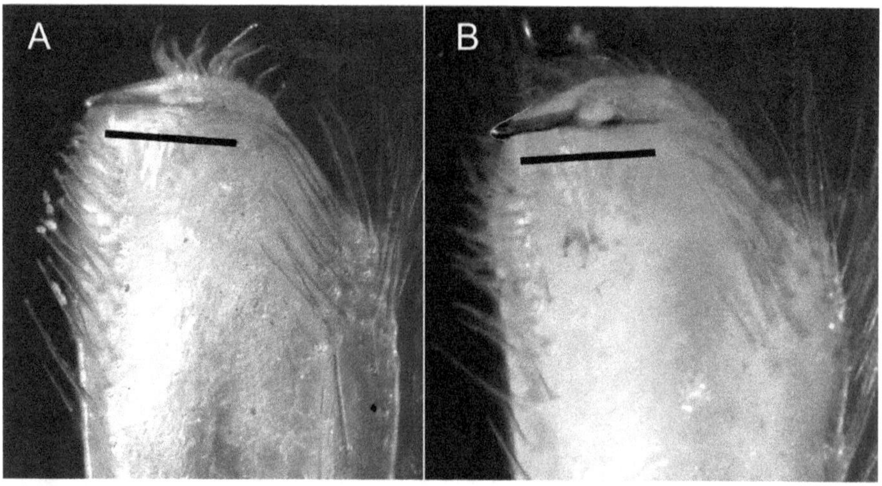

Fig. 2. Ventral view of distal portion of male G1: A, *Mictyris guinotae* sp. nov. from Amami Ohsima, Kagoshima Prefecture (QM-W28923); B, *Mictyris brevidactylus* from Taiwan (ZRC1998).

G1 (fig. 2A) slender, moderately setose, setae densest and longest distally; inner, well-developed, lateral subdistal shoulder, not produced into separate lobe; apex with conspicuous chitinous crest pointed sternolaterally and slightly upwards; opening of sperm tube clearly defined, placed medially near base of crest.

Colour. — *Mictyris guinotae* typically has pale and dark colour forms, though individuals can range from pale blue to a darker blue-grey to a dark ink-blue; occasional individuals may even be pale pinkish-fawn (see fig. 4A-F). The smallest ones tend to be the darkest, lightening as they grow and mature. The pale blue individuals are usually large, sexually mature males, and are more common in summer (T. Naruse, pers. comm.), which is perhaps related to mating. Walking legs similarly coloured to carapace but typically paler; never with red banding.

Remarks. — *Mictyris guinotae* sp. nov. is very similar morphologically to *M. brevidactylus* Stimpson, 1858. However, mature adults of both species can be easily separated by differences in colour pattern, with only *M. brevidactylus* exhibiting the broad red bands at the base of the walking legs.

There are also small but consistent morphological differences in the shape of the tip of the G1, and this is crucial for identifying preserved samples from which the colour has been lost. Even so, the differences in the G1 apex, while consistent, are not dramatic. In *M. guinotae* the chitinous apical crest is pointed sternolaterally and slightly upwards, whereas in *M. brevidactylus* it is a little more deflexed and pointed slightly downwards. This difference is emphasized by the placement of the black bars parallel with the apical crests in fig. 2A and B.

The shape of the G1 is proving consistently useful in separating known and putative new species of *Mictyris* (cf. P. J. F. Davie, work in progress), and it was through careful examination of the G1 that subtle but consistent differences between the Ryukyus and Hong Kong "*Mictyris brevidactylus*" were first noticed. While this, combined with the obvious colour differences, may have been enough to describe the new species, it was felt that genetic studies would be a useful test of the hypothesis since DNA differences would reinforce the value of the G1 as a useful diagnostic character among species. The present result is particularly useful in validating morphology as the differences between *M. guinotae* and *M. brevidactylus* are perhaps the most subtle of any of the future new species that will be recognized.

It also appears that individuals of *M. guinotae* are smaller than those of *M. brevidactylus*. There is commonly a disparity of size classes according to the amount of time since larval settlement events, and armies of smaller individuals commonly occur in *Mictyris*. Nevertheless, of more than 300 specimens examined from a number of localities, the largest *M. guinotae* individuals attained only 15.7 mm CL, and crabs of this size were rare. Sakai (1976) gave measurements of only 11 mm CL. On the other hand, we examined a number of specimens of *M. brevidactylus* that reached 17-18.9 mm CL, Stimpson (1858) himself gave a size of 17.3 mm (0.683 inches) CL, and Dai & Yang (1991: 453) recorded specimens up to 17.8 mm CL for that species. Pictures of marching armies (fig. 5B) generally seem to show smaller crabs than those of *M. brevidactylus* commonly seen around Hong Kong and on other continental beaches.

Ecology. — Sandy-mud intertidal flats, including estuaries; forming droving "armies" at low tide (Yamaguchi, 1976). Takeda & Murai (2004) discussed microhabitat use, and surface and subsurface feeding through burrow structure alteration. Fig. 5 shows a typical surface view of burrows in a sandy-mud flat in Ohura Bay, Okinawa, and a "marching troop" (dark colour form) at the same locality.

Reproduction and larval development. — Takeda (2005) reported on the sexual differences in the behaviour of *M. guinotae* during the breeding season. Nakasone & Akamine (1981) documented the reproductive cycle and early growth, and Fukuda (1990) described the early larval and postlarval morphology (under the name *Mictyris brevidactylus*), and made comparisons with the Australian *M. longicarpus* (see Fielder et al., 1984).

Geographical distribution. — Only known from the Ryukyu Islands, southern Japan, north to Tanega I. and south to Iriomote I. More precise localities include: Okinawa (Stimpson, 1907), Amami-Oshima and Okinawa (Sakai, 1976), Tanega I. (Takeda, 1976), Ishigaki I. (Miyake, 1983), Okinoerabu I. and Yoron I. (Muraoka, 1998).

Mictyris brevidactylus Stimpson, 1858 (figs. 2B, 3, 4G, H)

Myctiris brevidactylus Stimpson, 1858: 99; 1907: 103, pl. 13 fig. 4 (specimens from Hong Kong and China only).

Mictyris brevidactylus — Huang, Yu & Takeda, 1992: 155, fig. 18, pl. 2F; Kosuge et al., 1997: 180, 182; Ng et al., 2001: 39.

Not *Mictyris longicarpus* var. *brevidactylus* — McNeill, 1926: 116-119, text-fig. 2, pl. 9 figs. 3, 4 [undescribed species from the Philippines; P. J. F. Davie, in prep.].

Myctiris longicarpus — White, 1847: 34 (specimen from China only); Miers, 1884: 248 (specimens from South China Sea only); Ortmann, 1894: 748 (specimens from Hong Kong only); Parisi, 1918: 100 (specimen from Formosa [=Taiwan] only); Balss, 1922: 144 (specimens from Annam [Vietnam] only). [Not *M. longicarpus* Latreille, 1806.]

Mictyris longicarpus — Dai, Yang, Song & Chen, 1986: 415; Dai & Yang, 1991: 452, fig. 229B, pl. 58(1). [Not *M. longicarpus* Latreille, 1806.]

Material examined. — Neotype: WAM-237-80, male (CL 18.9, CW 17.1), Tong Fuk, Lantau, Hong Kong, coll. L. Bones, 29.iv.1980. Other material: WAM-239-80, 2 females (8.6 × 9.9, 8.1 × 9.6), male (12.2 × 14.3), Starfish Bay, Hong Kong, 21.iv.1980, coll. R. W. George. QM-W21355, 6 males (9.8 × 11.8 to 13.6 × 16.4), 3 females (9.2 × 11.3 to 11.3 × 12.3), sand flat, in pools, Three Fathom Cove, Hong Kong, 22°25′N 114°16′E, 20.iv.1989, P. Davie. QM-W27378, 2 males (8.7 × 10.1, 6.5 × 7.5), Starfish Bay, Hong Kong, 22°43′N 114°23′E, 12.iv.1989, coll. P. Davie. QM-W27379, 4 males (8.0 × 10.0 to 10.7 × 11.8), Hong Kong, 15.iv.1989, coll. P. Davie. QM-W27380, 2 females (8.1 × 10.1, 7.25 × 9.1), Starfish Bay, Wu Kwai Sha, Hong Kong, 22°43′N 114°23′E, 13.iv.1989, coll. R. Choi. ZRC-1999.1485, 2 males (8.4 × 9.4 to 9.4 × 10.7), 15 km from Halong City, Gulf of Tonkin, northern Vietnam, muddy sand, intertidal, 29.x.1999, coll. P. S. Tan. QM-W21919, 3 males (13.8 × 15.8, 13.9 × 15.5, 14.2 × 15.6), western Taiwan, 1996. QM W28888, 3 females (10.4 × 11.9, 11.0 × 11.7, 11.0 × 12.2), ovig. female (9.4 × 10.7), Anping, Tainan, Taiwan, 22°59′N 120°09′E, 26.iii.1998. QM-W28889, 6 males (12.1 × 13.6 to 13.6 × 15.2), Haomeiliao, Chiayi, Taiwan, 23°28′N 120°26′E, 06.ii.1996. ZRC-1995.619 (TMCD-2293), 2 males (15.6 × 17.5, 16.5 × 17.4), Yunlin County, Taiwan, 18.x.1987, coll. C. H. Wong. ZRC-1998.499 (CHC D388), 10 females (5.8 × 7.5 to 12.7 x 14.2), 11 males (8.0 × 10.0 to 15 × 16.4), Santiaolun, Yunlin County, Taiwan, 24.v.1994, coll. H. C. Lim. ZRC-1999.0705, 6 males (6.3 × 7.4 to 9.1 × 10.8), 7 females (7.0 × 8.0 to 8.7 × 10.0), Danshuei mangroves, Taipei County, Taiwan, 30.v.1999, coll. P. K. L. Ng. SMF-8786, male, near Mailiao, Yunlin County, western Taiwan, 27°47.02′N 120°10.80′E, silt/clay, fine sand, 29.viii.1977. SMF-8787, 5 males, 5 females, near Mailiao, Yunlin County, western Taiwan, 23°47.25′N 120°10.95′E, fine sand, silt/clay, 29.viii.1977. SMF-8788, female (7.7 × 9.1), stn 5, near Mailiao, west coast of Taiwan, 23°47.55′N 120°11.5′E, silt/clay, fine sand, 02.viii.1977. SMF-8789, 1 female (7.2 × 8.2), near Mailiao, Yunlin County, western Taiwan, 23°47.9′N 120°11.1′E, fine and medium grained

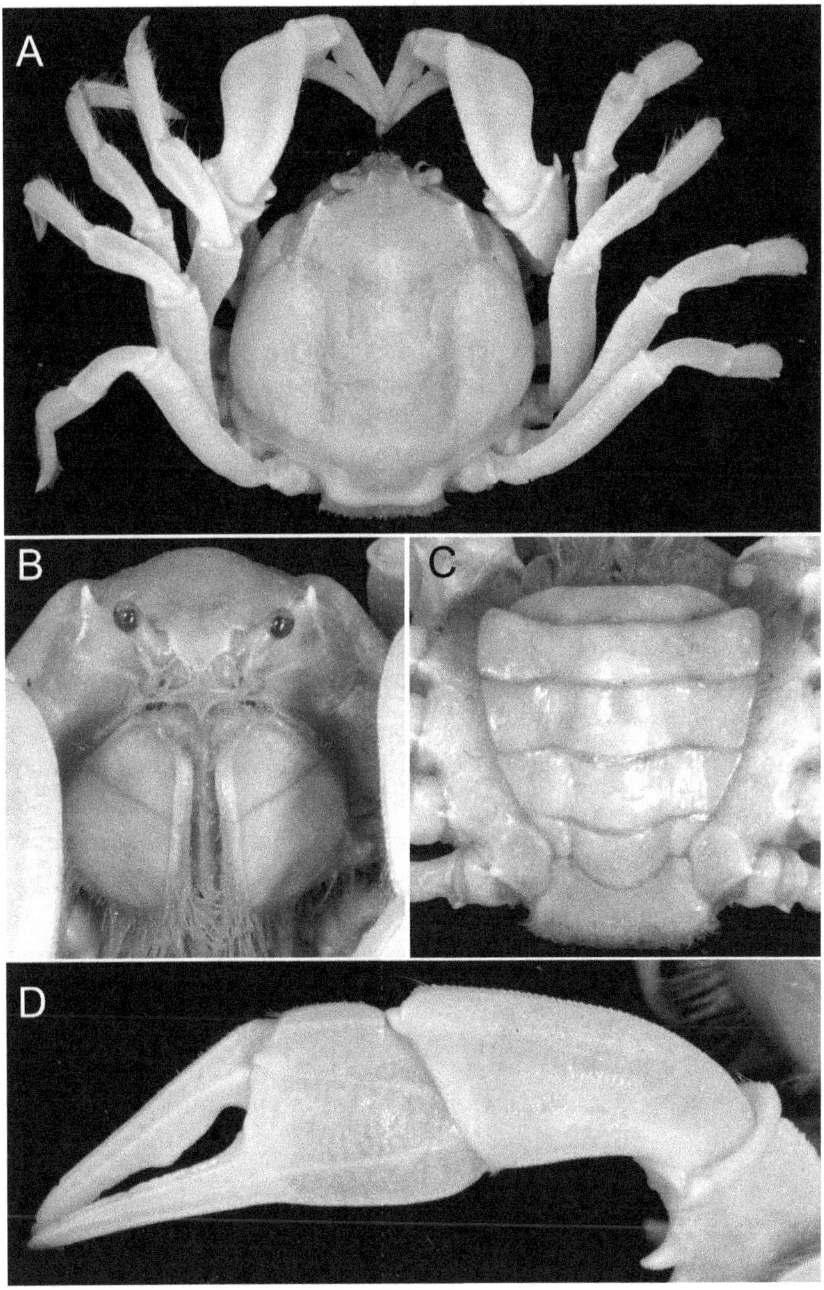

Fig. 3. *Mictyris brevidactylus* Stimpson, 1858. Neotype male (CL 18.9 mm), Tong Fuk, Lantau, Hong Kong (WAM-237-80). A, dorsal view; B, mouth frame, ventral view; C, abdomen; D, cheliped.

sand, 3.viii.1977. SMF-8790, 11 males (10.7 × 12.2 to 15.1 × 17.0), Taiwan, 1977. SMF-10675, 2 males (9.5 × 10.8, 9.5 × 9.6), female (9.0 × 9.9), east coast Taiwan, 1977.

Diagnosis. — See description of *M. guinotae* sp. nov. from which it is almost morphologically indistinguishable; apex of G1 (fig. 2B) has conspicuous chitinous crest that is more deflexed than *M. guinotae* sp. nov. and pointed slightly downwards.

Colour. — Adult males have a characteristic pale blue carapace, with wide bright red bands at the base of the walking legs, and pale creamy-yellow legs and claws (fig. 4G, H). Juveniles and immature crabs can be more plainly, evenly, and darkly coloured without the banding on the legs.

Remarks. — A nomenclatural complication has been caused by the fact that Stimpson's types of *M. brevidactylus* were lost when the Chicago Academy of Sciences was destroyed by fire in 1871 (fide Rathbun in Stimpson, 1907: 4). Because species differences in *Mictyris* can be subtle and confusing, it is necessary to establish a neotype for *M. brevidactylus* to stabilize the name. As Stimpson (1858) listed specimens from Hong Kong, China Sea, and "Loo Choo Islands" [=Ryukyus] which have two species among them, it is necessary to decide which species should bear Stimpson's name. Because Stimpson (1858) made extensive collections from Hong Kong, and because the Chinese and Taiwanese species is larger in size, more conspicuous, and more colourful than that from the Ryukyus, we believe that this was probably the species that Stimpson would have been most familiar with, and the one that would have most caught his attention. We have thus selected a specimen from Hong Kong (WAM-237-80) as the neotype of *M. brevidactylus* (fig. 3), and the soldier crab from the Ryukyus must have the new name, *M. guinotae*.

Differences between *M. brevidactylus* and *M. guinotae* sp. nov. are provided under the account of the latter species. In having broad red bands at the base of the walking legs, *M. brevidactylus* is similar to *M. longicarpus* from eastern Australia, and some earlier authors have considered it merely a subspecific variant of that species. *M. longicarpus* however has a much narrower band

Fig. 4. Live colour patterns. A-F, *Mictyris guinotae* sp. nov.: A, dark form male, from Ohura Bay, Okinawa I. (CW 15 mm); B, unusual pale colour form from Awase, Okinawa I.; C, intermediate darker form from Funaura Bay, Iriomote I. (paratype, QM-W28920, CW 14.2 mm); D, blue form from Ohura Bay (CW 20 mm); E, F, pale blue adult male from Funaura Bay, Iriomote I., Ryukyus (holotype, QM-W28919, CW 13.7 mm); G, H, *Mictyris brevidactylus* Stimpson, 1858 from Haomeiliao, Chiayi, Taiwan: G, dorsal view; H, frontal view; note characteristic pale blue carapace of adult males, with wide bright red bands at base of walking legs, and pale creamy-yellow legs and claws.

A
B
C
D
E
F
G
H

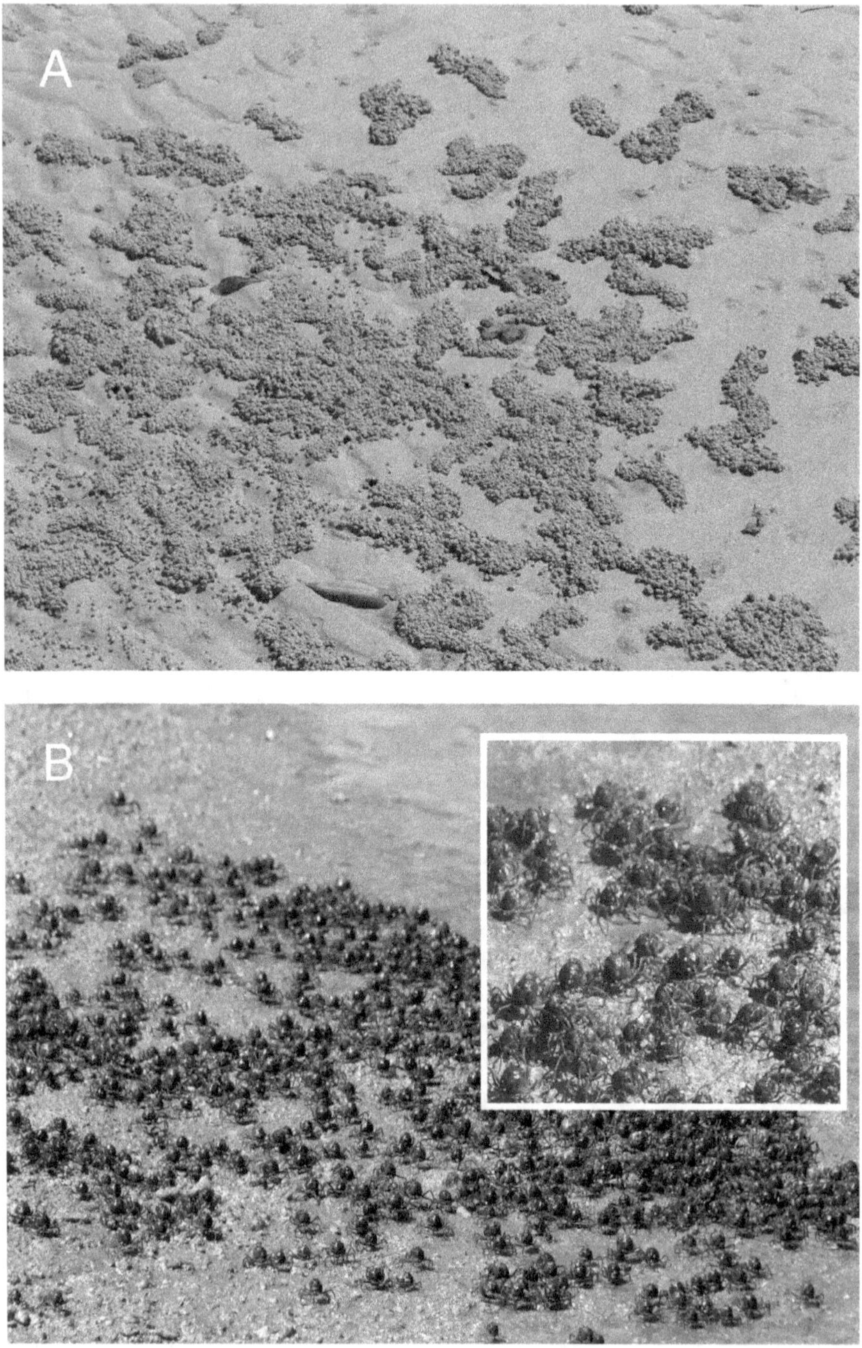

Fig. 5. A, typical feeding burrows of *Mictyris guinotae* sp. nov.; B, a marching troop (dark colour form). Both photographs taken on the sandy-mud flat of Ohura Bay, Okinawa.

at the base of the legs, and also another similar band over the meral-carpal articulation on each leg and on the cheliped (also see Takeda, 1978, for other morphological differences). Some *Mictyris* species, such as *M. livingstonei* and a number of the new species to be described, are naturally drab and inconspicuous in colour, but where a *Mictyris* species is brightly coloured, the colour and pattern appear to be useful diagnostic features.

Ecology. — Typically found at mid- to low tidal level on cleaner, sandy, sheltered, intertidal flats, where they form droving "armies" on the surface at low tide (Morton & Morton, 1983). Surface "armies" typically emerge in warm and sunny conditions, and mostly consist of adult or sub-adult males. Characteristic feeding pellets are deposited on the substrate, especially near the water's edge. When marching, the troop seems to wander with no purposeful direction. Females and juveniles tend to feed by tunnelling just below the surface forming raised meandering trails of pellets. Predators include shore birds and crabs of the family Matutidae, notably *Ashtoret lunaris* (Forskål, 1775) in Hong Kong (see Morton & Morton, 1983: 178), and *Matuta victor* (Fabricius, 1781) in Taiwan. Other predatory shore crabs such as species of *Ocypode*, *Helice*, and *Helicana* may also opportunistically prey on *Mictyris* (cf. Ho, 1995; H.-T. Shih, pers. obs.). See also Shih (1995) for a discussion of population densities and annual activity, and Shih et al. (1997) on the role of the hepatopancreas in young female crabs.

Geographical distribution. — Taiwan: Danshuei (Sakai, 1939), Danshuei River, Lugang, Wengang, and Cigu (Huang et al., 1992), Fukui et al. (1989), and present records. China: Hong Kong (Stimpson, 1858; Ortmann, 1894), Guangxi, Guangdong, including Hainan Island and Fujian (Dai & Yang, 1991). North Vietnam (Balss, 1922; Kosuge et al., 1997; present record).

DISCUSSION

This work has combined morphological and DNA approaches to understand a relatively subtle speciation event, and in so doing, provides further evidence to understand regional biogeography.

Different groups of crustaceans show different and varying degrees of divergence between species (see Lefébure et al., 2006; Costa et al., 2007) obviously reflecting the length of time since separate lineages have diverged. There is no simple rule that can be applied to DNA results that will necessarily guarantee that two putative species are reproductively isolated and thus biological species, however, comparisons with other taxa provide a rule-of-thumb approach, and a discontinuity in genetic variation between intra- and interspecific populations must also be expected. In the present study, *Mictyris brevidactylus* and *M. guinotae* have mean divergences (K2P distance) of 1.08% for 16S rRNA and 4.39% for COI (tables II and III), and these values, while low, are much larger (2.0 and 3.5 times respectively) than the intraspecific values. Thus the evidence from both genes supports the hypothesis that speciation has occurred. Similar results have also been recently reported for intertidal varunid crabs in the *Helice/Chasmagnathus* complex from East Asia (Shih & Suzuki, 2008). In that case minimum divergence between species of the *Helice latimera* Parisi, 1918 clade and *H. tridens* (De Haan, 1835) is 1.3% for 16S rRNA and 4.7% for COI (re-calculated from Shih & Suzuki, 2008).

Biogeographically, *M. guinotae* appears limited to the Ryukyus, while *M. brevidactylus* is more widely distributed from Taiwan and along the coast of China to Hainan Island and northern Vietnam. This distribution pattern closely mirrors that of another pair of intertidal crab sister-species, *Tmethypocoelis ceratophora* (Koelbel, 1897) and *T. choreutes* Davie & Kosuge, 1995 (Dotillidae), that are also separated by the strait between Taiwan and the Yaeyama Islands (which include Ishigaki and Iriomote Islands) (Davie & Kosuge, 1995). Naruse & Ng (2008) have also found the same distribution pattern for two sesarmid species, *Chiromantes haematocheir* (De Haan, 1833) and *C. ryukyuanum* Naruse & Ng, 2008. Similarly, *Scopimera ryukyuensis* Wong, Chan & Shih, 2010 has been described as endemic to the Ryukyu Islands, while its sister species *Scopimera intermedia* Balss, 1934 is found in Taiwan and mainland Asia south to Singapore and northern Indonesia. A further similar case concerns the mitten crab *Eriocheir japonica* (De Haan, 1835), with crabs from Okinawa proving to be genetically distinct from Taiwanese and Chinese populations, so also suggesting that there is a barrier to gene flow between the Ryukyus and waters to the south (Xu et al., 2009). This conjecture is strengthened by the floral composition of the mangrove communities between the two regions. Although Iriomote I. and north-eastern Taiwan are very close, the mangroves in Iromote I. are dominated by *Rhizophora* and *Bruguiera*, which are never found in NE and E Taiwan and are replaced by *Kandelia* in western Taiwan (Hosokawa et al., 1977). *Rhizophora*, *Bruguiera*, and *Kandelia* all produce dropper propagules that are dispersed by currents, so evidently the currents between the two areas are not intermixing. Contrary to this, an intertidal barnacle, *Tetraclita kuroshioensis* Chan, Tsang & Chu, 2007, that is common to mainland Japan and the Ryukyus, does also occur in northern Taiwan, though absent from the coast of mainland China (Chan et al., 2007a, b).

While there may be no universally consistent biogeographic pattern, nevertheless the Ryukyus appear to be much more influenced by the main Kuroshio Current in contrast to the continental coastline, which is impacted mainly by the South China Sea Current and westerly flowing Kuroshio Branch Current (Jan et al., 2002). Perhaps also the deep-water strait between Taiwan and the Yaeyama Islands plays an important role in the local circulation patterns of the region, and becomes an effective barrier for species that may have rapid larval development and/or abbreviated life cycles. In this respect, it would be particularly interesting to test the importance of the strait as a dispersal barrier by undertaking genetic studies on species pairs in groups such as benthic peracarids that brood their young and thus have much more limited dispersal capability.

ACKNOWLEDGEMENTS

We thank Tohru Naruse, Takashi Nagai, and T. Maenosoho (University of the Ryukyus, Japan), for help in collecting fresh specimens of *M. guinotae*. The first author is grateful for the support of an Australian Biological Resources Study Grant nos. 207-50 and 208-72, and to Lisa Pabriks (QM) and Tohru Naruse (University of the Ryukyus) for photography. The second author acknowledges the support of a grant from the National Science Council, Taiwan (NSC96-2621-B-005-005-MY3), and thanks I-Chu Chien and Kuo-Chih Wu for help with the molecular work. The third author thanks the National Science Council, Taiwan for grant support (NSC 96-2621-B-00I-007-MY3), Hsin-Ying Lee (Academia Sinica) for assistance with molecular work, and Kingsley Wong (National Taiwan University) for help with sampling in Hong Kong and Hainan Island.

REFERENCES

ALCOCK, A., 1900. Materials for a carcinological fauna of India. No. 6. The Brachyura Catametopa, or Grapsoidea. Journal of the Asiatic Society of Bengal, **69** (2) (3): 279-456.

BALSS, H., 1922. Ostasiatische Decapoden. IV. Die Brachyrhynchen (Cancridea). Archiv für Naturgeschichte, Berlin, (A) **88** (11): 94-166, text-figs. 1, 2, pls. 1, 2.

CHAN, B. K. K., L. M. TSANG & K. H. CHU, 2007a. Cryptic diversity of *Tetraclita squamosa* complex (Crustacea, Cirripedia) in Asia: description of a new species from Singapore. Zoological Studies, **46**: 46-56.

— —, — — & — —, 2007b. Morphological and genetic differentiation of the acorn barnacle *Tetraclita squamosa* (Crustacea, Cirripedia) in East Asia and description of a new species of *Tetraclita*. Zoologica Scripta, Stockholm, **36**: 79-91.

COSTA, F. O., J. R. D. WAARD, J. BOUTILLIER, S. RATNASINGHAM, R. T. DOOH, M. HAJIBABAEI & P. D. N. HEBERT, 2007. Biological identifications through DNA barcodes: the case of the Crustacea. Canadian Journal of Fisheries and Aquatic Sciences, **64**: 272-295.

CRANDALL, K. A. & J. F. FITZPATRICK, JR., 1996. Crayfish molecular systematics: using a combination of procedures to estimate phylogeny. Systematic Biology, **45**: 1-26.

DAI, A.-Y. & S.-L. YANG, 1991. Crabs of the China Seas: 1-682. (China Ocean Press & Springer Verlag, Beijing and Berlin).

DAI, A.-Y., S.-L. YANG, Y. Z. SONG & G. X. CHEN, 1986. Crabs of the China seas: 1-11, 1-642. (China Ocean Press, Beijing, China). [In Chinese.]

DANA, J. D., 1851. Conspectus Crustacearum quae in Orbis Terrarum circumnavigatione, Carolo Wilkes e classe Reipublicae Foederatae Duce, lexit et descriptsit J. D. Dana. Proceedings of the Academy of Natural Sciences of Philadelphia, **5**: 247-254.

DAVIE, P. J. F., 2002. Crustacea: Malacostraca: Eucarida (Part 2: Anomura, Brachyura): 1-630. (CSIRO Publishing, Melbourne).

DAVIE, P. J. F. & T. KOSUGE, 1995. A new species of *Tmethypocoelis* (Crustacea: Brachyura: Ocypodidae) from Japan. Raffles Bulletin of Zoology, **43** (1): 207-215.

DITTMANN, S., 1994. Impact of foraging soldier crabs (Decapoda: Mictyridae) on meiofauna in a tropical tidal flat. Revista de Biologia Tropical, **41** (3A): 627-637.

FIELDER, D. R., J. G. GREENWOOD & R. H. QUINN, 1984. Zoeal stages, reared in the laboratory, and megalopa of the soldier crab *Mictyris longicarpus* Latreille, 1806 (Decapoda, Mictyridae). Bulletin of Marine Science, **35** (1): 20-31.

FOLMER, O., M. BLACK, W. HOEH, R. LUTZ & R. VRIJENHOEK, 1994. DNA primers for amplification of mitochondrial cytochrome c oxidase subunit I from diverse metazoan invertebrates. Molecular Marine Biology and Biotechnology, **3**: 294-299.

FUKUDA, Y., 1990. Early larval and postlarval morphology of the soldier crab, *Mictyris brevidactylus* Stimpson (Crustacea: Brachyura: Mictyridae). Zoological Science, Tokyo, **7** (2): 303-309.

FUKUI, Y., K. WADA & C.-H. WANG, 1989. Ocypodidae, Mictyridae and Grapsidae (Crustacea: Brachyura) from some coasts of Taiwan. Journal of the Taiwan Museum, **42** (1): 225-238.

HALL, T. A., 2001. BioEdit: A user-friendly Biological Sequence Alignment Editor and Analysis, version 5.09. (Department of Microbiology, North Carolina State University, North Carolina).

HO, P.-H., 1995. Soldier crabs (3). Fisheries Extension, **108**: 17-20. [In Chinese.]

HOSOKAWA, T., H. TAGAWA & V. J. CHAPMAN, 1977. Mangals of Micronesia, Taiwan, Japan, the Philippines and Oceania. In: V. J. CHAPMAN (ed.), Wet coastal ecosystems: 271-291. (Elsevier, Amsterdam, Netherlands).

HUANG, J.-F., H.-P. YU & M. TAKEDA, 1992. A review of the ocypodid and mictyrid crabs (Crustacea: Decapoda: Brachyura) in Taiwan. Bulletin of the Institute of Zoology, Academia Sinica, **31** (3): 141-161, figs. 1-18, pls. 1, 2.

HUBERT, N., R. HANNER, E. HOLM, N. E. MANDRAK, E. TAYLOR, M. BURRIDGE, D. WATKINSON, P. DUMONT, A. CURRY, P. BENTZEN, J. ZHANG, J. APRIL & L. BERNATCHEZ, 2008. Identifying Canadian freshwater fishes through DNA barcodes. PLoS ONE, **3**: e2490.

JAN, S., J. WANG, C.-H. CHERN & S.-Y. CHAO, 2002. Seasonal variations of circulation in the Taiwan Strait. Journal of Marine Systems, **35**: 249-268.

KIMURA, M., 1980. A simple method for estimating evolutionary rates of base substitutions through comparative studies of nucleotide sequences. Journal of Molecular Evolution, **16**: 111-120.

KOSUGE, T., K. WADA & P. D. TRONG, 1997. Crab distribution in the Cam River estuary, Haiphong, northern Vietnam. In: P. N. ANH, J. T. BRANDS & P. N. HONG (eds.), National Workshop on the Relationship between Mangrove Rehabilitation and Coastal Aquaculture in Vietnam: 178-184. (Cres & Actmang, Hanoi).

LATREILLE, P. A., 1806. Genera Crustaceorum et Insectorum secundum ordinem naturalem in familias disposita, **I**: i-xviii, 1-302, pls. 1-16. (Parisiis et Argentorati).

LEFÉBURE, T., C. J. DOUDY, M. GOUY & J. GIBERT, 2006. Relationship between morphological taxonomy and molecular divergence within Crustacea: proposal of a molecular threshold to help species delimitation. Molecular Phylogenetics and Evolution, **40**: 435-447.

MCNEILL, F. A., 1926. Studies in Australian Carcinology. No 2. A revision of the family Mictyridae, the status and synonymy of the genus *Megametope* and its contained species. Records of the Australian Museum, **15** (1): 100-131, figs. 1-4, pls. 9, 10.

MIERS, E. J., 1884. Crustacea: 178-322, 513-575, pls. 18-32, 46-51, table. In: Report on the Zoological Collections made in the Indo-Pacific Ocean during the Voyage of H.M.S. 'Alert' 1881-2. Part I. The collections from Melanesia. Part II. Collections from the western Indian Ocean. (British Museum, London).

MILNE EDWARDS, H., 1852. De la famille des Ocypodides (Ocypodidae). Second mémoire. In: Observations sur les affinités zoologiques et la classification naturelle des Crustacés. Annales des Sciences Naturelles, Paris, (Zoology) **18**: 128-166, pls. 3, 4.

MIYAKE, S., 1983. Japanese crustacean decapods and stomatopods in color, **2**, Brachyura (crabs): i-viii, 1-277, pls. 1-64. (Hoikusha, Osaka). [In Japanese.]

MORTON, B. & J. MORTON, 1983. The sea shore ecology of Hong Kong: i-xiv, 1-350. (Hong Kong University Press, Hong Kong).

MURAOKA, K., 1998. Catalogue of the brachyuran and anomuran crabs donated by Prof. Dr. Tune Sakai to the Kanagawa Prefectural Museum. Catalogue of the Collection in the Kanagawa Prefectural Museum of Natural History, **11**: 5-67, pls. 1-16.

NAKASONE, Y. & T. AKAMINE, 1981. The reproductive cycle and young crab's growth of the soldier crab *Mictyris brevidactylus* Stimpson, 1858. The Biological Magazine, Okinawa, **19**: 17-23.

NARUSE, T. & P. K. L. NG, 2008. A new species of *Chiromantes* s. str. (Decapoda: Brachyura: Sesarmidae) from the Ryukyu Islands, Japan, with a note on the identity of *Holometopus serenei* Soh, 1978. Crustacean Research, **37**: 1-13.

NG, P. K. L., C.-H. WANG, P.-H. HO & H.-T. SHIH, 2001. An annotated checklist of brachyuran crabs from Taiwan (Crustacea: Decapoda). National Taiwan Museum Special Publication Series, **11**: 1-86.

ORTMANN, A., 1894. Die Decapoden-Krebse des Strassburger Museums. VIII. Theil. Abtheilung; Brachyura (Brachyura genuina Boas) III. Unterabtheilung: Cancroidea, 2. Section: Cancrinea, 2. Grupe: Catometopa. Zoologische Jahrbücher, Jena, (Abteilung für Systematik) **7**: 683-772, pl. 23.

PARISI, B., 1918. Decapodi gipponesi del Museo di Milano VI. Catometopa e Paguridae. Atti della Società Italiana di Science Naturale e del Museo Civile di Storia Naturale. Milano, **57**: 90-115, pl. 8 figs. 1-5.

QUINN, R. H., 1986. Experimental studies of food ingestion and assimilation of the soldier crab, *Mictyris longicarpus* Latreille (Decapoda, Mictyridae). Journal of Experimental Marine Biology and Ecology, **102** (2-3): 167-181.

ROHWEDER, D. A. & B. D. LEWIS, 2004. Day-night foraging behaviour of banded dotterels (*Charadrius bicinctus*) in the Richmond River estuary, northern New South Wales, Australia. Notornis, **51** (3): 141-146.

SAKAI, T., 1934. Brachyura from the coast of Kyushu, Japan. Science Reports of the Tokyo Bunrika Daigaku, Tokyo, (B) **1** (25): 281-330, text-figs. 1-26, pls. 17, 18.

— —, 1936. Crabs of Japan: i-xxvii, 1-239, pls. 1-66. (Sanseido, Tokyo).

— —, 1939. Studies on the crabs of Japan. IV. Brachygnatha, Brachyrhyncha: 365-741, text-figs. 1-129, pls. 42-111, tables 1-6. (Yokendo, Tokyo).

— —, 1976. Crabs of Japan and the adjacent seas: i-xxix, 1-773 [English vol.]; 1-16, pls. 1-251 [Plates vol.]; 1-461 [Japanese vol.]. (Kodansha, Tokyo).

SHIH, H.-T. & H. SUZUKI, 2008. Taxonomy, phylogeny, and biogeography of the endemic mudflat crab *Helice/Chasmagnathus* complex (Crustacea: Brachyura: Varunidae) from East Asia. Zoological Studies, **47**: 114-125.

SHIH, J.-T., 1995. Population densities and annual activities of *Mictyris brevidactylus* (Stimpson, 1858) in the Tanshui Mangrove Swamp of northern Taiwan. Zoological Studies, **34** (2): 96-106.

SHIH, J.-T., J.-W. CHANG & W.-M. LI, 1997. The role of the hepatopancreas of young female *Mictyris brevidactylus*. Biological Bulletin National Taiwan Normal University, **32** (2): 53-58.

STIMPSON, W., 1858. Prodromus descriptionis animalium evertebratorum quæin Expeditione ad Oceanum Pacificum Septentrionalem, a Republica Federata missa, Cadwaladaro Ringgold et Johanne Rodgers Ducibus, observavit et descripsit. Pars. V. Crustacea Ocypodoidea. Proceedings of the Academy of Natural Sciences of Philadelphia, **10**: 93-110. [Pp. 39-56 on original separate.]

— —, 1907. Report on the Crustacea (Brachyura and Anomura) collected by the North Pacific Exploring Expedition, 1853-1856 (with introductory note, and edited by Mary J. Rathbun). Smithsonian Miscellaneous Collections, **49** (1717): 1-240.

TAKEDA, M., 1976. Littoral and inshore decapod crustaceans, of Tanega-Shima Island, Southwest Japan. Memoirs of the National Science Museum, Tokyo, **9**: 151-161, figs. 1-5. [In Japanese with an English summary.]

— —, 1978. Soldier crabs from Australia and Japan. Bulletin of the National Science Museum, Tokyo, (A, Zoology) **4** (1): 31-38.

TAKEDA, S., 2005. Sexual differences in behaviour during the breeding season in the soldier crab (*Mictyris brevidactylus*). Journal of Zoology, **266** (2): 197-204.

TAKEDA, S. & M. MURAI, 2004. Microhabitat use by the soldier crab *Mictyris brevidactylus* (Brachyura: Mictyridae): interchangeability of surface and subsurface feeding through burrow structure alteration. Journal of Crustacean Biology, **24** (2): 327-339.

TAMURA, K., J. DUDLEY, M. NEI & S. KUMAR, 2007. MEGA4: Molecular Evolutionary Genetics Analysis (MEGA) software version 4.0. Molecular Biology and Evolution, **24**: 1596-1599.

THOMPSON, J. D., D. G. HIGGINS & T. J. GIBSON, 1994. CLUSTAL W: improving the sensitivity of progressive multiple sequence alignment through sequence weighting, position specific gap penalties and weight matrix choice. Nucleic Acids Research, **22**: 4673-4680.

UNNO, J., 2008. A new species of soldier crab, *Mictyris occidentalis* (Crustacea: Decapoda: Brachyura: Mictyridae) from Western Australia, with congener comparisons. Journal of the Royal Society of Western Australia, **91** (1): 31-50.

WADA, K., 1995. Brachyura. In: S. NISHIMURA (ed.), Guide to seashore animals of Japan with color pictures and keys, **2**: 379-418, pls. 101-118. (Hoikusha, Osaka). [In Japanese.]

WEBB, A. P. & B. D. EYRE, 2004. The effect of natural populations of the burrowing and grazing soldier crab (*Mictyris longicarpus*) on sediment irrigation, benthic metabolism and nitrogen fluxes. Journal of Experimental Marine Biology and Ecology, **309** (1): 1-19.

WHITE, A., 1847. List of the specimens of Crustacea in the collection of the British Museum: i-viii, 1-143. (British Museum, London).

WONG, K. J. H., B. K. K. CHAN & H.-T. SHIH, 2010. Taxonomy of the sand bubbler crabs *Scopimera globosa* De Haan, 1835, and *S. tuberculata* Stimpson, 1858 (Crustacea: Decapoda: Dotillidae) in East Asia, with description of a new species from the Ryukyus, Japan. Zootaxa, **2345**: 43-59.

XU, J., T.-Y. CHAN, L. M. TSANG & K. H. CHU, 2009. Phylogeography of the mitten crab *Eriocheir* sensu stricto in East Asia: Pleistocene isolation, population expansion and secondary contact. Molecular Phylogeny and Evolution, **52**: 45-56.

YAMAGUCHI, T., 1976. A preliminary report on the ecology of the sand bubbler crab, *Mictyris longicarpus* Latreille. Benthos Research, **11/12**: 22-34. [In Japanese.]

ZHARIKOV, Y. & G. A. SKILLETER, 2002. Sex-specific intertidal habitat use in subtropically wintering bar-tailed godwits. Canadian Journal of Zoology, **80** (11): 1918-1929.

— — & — —, 2003. Depletion of benthic invertebrates by bar-tailed godwits *Limosa lapponica* in a subtropical estuary. Marine Ecology Progress Series, **254**: 151-162.

— — & — —, 2004. Why do eastern curlews *Numenius madagascariensis* feed on prey that lowers intake rate before migration? Journal of Avian Biology, **35** (6): 533-542.

First received 20 July 2009.
Final version accepted 30 July 2009.

EOGARTHAMBRUS GUINOTAE N. GEN. AND N. SP. (DECAPODA, BRACHYURA, PARTHENOPIDAE) FROM THE EOCENE OF VICENZA, ITALY

BY

ANTONIO DE ANGELI[1,4]), ALESSANDRO GARASSINO[2,5]) and RICCARDO ALBERTI[3,6])

[1]) Piazzetta Nostro Tetto, 9, I-36100 Vicenza, Italy; Museo Civico G. Zannato, Piazza Marconi, 15, 36075 Montecchio Maggiore (Vicenza), Italy

[2]) Museo Civico di Storia Naturale di Milano, Corso Venezia, 55, I-20121 Milano, Italy

[3]) Associazione Amici del Museo Civico G. Zannato, Piazza Marconi, 15, I-36075 Montecchio Maggiore (Vicenza), Italy

ABSTRACT

Eogarthambrus guinotae n. gen., n. sp. (Parthenopidae MacLeay, 1838) is described from the Eocene (Ypresian and Priabonian) of Vicenza, northern Italy.

The shape of the carapace shows affinities with the extant representatives of *Garthambrus* Ng, 1996, but differs in the shape of the front and the hepatic margin. Even if strong affinities also exist with the fossil *Mesolambrus* Müller & Collins, 1991, from the late Eocene (Priabonian) of Hungary, the new genus differs in having a narrower front with two median teeth, anterolateral margins that form an acute angle and have five wide, flat spiny teeth.

Eogarthambrus n. gen. and *Mesolambrus* were widespread in the Eocene seas of Europe, having preference for coral reef environments.

RÉSUMÉ

Eogarthambrus guinotae n. gen., n. sp. (Parthenopidae MacLeay, 1838) est décrit de l'Éocène (Yprésien et Priabonien) de Vicenza, Italie du nord. La forme de la carapace montre des affinités avec les espèces éteintes de *Garthambrus* Ng, 1996, mais en diffère par la forme du front et du bord hépatique. De fortes affinités existent également avec les fossiles de *Mesolambrus* Müller & Collins, 1991 de l'Éocène supérieur (Priabonien) de Hongrie. Le nouveau genre en diffère par un front plus étroit avec deux dents médianes, bord antérolatéral

[4]) Corresponding author; e-mail: antonio.deangeli@alice.it

[5]) e-mail: agarassino63@gmail.com

[6]) e-mail: riccardoalberti@libero.it

© Koninklijke Brill NV, Leiden, 2010

formant un angle aigu et avec cinq dents aplaties, larges et pointues. *Eogarthambrus* n. gen. et *Mesolambrus* étaient largement répartis dans les mers d'Europe à l'Éocène, de préférence dans les environnements corallines.

INTRODUCTION

The present specimens come from the early Eocene (Ypresian) of the Rossi quarry of Monte di Malo and from the late Eocene (Priabonian) of San Feliciano, Orgiano (Vicenza, northern Italy).

The quarry, located on the eastern side of Monti Lessini Vicentini, represents one of many quarries where nummulitic limestones are extracted. Many decapod crustaceans have been discovered and described by Beschin et al. (1988, 1996, 1998, 2000, 2007b).

The micropalaeontologic study of the quarry was carried by Beschin et al. (1998), Beccaro (2003) and Beccaro et al. (2001). These authors recognized three different levels containing decapod crustaceans. The first two levels, located in the lower part of the quarry, are middle Ypresian (*Nummulites nitidus* zone) and late Ypresian (*Nummulites campesinus* zone), both composed of an alternation of grey-green volcanodetritic materials, micritic limestone, and calcarenite; the third level, at the top of the quarry, and middle Lutetian in age, is composed of micrite and tufaceous micrite (*Morozovella lehneri* zone).

The quarry is also known to have yielded small fragments of Eocene amber (Boscardin & Violati Tescari, 1996; Ragazzi, 1998; Trevisani et al., 2005).

One of the present specimens was discovered in the layers located below the levels containing previously known decapods (Beschin et al., 1998). These layers, perhaps from patch reefs, are composed of stratified calcarenites, are rich of coralligenous algae and corals, associated with natural moulds of molluscs, decapod crustaceans, and macroforaminifers. One similar formation was identified from the middle Ypresian of Contrada Gecchelina of Monte di Malo (Vicenza, Italy), where 350 fossil decapods belonging to 48 species were collected (Beschin et al., 2007a).

The second specimen, from the quarry of San Feliciano Hill, Orgiano (south-western Monti Berici, Vicenza), was discovered in calcarenites rich in corals of late Eocene (Priabonian) age. The holotype of *Phlyctenodes dalpiazi* studied and described by Fabiani (1911), and housed in the collections of the University of Padova, comes from this level. The discovery of many decapods from this calcarenitic level has resulted in the description of many galatheid,

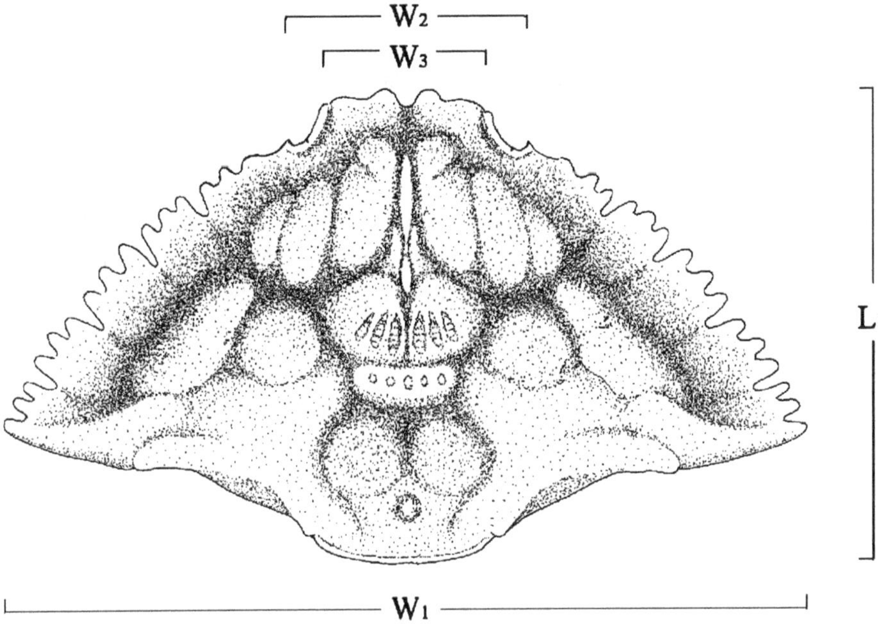

Fig. 1. *Eogarthambrus guinotae* n. gen., n. sp., carapace reconstruction; W_1 = maximum width of the carapace; W_2 = orbito-frontal width; W_3 = frontal width; L_1 = maximum length of the carapace.

chirostylid, and porcellanid decapods (De Angeli & Garassino, 2002) and one sphaeromatid isopod (De Angeli & Lovato, 2009).

The specimens have been deposited in the palaeontological collection of Museo Civico G. Zannato of Montecchio Maggiore (Vicenza, Italy) (MCZ); the measurements (fig. 1) of the carapace are given in millimetres (mm); abbreviations: W_1 = maximum width of the carapace, W_2 = orbito-frontal width, W_3 = frontal width, L_1 = maximum length of the carapace.

SYSTEMATIC PALAEONTOLOGY

Infraorder BRACHYURA Linnaeus, 1758

Section EUBRACHYURA de Saint Laurent, 1980

Subsection HETEROTREMATA Guinot, 1977

Superfamily PARTHENOPOIDEA MacLeay, 1838

Family PARTHENOPIDAE MacLeay, 1838

Subfamily PARTHENOPINAE MacLeay, 1838

Eogarthambrus n. gen.

Type species. — *Eogarthambrus guinotae* n. sp.

Diagnosis. — Carapace wider than long, steeply downturned anteriorly; front sinuous with two median lobes; orbits small, rounded, oriented obliquely; supraorbital margin well marked, raised; anterolateral margin with 5 broad, flat teeth, each with 3 spines; anterolateral tooth acute; posterolateral margins strongly convergent with 2 tubercles; regions well marked with distinct epibranchial lobes; dorsal surface finely granular.

Etymology. — *Eogarthambrus* (m.) is a compound name derived from Eocene, the geological age of the studied specimens, and *Garthambrus* Ng, 1996, to refer to its morphological affinities with the new genus.

Remarks. — The available specimens consist of well preserved carapaces and show the typical subtriangular shape of parthenopid crabs. The morphological characters of Parthenopidae have been recently reviewed by Tan & Ng (2007a, b). The genus *Garthambrus* has recently been revised by McLay & Tan (2009) and some species transferred to two new genera, *Zarenkolambrus* McLay & Tan, 2009 and *Hispidolambrus* McLay & Tan, 2009. As a result, *Garthambrus* now contains 11 species from the Indo-West Pacific. Those authors also redefined some morphological terms used to describe these parthenopid species, and we follow their terminology herein.

Eogarthambrus n. gen. shows evident affinities with the Recent *Garthambrus* Ng, 1996, in particular with *G. complanatus* (Rathbun, 1906) having a carapace with an acute, toothed dorsal angle and anterolateral margins with many flat, spiny teeth (Ng & Tan, 1999). *Garthambrus*, however, has a narrower front extending beyond the orbits, a median protuberance, well marked hepatic margins, and weaker defined and less convex dorsal regions.

The shape of the carapace also shows affinities with the fossil *Mesolambrus* Müller & Collins, 1991, known only by the type species *M. declinatus* Müller & Collins, 1991 (cf. their pl. 3 figs. 11, 12; text-fig. 3f), from the late Eocene of Hungary.

Comparison of the new specimens with a cast of the holotype of *M. declinatus* allowed the following observations: *M. declinatus* differs from *Eogarthambrus guinotae* n. gen., n. sp. in having a carapace shorter than wide; its maximum length is found on 3/5 of the width of the carapace, while in *E. guinotae* the maximum width is found on 4/5 of the length of the carapace; the lateral angle teeth are convex and do not form an acute angle as in the new species, anterolateral and posterolateral margins have spines or lobes in the Hungarian species (*E. guinotae* n. sp. has the anterolateral margins with flat teeth, each with three flat spines, and the posterolateral margins lack spines, but bear only two small tubercles), and the dorsal surface of the carapace has more tubercles than in the new species.

The specimen ascribed to *Mesolambrus declinatus* reported by Beschin et al. (2007a) from the early Eocene (Ypresian) of Contrada Gecchelina of Monte di Malo (Vicenza) shows the shape of its carapace to be similar to that of *E. guinotae* n. sp. The front, however, is wider, the frontal margin is more sinuous and entirely lobate, the anterolateral margins form a convex angle with many teeth, and the dorsal surface of the carapace has more tubercles.

Eogarthambrus n. gen. and *Mesolambrus* represent extinct parthenopids with very large carapaces for the family, lobate fronts slightly extended beyond the orbits, protogastric regions and epibranchial lobes strongly raised and strongly inclined downward anteriorly.

Eogarthambrus guinotae n. sp. (figs. 1-3)

Material examined. — Holotype, MCZ 2992, from the early Eocene (middle Ypresian) of the Rossi quarry of Monte di Malo (Vicenza); measurements: $W_1 = 44.2$; $W_2 = 15.7$; $W_3 = 9.6$; $L_1 = 26.9$. Paratype, MCZ 2993, from the late Eocene (Priabonian), San Feliciano, Orgiano (Monti Berici, Vicenza), an incomplete carapace; measurements: $L_1 = 11.2$.

Etymology. — The specific name honours Danièle Guinot, Département Milieux et Peuplements aquatiques, Muséum national d'Histoire naturelle, Paris, for her important contribution to the study of Recent and fossil brachyurans.

Diagnosis. — As for the new genus.

Description. — Carapace subhexagonal, wider than long ($W_1/L_1 = 1.63$), with maximum width at 4/5 of its length; dorsal surface longitudinally convex, steeply downturned anteriorly; orbito-frontal margin moderately wide ($W_2/W_1 = 0.35$); front narrow ($W_3/W_1 = 0.22$), slightly extended beyond orbits; frontal margin slightly sinuous; median part with 2 lobes directed downward, well marked by median incision; orbits small, rounded, oriented obliquely; supraorbital margin well marked, raised, with short, narrow supraorbital fissure; anterolateral margins elongate, convex, diverging, connecting to posterolateral margins by acute angle; anterolateral margins with 5 flat teeth, separated by narrow, deep fissures; each anterolateral tooth with 3 flat spines on outer margin; the first tooth is found on the hepatic margin and the other four teeth on the epibranchial margin; posterolateral margins shorter

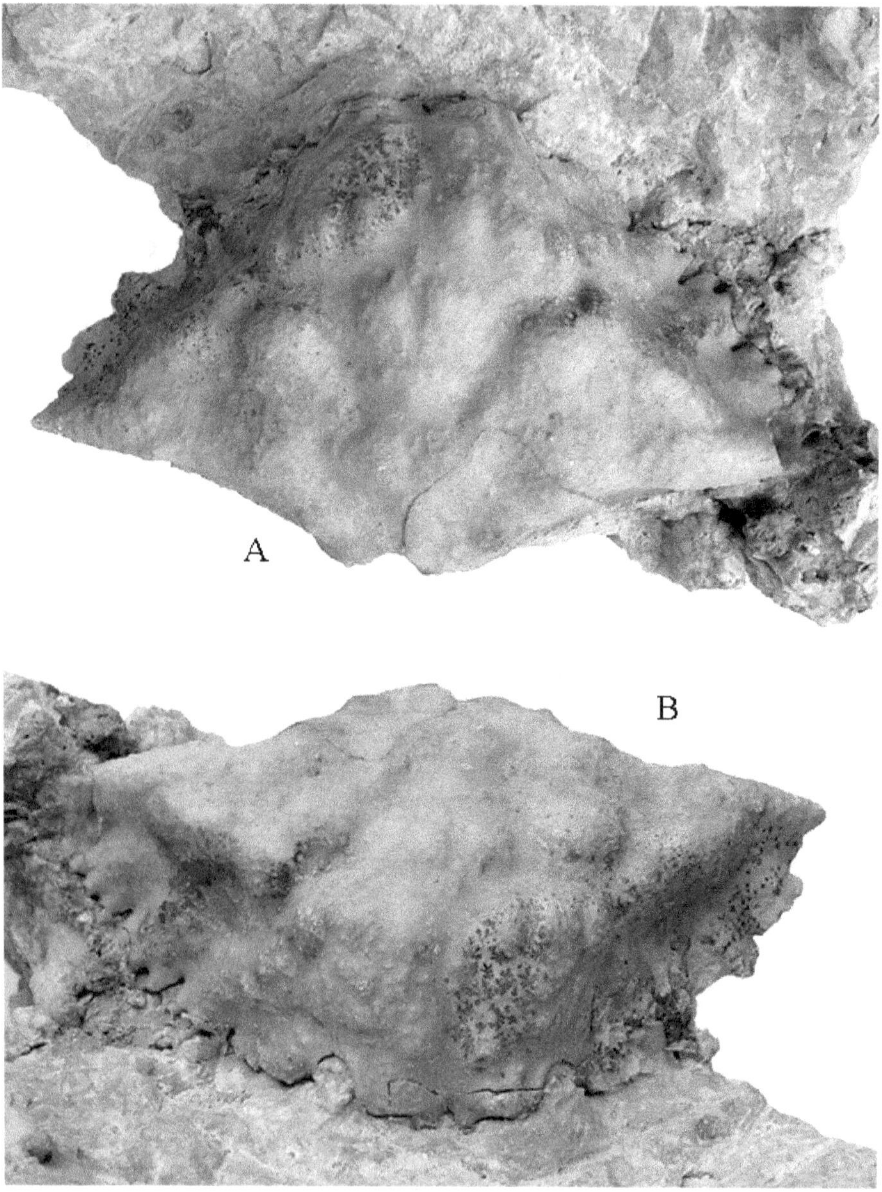

Fig. 2. *Eogarthambrus guinotae* n. gen., n. sp., holotype MCZ 2992; A, dorsal view; B, frontal view (×2.2).

than anterolateral margin, strongly convergent, slightly concave; 2 small spiny protuberances, one epibranchial and one metabranchial located before posterior angle; posterior margin narrow, slightly curved; regions of carapace well marked; frontal region marked by median longitudinal depression; epigastric

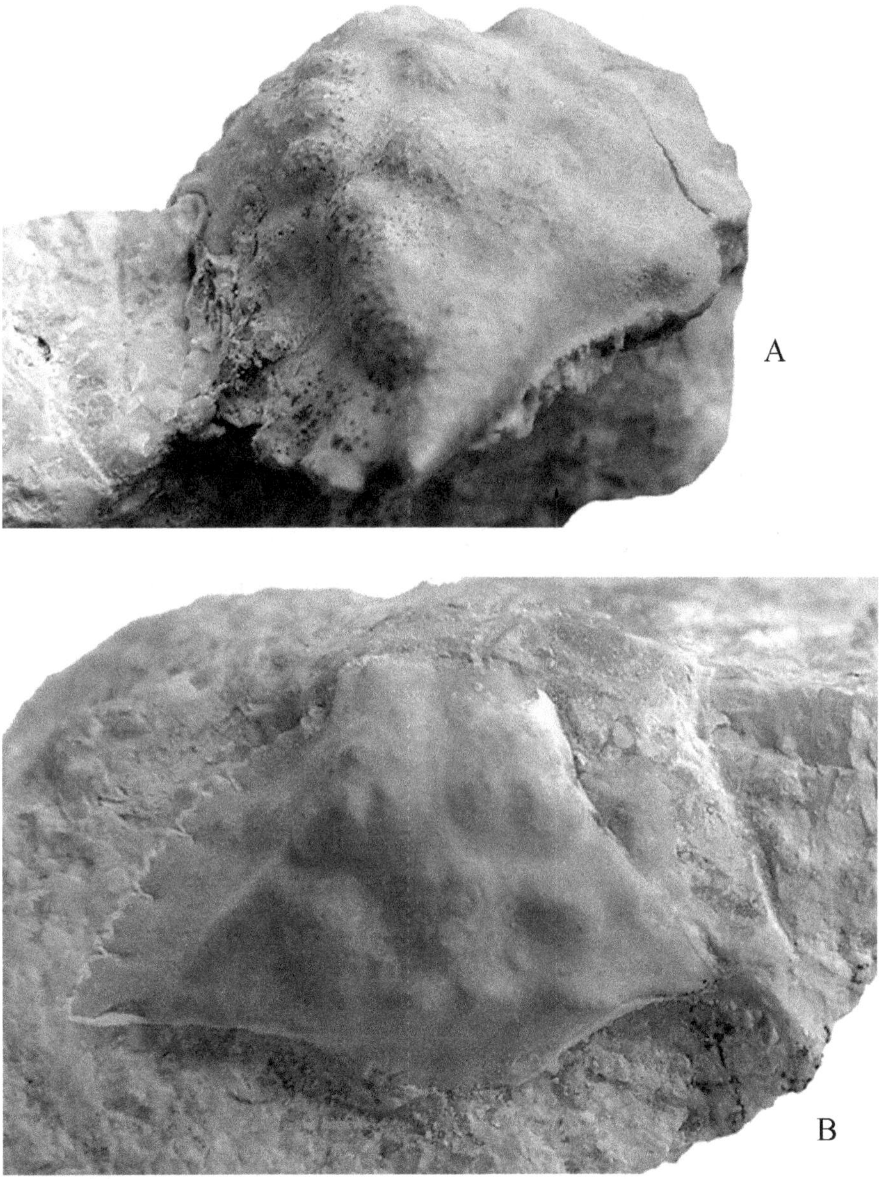

A

B

Fig. 3. *Eogarthambrus guinotae* n. gen., n. sp.; A, holotype MCZ 2992, oblique lateral view (×2.6); B, paratype MCZ 2293, dorsal view (×4.6).

lobes represented by 2 small, rounded nodes; protogastric regions strongly raised, trilobate; mesogastric region well marked by cervical groove; mesogastric region subpentagonal with 2 wide elevations separated by median groove; anterior part of mesogastric region with elongated, narrow process, wedged between protogastric regions; metagastric region well marked with transverse, tuberculated elevation; cardiac region inflated, well marked by branchiocardiac grooves; cardiac region wider anteriorly with 2 wide circular elevations, restricted posteriorly to small node; intestinal region small, depressed; hepatic regions relatively well developed, inclined, marked by cervical groove; epibranchial regions wide with elongated, oblique elevation, extending to last tooth of anterolateral margin; additional rounded elevation present between oblique relief and gastric regions; mesobranchial regions with one weak protuberance; metabranchial regions small, with spiny tubercle; dorsal ornamentation with small irregular granulations located primarily on swellings of regions.

Remarks. — The Parthenopidae include sixteen genera represented in the fossil record: *Acantholambrus* Blow & Manning, 1996 (fossil); *Branchiolambrus* Rathbun, 1808 (fossil); *Daldorfia* Rathbun, 1904 (fossil and extant); *Derilambrus* Tan & Ng, 2007 (fossil and extant); *Enoplolambrus* A. Milne-Edwards, 1878 (fossil and extant); *Eogarthambrus* n. gen. (fossil); *Latolambrus* Tan & Ng, 2007 (fossil and extant); *Heterocrypta* Stimpson, 1871 (fossil and extant); *Leiolambrus* A. Milne-Edwards, 1879 (fossil and extant); *Mesolambrus* Müller & Collins (fossil); *Mesorhoea* Stimpson, 1871 (fossil and extant); *Parthenope* Weber, 1795 (fossil and extant); *Phrynolambrus* Bittner, 1893 (fossil); *Platylambrus* Stimpson, 1871 (fossil and extant); *Rhinolambrus* A. Milne-Edwards, 1879 (fossil and extant); *Tutankhamen* Rathbun, 1925 (fossil and extant). Glaessner (1969) has considered *Phrynolambrus* a synonym of *Pseudolambrus*. In the recent study of new Italian fossil decapods, De Angeli & Beschin (2008) have seen the characteristics of *Phrynolambrus* and considered it a valid genus.

Eogarthambrus n. gen. and *Mesolambrus*, its closest similar genus, have wide carapaces with dorsal surfaces steeply downturned anteriorly, fronts slightly extended beyond the orbits, and hepatic regions margins weakly marked. Both genera originated in the Eocene Tethyan Sea, spreading into environments rich of coralligenous algae and corals, and living associated with other crustaceans (Müller & Collins, 1991; De Angeli & Garassino, 2002; Beschin et al., 2007a).

ACKNOWLEDGEMENTS

We wish to thank R. M. Feldmann, Kent State University, Ohio, for his careful review of and criticism on the manuscript.

REFERENCES

BECCARO, L., 2003. Revisioni stratigrafiche nel Paleogene del Veneto occidentale. (Dottorato di Ricerca in Scienze della Terra Ciclo XVI°, Università degli studi di Padova, Dipartimento di Geologia, Paleontologia e Geofisica). [Unpublished Ph.D. Thesis.]

BECCARO, L., E. FORNACIARI, P. MIETTO & N. PRETO, 2001. Analisi delle facies e ricostruzione ambientale dei "Calcari nummulitici" (Eocene; Monti Lessini orientali — Vicenza): dati preliminari. Studi Trentini di Scienze Naturali, (Acta Geologica) **76**: 3-16.

BESCHIN, C., A. BUSULINI, A. DE ANGELI & G. TESSIER, 1988. Raninidae del Terziario berico-lessineo (Italia settentrionale). Lavori — Società Veneziana di Scienze Naturali, **13**: 155-215.

— —, — —, — — & — —, 1996. Retroplumoidea (Crustacea, Brachyura) nel Terziario del Vicentino (Italia settentrionale). Lavori — Società Veneziana di Scienze Naturali, **21**: 83-102.

— —, — —, — — & — —, 2007 (cf. a). I decapodi dell'Eocene inferiore di Contrada Gecchelina (Vicenza, Italia settentrionale) (Anomura e Brachyura). Museo di Archeologia e Scienze Naturali "G. Zannato", **2007**: 1-76.

BESCHIN, C., A. BUSULINI, A. DE ANGELI, G. TESSIER & S. UNGARO, 1998. Crostacei eocenici di "Cava Rossi" presso Monte di Malo (Vicenza — Italia settentrionale). Studi Trentini di Scienze Naturali, (Acta Geologica) **73**: 7-34.

BESCHIN, C., A. DE ANGELI & R. ALBERTI, 2000. *Zanthopsis bruckmanni* (Meyer) (Crustacea, Decapoda) dell'Eocene del Vicentino (Italia settentrionale). Studi e Ricerche, Associazione Amici del Museo, Museo Civico "G. Zannato", **2000**: 13-16.

BESCHIN, C., A. DE ANGELI & A. CHECCHI, 2007 (cf. b). Crostacei dell'Eocene inferiore di Cava "Rossi" di Monte di Malo (Vicenza — Italia settentrionale). Studi e Ricerche, Associazione Amici del Museo, Museo Civico "G. Zannato", **14**: 11-24.

BOSCARDIN, M. & O. VIOLATI TESCARI, 1996. Gemme del Vicentino. Pubblicazione Museo Civico "G. Zannato", **1996**: 1-114.

DE ANGELI, A. & C. BESCHIN, 2008. Crostacei decapodi dell'Oligocene di Soghe e Valmarana (Monti Berici, Vicenza — Italia settentrionale). Studi e Ricerche, Associazione Amici del Museo Zannato, Museo Civico "G. Zannato", **15**: 15-39.

DE ANGELI, A. & A. GARASSINO, 2002. Galatheid, chirostylid and porcellanid decapods (Crustacea, Decapoda, Anomura) from the Eocene and Oligocene of Vicenza (N Italy). Memorie della Società Italiana di Scienze Naturali e del Museo Civico di Storia Naturale di Milano, **30** (3): 1-40.

DE ANGELI, A. & A. LOVATO, 2009. *Sphaeroma gasparellai* n. sp. (Isopoda, Flabellifera, Sphaeromatidae), nuova specie di isopode dell'Eocene superiore dei Monti Berici (Italia settentrionale). Lavori — Società Veneziana di Scienze Naturali, **34**: 101-104.

FABIANI, R., 1911. Di una nuova specie di *Phlyctenodes* (*Phl. Dalpiazi*) dell'Oligocene dei Berici. Bollettino del Museo Civico di Vicenza, **1** (3) (1910): 3-6.

GLAESSNER, M. F., 1969. Decapoda. In: R. C. MOORE (ed.), Treatise on Invertebrate Paleontology, **R** (Arthropoda 4) (2): R399-R533, R626-R628. (Geological Society of America and University Kansas Press, Lawrence).

MCLAY, C. L. & S. H. TAN, 2009. Revision of the crab genus *Garthambrus* Ng, 1996, with the description of two new genera and discussion of the status of *Tutankhamen* Rathbun, 1925 (Crustacea: Brachyura: Parthenopidae). Zootaxa, **2122**: 1-50.

MÜLLER, P. & J. S. H. COLLINS, 1991. Late Eocene coral-associated decapods (Crustacea) from Hungary. Contributions to Tertiary and Quaternary Geology, **28** (2-3): 47-92.

NG, P. K. L. & S. H. TAN, 1999. The Hawaiian parthenopid crabs of the genera *Garthambrus* Ng, 1996, and *Dairoides* Stebbing, 1920 (Crustacea: Decapoda: Brachyura). Proceedings of the Biological Society of Washington, **112** (1): 120-132.

RAGAZZI, E., 1998. Ambra: mito e realtà. Studi e Ricerche, Associazione Amici del Museo, Museo Civico "G. Zannato", **1998**: 7-16.

TAN, S. H. & P. K. L. NG, 2007a. Descriptions of new genera from the subfamily Parthenopinae (Crustacea: Decapoda: Brachyura: Parthenopidae). The Raffles Bulletin of Zoology, (Supplement) **16**: 95-119.

— — & — —, 2007b. Review of the subfamily Daldorfiinae Ng & Rodríguez, 1986 (Crustacea: Decapoda: Brachyura: Parthenopidae). The Raffles Bulletin of Zoology, (Supplement) **16**: 121-167.

TREVISANI, E., C. A. PAPAZZONI, E. RAGAZZI & G. ROGHI, 2005. Early Eocene amber from the "Pesciera di Bolca" (Lessini Mountains, Northern Italy). Palaeo, **223**: 260-274.

First received 5 May 2009.
Final version accepted 26 June 2009.

DESCRIPTION OF *ETISUS GUINOTAE* N. SP., AND DISCUSSION OF ITS RECENT DISCOVERY IN THE GULF OF MEXICO (BRACHYURA, DECAPODA, XANTHIDAE)

BY

DARRYL L. FELDER[1,2]) and BRENT P. THOMA[1,3])

[1]) Department of Biology and Laboratory for Crustacean Research, University of Louisiana at Lafayette, Lafayette, Louisiana 70504, U.S.A.

ABSTRACT

Recent collecting efforts on continental shelf banks in the Gulf of Mexico have produced material of a previously unknown but colorful and conspicuous xanthoid crab, herewith assigned to the species, *Etisus guinotae* n. sp. Initial collection of a single specimen from the eastern Gulf of Mexico in 2004 raised suspicion that it was an introduction or morphological anomaly of some known species, but exhaustive searches of world literature and consultations with colleagues produced no clear evidence of its identity. Subsequent discovery of additional specimens on banks in the northwestern Gulf of Mexico during August 2008 revealed presence of this species in habitats where it was not found during extensive sampling over 12 previous years and provided additional materials that confirmed its unique characters. Molecular genetic analyses based on 16S mt, 12S mt, and Histone 3 nuclear genes position this species among other Xanthidae sensu stricto, and morphological comparisons place it nearest Indo-West Pacific representatives of *Etisus*, the genus to which it is provisionally assigned.

RÉSUMÉ

L'échantillonnage récent des bancs situés sur le plateau continental du Golfe du Mexique a révélé des spécimens d'un crabe xanthoide inconnu jusqu'à présent, bien que très coloré et facilement visible, ici nommé *Etisus guinotae* n. sp. Un unique spécimen, prélevé dans l'est du Golfe du Mexique en 2004, fut initialement suspecté d'appartenir à une espèce introduite ou d'être une forme morphologiquement anormale d`une espèce connue. La découverte de plusieurs spécimens de cette nouvelle espèce sur les bancs du Golfe du Mexique en août 2008 confirme la validité des caractères clés et révèle sa présence dans des habitats ou de tels spécimens n'avaient jamais été trouvés durant 12 ans d'échantillonnage intensif. L'analyse

[2]) Corresponding author; e-mail: dlf4517@louisiana.edu

[3]) e-mail: bpt0706@louisiana.edu

© Koninklijke Brill NV, Leiden, 2010

des gènes mitochondriaux 16S et 12S, ainsi que du gène nucléaire Histone 3 place cette espèce parmi d'autres Xanthidae s. s. Morphologiquement, ses congénères les plus proches sont des représentants d'*Etisus* de l'Indo-Pacifique. Les spécimens examinés sont assignés à une nouvelle espèce dans ce genre.

INTRODUCTION

A series of cruises aboard the R/V "Pelican" was conducted beginning in 1996 to assess communities of macroalgae and decapod crustaceans on deep banks of the continental shelf in the Gulf of Mexico, primarily in waters greater than 50 m depth. Initially funded by the United States Department of Energy and thereafter by the United States National Science Foundation Biotic Surveys and Inventories Program, the primary intent was to evaluate regional endemism of benthic organisms, which have putatively disjunct distributions. While a number of new taxa were discovered and many remain under study, most were more or less the predictable outcome of carefully examining fresh collections of supposedly known species and generic complexes. Others appear to be previously unknown Gulf of Mexico occurrences for unexpected but well-known western Atlantic genera. There were in addition new records of known western Atlantic species. Most of the latter are included in a recent summary on regional biodiversity (Felder et al., 2009). A number of the identified regional taxa, and some of the questionable finds, were treated along with general overviews of decapod relationships in the course of molecular phylogenetic studies (Mantelatto et al., 2007, 2009; Robles et al., 2007, 2009; Bracken et al., 2009; Felder & Robles, 2009; Palacios-Theil et al., 2009; Thoma et al., 2009; Tshudy et al., 2009). As expected in any major collection and analysis effort, a few individuals that were damaged or appeared atypical could not be identified, and these were set aside for eventual study.

One such collection was a female brachyuran crab taken in 2004 from calcareous coralline and mollusk shell rubble off Dry Tortugas in the eastern Gulf of Mexico. As the specimen obviously represented an unfamiliar xanthoid, photographs were circulated to colleagues generally knowledgeable of brachyurans elsewhere in the world under suspicion that it represented a non-indigenous species that had been recently introduced. Neither this effort nor study of literature fixed identity of the specimen to a known species, as was also the case for initial sequence analyses. As we held only one female specimen and, thus, lacked knowledge of critically important gonopod morphology, the specimen was shelved pending inclusion of its tissues in further genetic analysis of xanthoids, but also with hopes of finding additional material in new samples or in the course of further work with our collection holdings. No additional specimens had been found, however, when field collections and lab identifications were completed under NSF BS&I support in 2006.

The opportunity arose in late summer of 2008 to join a cruise to collect mollusks from deep banks off Louisiana. Mr. William Cargile and colleagues specifically targeted those sites because our previous work there made the benthic assemblage well known and, thus, a predictable source of materials they sought. Remarkably, five specimens of a xanthoid crab matching the previously collected female from the eastern Gulf of Mexico were obtained from three different stations on these northwestern Gulf of Mexico deep banks. Juveniles, mature males, and mature females were included among these, which provided perspective on maturational change, gonopod structure, and sexual dimorphism.

The abrupt appearance of this new find in the extreme eastern Gulf of Mexico and its apparent rapid expansion to banks off Louisiana, strongly suggest that these discoveries represent an invasive species (see Remarks below). Because there are no published descriptions of species that apply to these materials, we are left with no choice but to provide a description and name for what is most probably a new species. We, thus, must also designate a possible non-native habitat as the type locality. We undertake these actions to facilitate additional studies of this animal, knowing that a previously named source population may yet come to light and require future synonymy. However, we welcome that development should it explain how this species has recently achieved wide distribution in the Gulf of Mexico.

MATERIAL AND METHODS

All specimens used in this study were collected aboard the R/V "Pelican" by dredge sampling. Gear and procedures were modeled after those used extensively in the eastern Gulf of Mexico during the "Hourglass" surveys conducted by the Florida Department of Natural Resources (Joyce & Williams, 1969), except that intended sampling tracts were first surveyed by submarine camera to assure that they did not transect hermatypic corals or other massive accumulations of live bottom habitat. The 1-m box dredge was towed at dead-slow vessel speed (estimated $\leqslant$1.0 knots), at a ratio of 3 : 1 cable length to water depth, for a period of 5-10 min. Upon retrieval, rubble and sediment were emptied onto a well-illuminated sorting table for complete processing, during which specimens were removed by hand, labeled, and frozen fully immersed in seawater for later laboratory photography, identification, removal of tissues for genetic analyses, preservation, and archival. Type material was archived in the Smithsonian Institution, National Museum of Natural History, Washington, D.C. (USNM) and the University of Louisiana at Lafayette Zoological Collections (ULLZ). Materials bearing USNM catalog numbers also retain originally issued cross-reference ULLZ numbers that correlate to a database for tissue samples used in genetic analyses. Sizes are stated as carapace width including lateral spines (cw) and carapace length including maximum extent of frontal lobes (cl), measured to the nearest 0.1 mm.

Sequences of two mitochondrial (12S and 16S) and one nuclear marker (Histone 3) from 44 species representing 41 genera, eight families, and five superfamilies were used to generally frame phylogenetic relationships of *Etisus guinotae* n. sp. (Appendix I). Genomic DNA extraction, as well as PCR and sequencing of 12S and 16S markers, were conducted following Thoma et al. (2009). Histone 3 was amplified and sequenced using the same parameters as for 12S and 16S sequencing, with the exception of the annealing temperature, which was 58°C using primers Hex-AF (ATG GCT CGT ACC AAG CAG ACG GC) and Hex-AR (ATA TCC TTG GGC ATG ATG GTG AC) (Svenson & Whiting, 2004). Sequences were aligned in MAFFT using the L-INS-i algorithm (Katoh et al., 2002). The model of evolution that best fit each of the datasets was determined by likelihood tests as implemented in MrModelTest 2.3 (Nylander, 2004) under the Akaike Information Criterion (AIC). The Bayesian analyses of the concatenated dataset were conducted in MrBayes (Huelsenbeck & Ronquist, 2001) with computations performed on the BioHPC computer cluster at the Cornell Computational Biology Service Unit, using parameters selected by MrModeltest. A Markov Chain Monte Carlo (MCMC) algorithm with four runs of four chains each was run for 15 000 000 generations, sampling one tree every 10 000 generations. A 50% majority rule consensus tree was obtained from the 34 000 trees (8500 per run) taken after the standard deviation between runs dropped below 0.01. Clade support was assessed with posterior probabilities (p*P*) and 1000 maximum-likelihood bootstrap replications (BS) performed in RAxML 7.0.4 on the computer cluster of the CyberInfrastructure for Phylogenetic RESearch project (CIPRES) at the San Diego Supercomputer Center.

SYSTEMATIC ACCOUNT

Family XANTHIDAE MacLeay, 1838, sensu Ng et al., 2008

Etisus guinotae n. sp. (figs. 1-3)

Material examined. — Holotype: male (cw 12.7, cl 9.9 mm) USNM 1129926 (ULLZ 11015 tissue sample), south-southwest of Isles Dernieres, Louisiana, near Ewing Bank or 29 Fathom Fish Haven (28°05.819′N 91°02.157′W), 57-58 m, Cargile Cruise stn 72, R/V "Pelican", 25 August 2008.

Paratypes: 1 female (cw 14.5, cl 11.5 mm) USNM 1129927 (ULLZ 6595 tissue sample), west-northwest of Dry Tortugas, Florida, near South Pulley Ridge (24°49.35′N 83°51.65′W), 88.3-85.2 m, NSF BS&I Cruise stn I-04, R/V "Pelican", 31 May 2004; 1 juvenile male (cw 5.5, cl 4.5), 1 female (cw 10.8, cl 8.0 mm) USNM 1129928 (ULLZ 10988 and 10925 tissue samples, respectively), south-southwest of Isles Dernieres, Louisiana, near Ewing Bank or 29 Fathom Fish Haven (28°05.849′N 91°01.539′W), 57-62 m, Cargile Cruise stn 17, R/V "Pelican", 21 August 2008; 1 male (cw 13.4, cl 10.5 mm) ULLZ 10936, south-southwest of Isles Dernieres, Louisiana, near Ewing Bank or 29 Fathom Fish Haven (28°05.849′N 91°01.539′W), 57-62 m, Cargile Cruise stn 17, R/V "Pelican", 21 August 2008; 1 male (cw 12.9, cl 10.3 mm) ULLZ 10944, south of Marsh Island, Louisiana, near Alderdice Bank (27°57.702′N 92°02.784′W), 64-68 m, Cargile Cruise stn 35, R/V "Pelican", 22 August 2008.

Diagnosis. — Carapace rounded in outline, dorsally elevated; regions strongly marked anteriorly, protogastric lobes prominent. Frontal lobes strongly produced, anteriorly truncate, widely separated from orbit by deep, rounded, antennal sulcus; frontal width between orbits exceeding 1/3 carapace width. Anterolateral teeth at least five; third to fifth terminating in sharp corneous spines, arched anteriorly. Basal antennal article with lateral flange extending well beyond articulation with flagellum. Endostome lacking branchial ridges. Chelipeds unequal, robust, major manus mostly smooth, finger tips spooned. Ambulatory pereopods spinous, dactyls bearing paired superior rows of strong spines. Male first gonopods tapering to narrow tip, distal 1/3 curved ventrolaterally, ornamented with short spinules, largest on mesial side slightly hooked. Second abdominal somite in male covering eighth sternite, third to fifth somites fused, lateral lobe of third fully covering gonopore, male papilla on coxa of fifth pereopod.

Description of holotype. — Carapace outline distinctly rounded, subcircular, width slightly exceeding length, dorsally elevated, bearing strong, rounded protogastric elevation to either side of midline (fig. 1a-d); mesogastric elevations broad, each subdivided anteriorly by longitudinal depression to form paired lobes bearing short transverse granulate crest, posteriorly with less defined lobes, crests; weak hepatic lobe, pair of strong epibranchial lobes also bearing short granulate crests, small outer mesobranchial lobe bearing dis-

Fig. 1. *Etisus guinotae* n. sp.: a, holotype male, USNM 1129926 (ULLZ 11015 tissue sample), dorsal surfaces, cw 12.7 mm; b, paratype male, ULLZ 10944, dorsal surfaces, cw 12.9 mm; c, paratype female, USNM 1129927 (ULLZ 6595 tissue sample), dorsal surfaces, cw 14.5 mm; d, same specimen, preserved, posterior oblique perspective; e, same specimen, major chela, upper external surfaces; f, same specimen, left third pereopod, posterior surface.

tinctly produced granular prominence mesial to fourth anterolateral tooth, broad inner mesobranchial lobe surmounted by long, broken granular crest running obliquely to cervical groove. Frontal lobes strongly produced, subquadrate, distinctly truncate, anterior margins bearing strong granules, separated from inner orbital angle by deep, broad, rounded antennal sulcus; width of front more than 1/3 carapace width. Anterolateral margin bearing 5 distinct teeth, including triangular postorbital angle as first, with weak evidence of sixth as small swelling of spinous tubercles, radius of anterolateral arch decreasing slightly to posterior, where margin continues onto posterolateral region of carapace; second tooth broadly triangular, terminating in rounded granular tubercle, small subhepatic cluster or tubercle of sharp granules immediately beneath; third to fifth teeth terminating in corneous spine or spines, arched or set anteriorly, tooth margins bearing sharp granules and small spines, larger of which terminate in corneous tips.

Eyestalk bearing cluster of 2 or 3 strong, thick, anteriorly directed spines with corneous tips, these proximal to line of smaller sharp granules extending dorsally, distally onto dilation of cornea (fig. 2a). Orbit with thick, strongly produced, triangular inner orbital tooth; convex superior orbital lobe between 2 dorsal fissures, less than half orbital width, deep lateral fissure separating it from strong, coarsely granular, triangular outer angle forming narrow first anterolateral tooth; inner, outer suborbital lobes strong, rounded, orbital margin between suborbital lobes regularly concave, granular.

Antennules folded obliquely. Basal article of antenna not reaching margin of carapace front, strong lateral flange greatly exceeding articulation with flagellum, abutting inner suborbital lobe blocking orbital hiatus, flagellum centered on broad antennal sulcus, excluded from orbit.

Endostome smooth, lacking longitudinal branchial ridges anterior to mandibles, strong oblique crest in anterolateral corner defining upturned posteromesial border of concave plate abutting basal article of antenna, overreached laterally by strongly tapered, granular, pterygostomial projection (fig. 2a). Third maxilliped coxa branchial arm bearing 2 superior teeth, distal podobranch; elongated, sparsely setose uncalcified epipod (fig. 2b); endopod with robust carpus, dense field of stiff mesially directed setae on internal carina of merus, ischium distal margin arched, bearing regularly set triangular teeth overlain by row of uniform setae on mesial margin.

Chelipeds massive, robust, unequal. Carpus in both bearing 2 or more strong corneous tipped spines, one at inner angle and another on distomesial margin, with adjacent smaller teeth, sharp granules, low, heavy tooth on distal superior surface just proximal to condyle, distolateral depression shallow, fixed finger in both with deep, punctuate external groove, finger color extended slightly onto palm (figs. 1e, 2c); major chela manus deep, swollen, surfaces generally smooth except for weak evidence of granular ridging on proximal superior margin, fixed finger short, thick, spooned tip evident, weakly produced, opposable surface bearing 3 well-coalesced molar teeth, distalmost rounded,

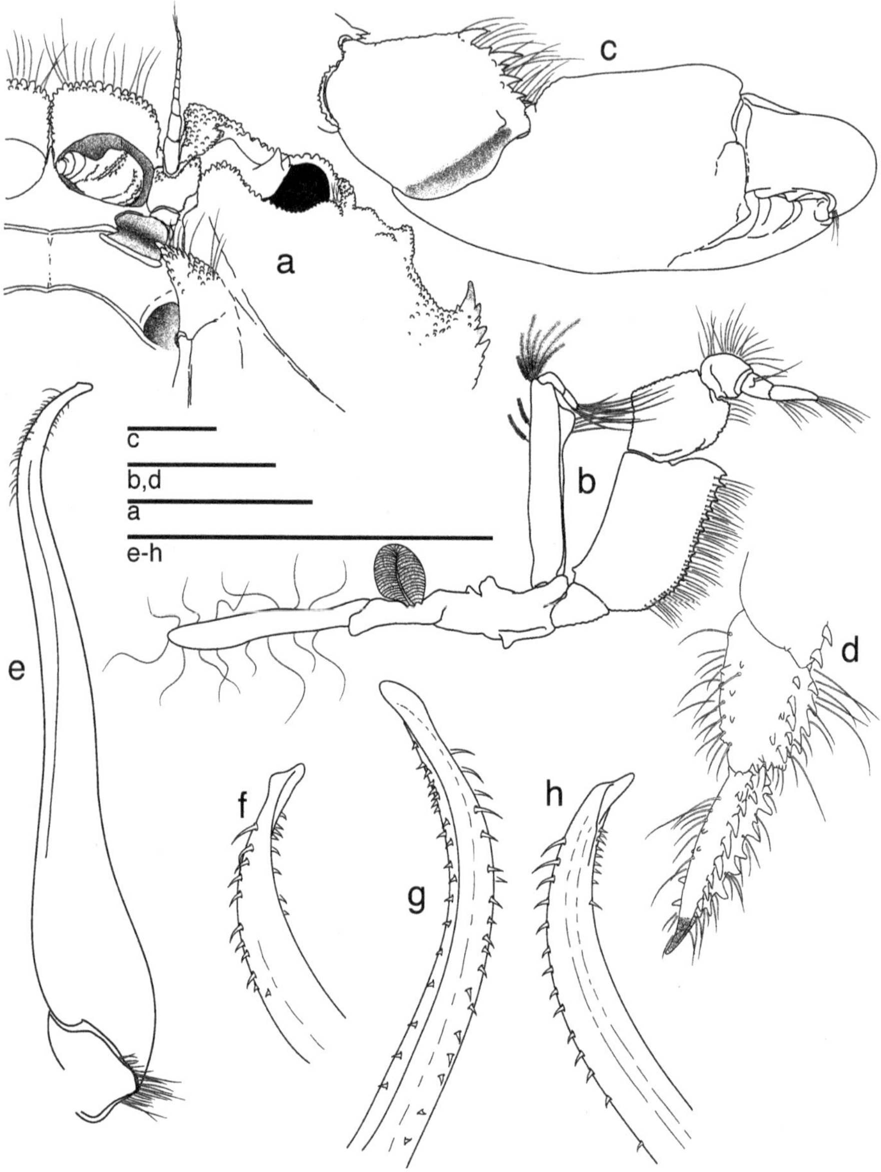

Fig. 2. *Etisus guinotae* n. sp.: a, d, paratype female, USNM 1129927 (ULLZ 6595 tissue sample): a, left anteroventral surface of carapace (antennular flagellum and mouth appendages not shown); d, superoposterior surfaces of dactylus and propodus of right fifth pereopod; b, e-g, holotype male, USNM 1129926 (ULLZ 11015 tissue sample): b, external surface of right third maxilliped; e, sternal side of right first gonopod; f, same appendage and perspective, enlarged apex; g, same appendage, abdominal side of enlarged apex; c, h, paratype male, ULLZ 10944: c, upper external surfaces of right chela; h, sternal side of right gonopod terminus. Scale bars: a = 3 mm; b-e = 2 mm; f-h = 1 mm.

decreasing in size, becoming transverse proximally; dactylus strongly arched, recurved to meet tip of fixed finger, terminally worn, rounded, weak evidence of spooned tip, opposable surface bearing 2 large, slightly coalesced teeth, distal slightly the larger; minor chela not as robust or deep as major, fixed finger and dactylus longer and straighter, less worn and arched, dentition less defined, finger tips spooned.

Ambulatory pereopods (walking legs) short, heavy, moderately setose, subequal to or shorter than carapace length, dorsal (superior) margins with sharp spines, becoming strongest on distal 3 articles, arranged in paired longitudinal lines either side of shallow furrow on dactylus; dactylus proximally bearing small, slightly elevated condyle abutting locking arch on distal margin of propodus, distally terminating in elongated corneous tip (fig. 2d); fourth and fifth pereopod dactylus bearing at least one auxiliary calcareous prehensile tooth on distal inferior margin.

Male abdomen narrow, elongated, retained in flexed position by press-button lock on fifth sternite (fig. 3a-d); first somite bearing strong transverse, offset, granular ridge at intersection with posterior of carapace; second somite broadly rectangular, laterally granular; third to fifth somites fused, lateral lobes of fused third somite weakly sculptured, granular, somite suture limited to slight indentations in lateral margin; sixth somite broadest distally, greatest width about equal to length, telson rounded, subtriangular, slightly wider than long.

Thoracic sternum narrow, elongated; mesial corners of fourth to sixth sternites converging in star-shaped depression on sternal midline. Large rounded opening for fleshy male papilla on fifth pereopod coxa, just anterior to sternal condyle, fully covered by lateral lobe of fused third abdominal somite (fig. 3f); eighth sternite completely concealed from ventral view by second somite of flexed abdomen.

First gonopod narrow, tapered over full length, distal 1/3 arched ventrolaterally, subterminally bearing longitudinal rows of thick spinules including elongated, hooked, spinules on distomesial margins, scant evidence of setae, narrowed terminally to compressed, slightly twisted translucent blade (fig. 2e-h). Second gonopod small, short, extending about 1/4 length of first gonopod, slightly swollen terminal bulb narrowing to sharp spine (fig. 3d).

Variations. — The mature female paratypes closely conform to the described morphology of the male holotype except for sexual structures. The second abdominal somite through the telson collectively form an elongated oval in outline in females, with the second and fifth somites being the longest

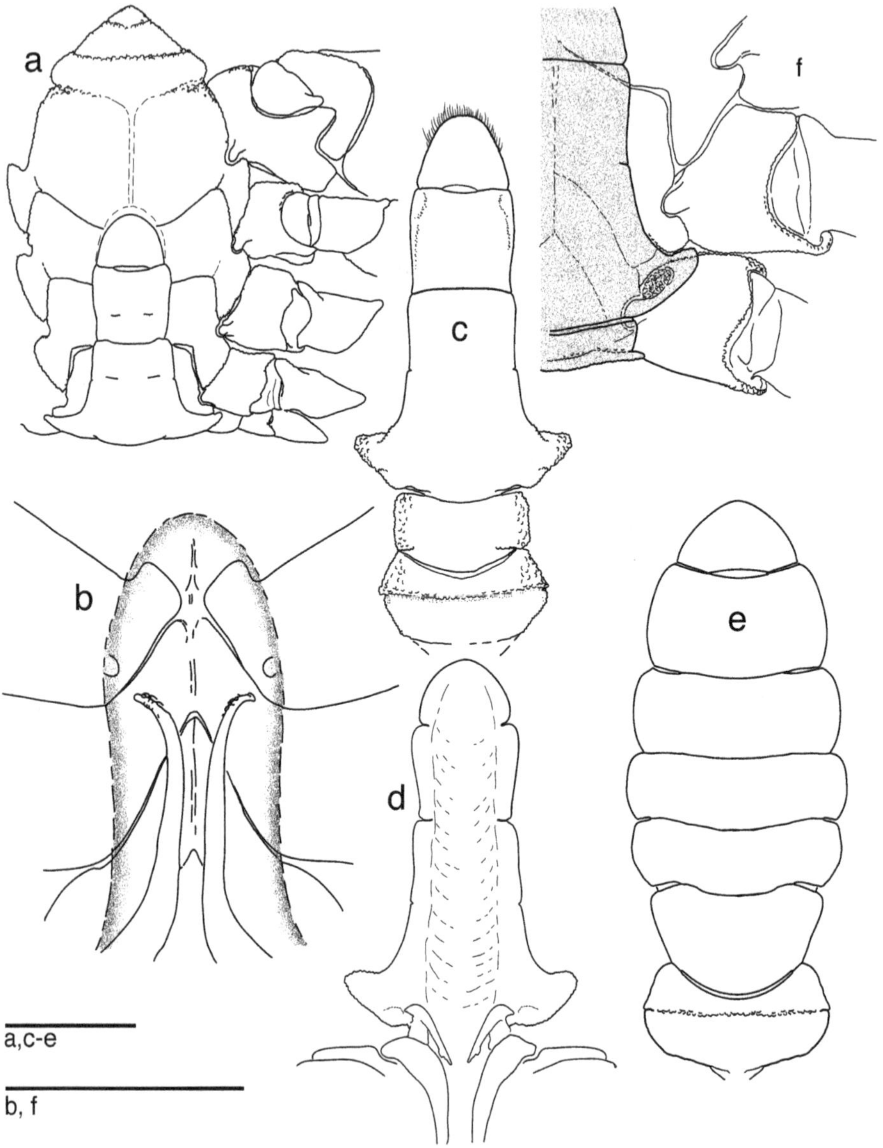

Fig. 3. *Etisus guinotae* n. sp.: a-d, f, holotype male, USNM 1129926 (ULLZ 11015 tissue sample): a, ventral surfaces of exposed sternum and abdomen, with articulation to proximal left appendage articles; b, diagrammatic anterior oblique perspective, sternal depression and gonopods; c, ventral (external) surface, extended abdomen, disarticulated from thorax; d, sternal (internal) surface, extended abdomen with gonopods; f, posteroventral perspective, diagrammatic overlay of flexed abdomen on sternites and fifth pereopod coxa; e, paratype female, USNM 1129927 (ULLZ 6595 tissue sample): e, ventral (external) surface of extended abdomen, disarticulated from thorax. Scale bars: a = 2.5 mm; b-f = 2 mm.

and subequal. The female gonopores are positioned near the sternal midline, set against the anteromedial margin of the sixth sternite, and reach halfway or less to the anterior margin of the seventh sternite, anteromedial margin of the fifth sternite is more sinuous than in the male, and converging sutures between the fourth through sixth sternite widen mesially to form four arms of a medial 5-armed star-like depression, the fifth arm being a small anteromedial notch. A similar star-shaped depression in the male (fig. 3b) is not as deep.

In the single juvenile specimen, a male, the third to fifth abdominal somites are incompletely fused, with clear evidence of suture lines extending across the full width of the abdomen but without obvious ability for flexure along these lines. In contrast to adults, the major cheliped manus is less conspicuously larger than the minor, and both show more evidence of granulation and surface irregularities on the dorsal (superior) and outer surfaces. The tips of the fingers of the major chela are more spooned than in larger specimens, some of which may be worn. The produced, quadrate frontal lobes of the carapace are truncate, but with more rounded mesial and lateral corners than in mature specimens. The second anterolateral tooth is terminally more acute and anteriorly directed than in adults, being surmounted by sharp granules.

Color. — The dorsal pattern on the carapace is highly variable (fig. 1a-c). A pale horn background is usually mottled with some pattern of yellow-orange, deep orange, rose, and white, sometimes with a solid patch or patches of deep orange or rust centered on gastric and hepatic regions, and commonly with a reddish orange patch on each posterolateral region. The carapace of mature males may be pigmented overall in moderate rose to pinkish red, with little other pattern except for a possible rust colored spot on the gastric region and the aforementioned red spots on each posterolateral region. Similarly, the chelae commonly match the carapace dorsally, varying from mottled with orange patches to near solid orange or moderate rose (fig. 1e). The fingers are of deeper orange to horn brown coloration, fading to whitish tips. The color patterns of the ambulatory pereopods also vary, one consistency being a sometimes ill-defined reddish band in the proximal 1/4 of an otherwise lightly pigmented dactylus (fig. 1a-c, f). The propodus and carpus can be fully covered or having very broad bands of reddish rose, orange, or yellow-orange, and the meri are typically marked by two highly variable bands of reddish or yellow-orange, the latter of which can sometimes cover the whole surface. All of these bands can be very diffuse in lightly pigmented individuals, however, or not obvious as bands in those that tend to be overall solidly pigmented. Even in the least pigmented individuals, there is usually some evidence of moderate red to rose-red pigment on the carpi, either overall or as a broad band. The dorsal surface of the fifth pereopod coxa consistently bears a rhomboidal reddish orange patch, this being evident in all mature specimens except when the body is overall solidly colored.

In the only juvenile specimen, the carapace bears very little dorsal pigment over the pale horn background, except for the posterolateral reddish orange patches on either side. In contrast, the chelae are rather uniformly moderate rose to reddish orange. This matches pigment generally covering the carpi of the second to fourth pereopods, but most color is absent on the carpus of the fifth. Otherwise, the ambulatory pereopods bear the characteristic proximal reddish band on each dactylus, a broad diffuse band of the same on the propodi, and a pair of very light such bands on otherwise poorly pigmented meri. The dorsal reddish patch on the fifth pereopod coxa is clearly evident and subtriangular in shape.

Etymology. — The epithet *guinotae* honors our distinguished colleague, Danièle Guinot, for her magnificent contributions to studies of xanthoids and other brachyuran crabs.

Distribution. — Known only from banks on the outer continental shelves of the eastern and northwestern Gulf of Mexico, occurring among mollusk shells, dead coral fragments, and other calcareous rubble at depths of 57-88 m.

Remarks. — The initial examination of specimens belonging to the new species showed similarities in superficial features of general habitus with some species of the genus *Pilumnoides* Lucas in H. Milne Edwards & Lucas, 1844 (Pseudozoidea), *Actumnus* Dana, 1851 (Pilumnoidea), and four genera of Xanthoidea: *Pilodius* Dana, 1851, *Chlorodiella* Rathbun, 1897, *Cyclodius* Dana, 1851 [=*Phymodius* A. Milne-Edwards, 1863], and *Paretisus* Ward, 1933 (see Forest & Guinot, 1961; Guinot, 1964; Takeda & Kim, 1977; Serène, 1984; Guinot & Macpherson, 1987; Clark & Galil, 1993). The gonopods in particular resemble those of *Chlorodiella nigra* (Forskål, 1775) despite great differences in their carapaces and pereopods (see Forest & Guinot, 1961). Diagnostic characters nevertheless most closely approach Indo-West Pacific members of the xanthoidean genus *Etisus* H. Milne Edwards, 1834 (see Serène, 1984), though species assigned to that taxon may themselves not represent a monophyletic group. The new species shares with the majority of the 23 Indo-West Pacific congeners (Ng et al., 2008) an elongated lateral flange on the basal antennal article that excludes the flagellum from the orbit, the general configuration of the male abdomen relative to the sternum and gonopore, the general shape of the first male gonopods, and at least a somewhat projected front with relatively large antennal sulci. At least some also share paired superior rows of conspicuous spines or teeth on the walking leg dactyls, but the presence of a relatively obvious locking articulation between the propodus and dactylus of the walking legs in *E. guinotae* n. sp. contradicts one dactylar character previously used in definition of the genus (A. Milne-Edwards, 1873; Edmondson, 1935, 1962; Guinot, 1964; Serène, 1980, 1984). However, few species of the genus exhibit comparably produced truncate frontal lobes in combination with such broadly excavated, U-shaped antennal sulci. The antennal sulci in *E. anaglyptus* H. Milne Edwards, 1934, are certainly of comparable development, but size of the frontal lobes, carapace shape, and cheliped ornamentation differ markedly from those of *E. guinotae* n. sp. On the other hand, the general carapace shape and production of the frontal lobes in *E. godeffroyi* (A. Milne-Edwards, 1873) resemble those features in the new species, but the antennal sulci are relatively less deeply developed, especially in dorsal perspective, while both cheliped and carapace dentition also appear to differ in form. The presence of corneous spines on the posterior anterolateral teeth as seen in *E. guinotae* n. sp. is not specifically mentioned in descriptions of any congener, though similarly acute tips of anterolateral teeth are depicted in an unnamed "*Etisus* sp." (Serène, 1984, pl. XXXIII fig. A), one in which the carapace and front again differ markedly from those features in the new species.

There are no Gulf of Mexico species with which *E. guinotae* n. sp. can be confused on even casual inspection. It does not, remarkably, bear close resemblance to *E. maculatus* (Stimpson, 1860), the only western Atlantic species that is currently (perhaps questionably) assigned to that genus (Guinot, 1969, 1971; see molecular phylogenetic discussion below). While these species share an elongated lateral lobe on the basal article of the antenna, blocking the flagellum from the orbital hiatus, the overall shape and positioning of this article varies greatly between the species (fig. 4a). While the two species also exhibit similar spination on the dorsal (superior) surfaces of the walking leg dactyls, it is less strongly developed in *E. maculatus* (fig. 4b). The features of the carapace differ even more conspicuously. Rather than being quadrate and truncate, each of the produced frontal lobes of *E. maculatus* slopes laterally to a prominent, rounded tooth, lateral to which the antennal sulci are broadly V-shaped (fig. 4c). The overall carapace outline in *E. maculatus* is also much less rounded than in *E. guinotae* n. sp., its width greatly exceeding its length. The male first gonopods of *E. maculatus* (fig. 4d) are in their general form more like those of *Pilodius*

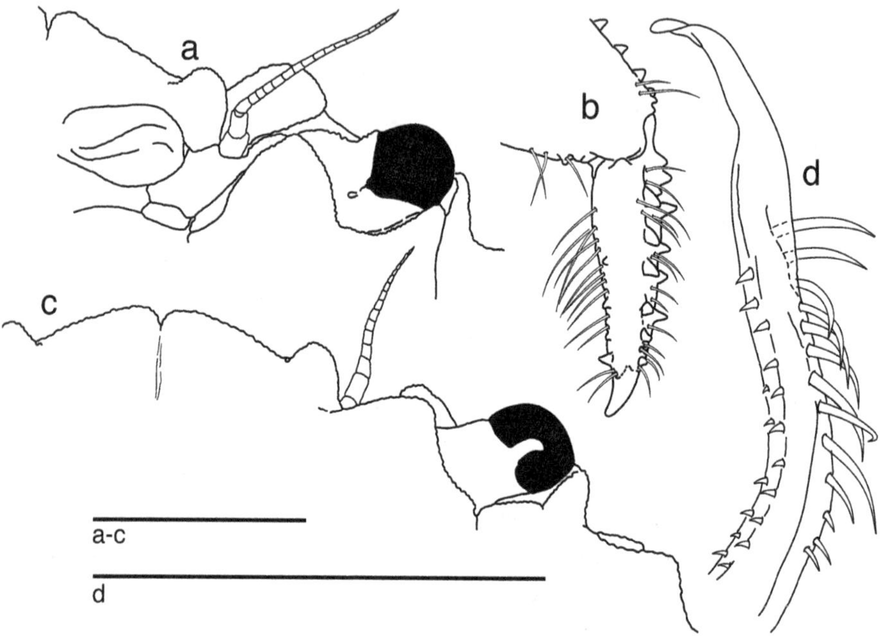

Fig. 4. *Etisus maculatus* (Stimpson, 1860), male, Carrie Bow Cay, Belize, ULLZ 10463: a, left anteroventral surface of carapace (antennule and setae not shown); b, superoposterior surface of dactylus of right fourth pereopod; carapace front and right orbital region, setae not shown; enlarged terminus of right first gonopod, abdominal surface. Scale bars :a-c = 3 mm; d = 1.0 mm.

pugil Heller, 1861 and *P. spinipes* Dana, 1852, than those of *E. guinotae* n. sp. or for that matter those of all of its congeners (see Serène, 1984; Clark & Galil, 1993). Finally, the two species differ in habitat, with *E. maculatus* occurring only in intertidal to shallow subtidal rubble settings to a maximum of about 3 m depth (Felder et al., 2009).

The general shape of the carapace, along with size and coloration, may at first glance suggest *E. guinotae* n. sp. to be one of the small calappoids or large pilumnoids that share its habitat. Likewise, the somewhat similarly proportioned xanthids *Carpoporus* Stimpson, 1871, and *Allactaea* Williams, 1974, occur in the same habitat, but all can be readily distinguished, even if only by inspection of the carapace front, anterolateral dentition, and spination of the pereopod dactyls.

If *E. guinotae* n. sp. was proved to be native to the Gulf of Mexico its discovery could perhaps be explained as the very belated detection of a Tethyan relict with close ties to modern Indo-West Pacific brachyuran groups. Other cases of single Atlantic species with such Indo-West Pacific ties are certainly known among the 109 xanthoid crabs currently reported from the Gulf of

Mexico (sensu "Xanthoidea" of Martin & Davis, 2001; Felder et al., 2009), *Ozius reticulatus* (Stimpson, 1860), and *Eriphia gonagra* (Fabricius, 1781), being two such examples (although their relationships to Indo-West Pacific congeners are currently under re-evaluation by J. C. Y. Lai, the present authors, and others, in prep.). This would mean that over 100 years of sampling effort in the western Atlantic had until now failed to produce the first specimen of a supposedly indigenous population, in which specimens are relatively large, colorful, and of remarkably unique morphology.

Prior to our own collection efforts on Gulf of Mexico deep banks during the last decade, which included over 50 dredge hauls in the very habitats off Louisiana where five specimens of *E. guinotae* n. sp. were finally found in 2008, intensive surveys at similar depths and with similar collecting equipment were conducted in the course of the "Hourglass Cruises" on the southwestern Florida continental shelf in the eastern Gulf of Mexico (Joyce & Williams, 1969). While the xanthoid crabs from those cruises have yet to be exhaustively studied and reported upon, our cursory inspection of archived collections revealed nothing that might represent *E. guinotae* n. sp. Our own dredge sampling aboard the R/V "Pelican" returned to many of those same areas on the Florida shelf and similar adjacent banks of the northeastern and north-northeastern Gulf of Mexico in 2004 and 2006, but only one of approximately 90 stations (restricting that count to those matching depth and habitat as now known for *E. guinotae* n. sp.) produced the original single specimen. Approximately 90 such stations sampled on the Campeche Banks off Yucatan in 2005 produced no specimens.

Failure of recent intensive Gulf of Mexico sampling efforts to produce specimens of such a conspicuous animal as *E. guinotae* n. sp. until now, along with the absent mention of any similar materials in older classic survey reports for the region (see review, Felder et al., 2009), suggests that this species is a probable invasive. However, unless a source population is identified, that is difficult to prove. If not an introduction, the recent appearance of this crab in our samples could alternatively represent a highly periodic bloom in abundance of some rare but indigenous species, or a recent expansion of range from inadequately sampled western Atlantic habitats adjacent to the Gulf of Mexico, perhaps under the influence of rising coastal ocean temperature. Regardless of the possible explanations, only more exhaustive compositional accounts for global brachyuran assemblages, along with continued monitoring of regional populations, can provide further insights.

MOLECULAR PHYLOGENETICS AND DISCUSSION

In this preliminary molecular phylogenetic analysis, *Etisus guinotae* n. sp. is nested within a clade containing representatives of the subfamilies Xanthinae, Euxanthinae, Actaeinae, Chlorodiellinae, and Etisinae, as the sister-taxon to a clade composed of two members of the subfamily Xanthinae, *Cataleptodius floridanus* (Gibbes, 1850) and *Xanthodius cooksoni* Stimpson, 1859 (fig. 5). Despite the suggestion of a sister-taxon relationship between *E. guinotae* n. sp. and the clade containing *C. floridanus* and *X. cooksoni*, branch lengths between them and poor support for this grouping (p*P* and BS values less than 50) suggest that *E. guinotae* n. sp. is not closely related to either of these taxa. However, other trees from the Bayesian and maximum-likelihood analyses likewise provide no evidence of a close relationship between *E. guinotae* n. sp. and any other xanthoid taxa included in our preliminary molecular genetic study. Somewhat surprisingly, this statement also applies to *Etisus maculatus*, raising question as to the appropriate generic assignment for either or both of these present congeners.

E. maculatus is the only western Atlantic species (and only American species) to be previously assigned to the genus *Etisus*, and the only congener of *E. guinotae* n. sp. included in the present molecular analysis. While morphological characters of the basal antennal article and general profile of the carapace front may for the time being justify assignment of *E. maculatus* to this genus, other features of its morphology do not appear typical of the genus. For example, as has been discussed, the male first gonopod is more like those seen in species of *Pilodius* than among the known species of *Etisus*. Also, the propodal-dactylar locking articulation is clearly evident in the walking legs, an exception to what is considered the typical mature state in *Etisus* spp. of the Indo-West Pacific (Serène, 1984). Thus, we concluded that its position in our present molecular phylogenetic analysis could not be used to argue for or against placement of our new species in *Etisus*. Rather, we for now retain a broad definition of that genus, accepting that it could later prove to be polyphyletic. Before concluding that restriction of *Etisus* is required, or erecting new genera to receive any of its present members, additional analysis must be undertaken with broad representation of Indo-West Pacific members and inclusion of the type species of the genus, *E. dentatus* (Herbst, 1785). Until such time, we recommend that the generic assignment of both *E. maculatus* and *E. guinotae* n. sp. be regarded as provisional.

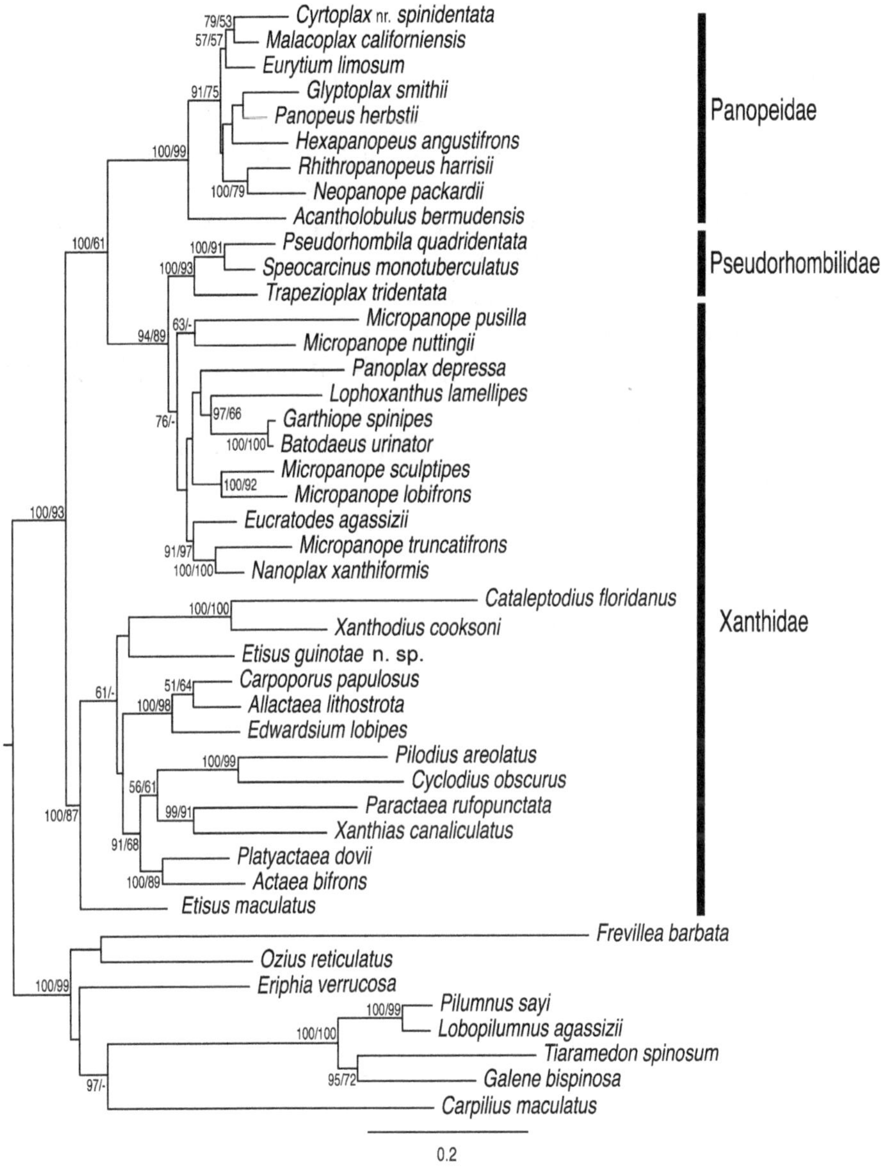

Fig. 5. Phylogenetic relationships among select representatives of Xanthoidea sensu Martin & Davis (2001), inferred by Bayesian analysis of 1322 basepairs of concatenated 12S, 16S, and Histone-3 dataset. Confidence values are Bayesian posterior probabilities followed by 1000 bootstrap maximum-likelihood values. Vertical bars indicate families of Xanthoidea sensu Ng et al. (2008). Values below 50 are indicated by "-".

ACKNOWLEDGEMENTS

For access to certain comparative materials for this analysis, we thank Robert Lasley, Arthur Anker, and Gustav Paulay, Florida Natural History Museum, University of Florida, Gainesville. Among many who in varied ways assisted in field sampling, analytical, and writing phases of this project, we thank F. Álvarez, H. Bracken, P. Castro, C. Ehrenhaus, E. Escobar, S. Fredericq, E. García, F. Gurgel, D. Krayesky, P. Ng, J. Mendoza, E. Pante, R. Robles, and W. Schmidt. Special thanks are also due to W. Cargile who allowed us to join his team of malacologists on a 2008 cruise of the R/V "Pelican" that he funded, herein termed the "Cargile Cruise". Other field work and laboratory analyses were supported by funding to D. Felder from U.S. National Science Foundation grants NSF/BS&I DEB-0315995 and NSF/AToL EF-0531603, along with U.S. Department of Energy grant no. DE-FG02-97ER1220. Participation of B. Thoma was supported under a Louisiana Board of Regents doctoral fellowship. This is contribution number 135 from the UL-Lafayette Laboratory for Crustacean Research.

Appendix I

Crab species used for phylogeny reconstruction, showing catalog number, collection locality, and GenBank accession numbers for partial sequences of 12S, 16S and Histone-3 respectively (ULLZ = University of Louisiana at Lafayette Zoological Collections, Lafayette, Louisiana; USNM = United States National Museum of Natural History, Smithsonian Institution, Washington D.C.)

Taxon name	Catalog number	Locality	12S	16S	H3
Carpiliidae					
Carpilius maculatus (Linnaeus, 1758)	ULLZ 7164	United States: Hawaii	GU144423	GU144450	GU144479
Eriphiidae					
Eriphia verrucosa (Forskål, 1775)	ULLZ 4275	Spain: Cadiz	EU863332	EU863398	GU144467
Ozius reticulatus (Desbonne, in Desbonne & Schramm, 1867)	ULLZ 8420	Mexico: Veracruz	GU144414	GU144446	GU144468
Euryplacidae					
Frevillea barbata A. Milne-Edwards, 1880	ULLZ 8369	United States: Florida	EU863333	EU863399	GU144484
Panopeidae					
Acantholobulus bermudensis (Benedict & Rathbun, 1891)	ULLZ 6924	United States: Florida	EU863306	EU863372	GU144464
Cyrtoplax nr. *spinidentata* (Benedict, 1892)	ULLZ 8423	United States: Florida	EU863303	EU863369	GU144501
Eurytium limosum (Say, 1818)	ULLZ 4012	Mexico: Veracruz	GU144421	GU144455	GU144477
Glyptoplax smithii A. Milne-Edwards, 1880	ULLZ 6793	Southwestern Gulf of Mexico	EU863276	EU863342	GU144463
Hexapanopeus angustifrons (Benedict & Rathbun, 1891)	ULLZ 8368	United States: Florida	EU863314	EU863380	GU144465
Malacoplax californiensis (Lockington, 1877)	ULLZ 10572	Mexico: Baja California	GU144433	GU144460	GU144506
Neopanope packardii Kingsley, 1879	ULLZ 3772	United States: Florida	EU863283	EU863349	GU144499
Panopeus herbstii H. Milne Edwards, 1834	ULLZ 8457	United States: South Carolina	EU863296	EU863362	GU144466
Panoplax depressa Stimpson, 1871	ULLZ 8056	Northern Gulf of Mexico	EU863281	EU863347	GU144493
Rhithropanopeus harrisii (Gould, 1841)	ULLZ 3995	United States: Texas	EU863280	EU863346	GU144498
Pilumnidae					
Galene bispinosa (Herbst, 1783)	ULLZ 4472	Thailand: Phuket	GU144417	GU144444	GU144470
Lobopilumnus agassizii (Stimpson, 1871)	ULLZ 7121	Southwestern Gulf of Mexico	EU863336	EU863402	GU144475
Pilumnus sayi Rathbun, 1923	ULLZ 6938	United States: Florida	GU144407	GU144435	GU144492
Tiaramedon spinosum (Miers, 1879)	ULLZ 4563	South Africa: Sodwana Bay	GU144413	GU144441	GU144476

APPENDIX I
(Continued)

Taxon name	Catalog number	Locality	12S	16S	H3
Pseudorhombilidae					
Pseudorhombila quadridentata (Latreille, 1828)	ULLZ 9326	Northern Gulf of Mexico	GU144426	GU144454	GU144483
Speocarcinus monotuberculatus Felder & Rabalais, 1986	ULLZ 7562	Southwestern Gulf of Mexico	EU863293	EU863359	GU144497
Trapezioplax tridentata (A. Milne-Edwards, 1880)	ULLZ 8054	Northern Gulf of Mexico	EU863278	EU863344	GU144500
Xanthidae					
Actaea bifrons Rathbun, 1898	ULLZ 7843	Northern Gulf of Mexico	GU144430	GU144458	GU144487
Allactaea lithostrota Williams, 1974	ULLZ 4613	United States: Louisiana	GU144428	GU144456	GU144485
Batodaeus urinator (A. Milne-Edwards, 1881)	ULLZ 8131	Southern Gulf of Mexico	EU863339	EU863405	GU144496
Carpoporus papulosus Stimpson, 1871	ULLZ 7355	Southern Gulf of Mexico	GU144429	GU144457	GU144486
Cataleptodius floridanus (Gibbes, 1850)	ULLZ 4806	United States: Florida	GU144422	GU144449	GU144478
Cyclodius obscurus (Hombron & Jacquinot, 1846)	ULLZ 9034	United States: Hawaii	GU144431	GU144459	GU144488
Edwardsium lobipes (Rathbun, 1898)	ULLZ 9321	Panama: Afuerita Channel	GU144427	GU144453	GU144482
Etisus guinotae n. sp.	ULLZ 6595 = USNM 1129926	Gulf of Mexico	GU144416	GU144443	GU144469
Etisus maculatus (Stimpson, 1860)	ULLZ 9983	Belize: Twin Cays	GU144432	GU144461	GU144489
Eucratodes agassizii A. Milne-Edwards, 1880	ULLZ 8400	United States: Louisiana	EU863323	EU863389	GU144505
Garthiope spinipes A. Milne-Edwards, 1880	ULLZ 7840	Northern Gulf of Mexico	GU144410	GU144438	GU144494
Lophoxanthus lamellipes (Stimpson, 1860)	ULLZ 9313	Panama: Cohiba	GU144425	GU144452	GU144481
Micropanope lobifrons A. Milne-Edwards, 1881	ULLZ 6678	Northern Gulf of Mexico	GU144412	GU144439	GU144504
Micropanope nuttingii (Rathbun, 1898)	ULLZ 8209	Northern Gulf of Mexico	GU144409	GU144437	GU144503
Micropanope pusilla A. Milne-Edwards, 1880	ULLZ 8139	Northern Gulf of Mexico	GU144411	GU144440	GU144495
Micropanope sculptipes Stimpson, 1871	ULLZ 8025	Northern Gulf of Mexico	EU863312	EU863378	GU144491
Micropanope truncatifrons Rathbun, 1898	ULLZ 7359	Northern Gulf of Mexico	GU144406	GU144434	GU144490
Nanoplax xanthiformis (A. Milne-Edwards, 1880)	ULLZ 6752	Northern Gulf of Mexico	GU144408	GU144436	GU144462
Paractaea rufopunctata (H. Milne Edwards, 1834)	ULLZ 8946	Panama: Bocas del Toro	GU144415	GU144442	GU144472
Pilodius areolatus (H. Milne Edwards, 1834)	ULLZ 9033	United States: Hawaii	GU144420	GU144448	GU144474
Platyactaea dovii (Stimpson, 1871)	ULLZ 9043	Panama	GU144424	GU144451	GU144480
Xanthias canaliculatus Rathbun, 1906	ULLZ 4381	South Africa: Sodwana Bay	EU863316	EU863382	GU144502
Xanthodius cooksoni (Miers, 1877)	ULLZ 5936	Mexico: Baja California	GU144418	GU144445	GU144471

REFERENCES

BRACKEN, H. D., S. DE GRAVE & D. L. FELDER, 2009. Phylogeny of the infraorder Caridea based on mitochondrial and nuclear genes (Crustacea: Decapoda). In: J. W. MARTIN, K. A. CRANDALL & D. L. FELDER (eds.), Decapod crustacean phylogenetics. Crustacean Issues, **18**: 281-305. (Taylor & Francis/CRC Press, Boca Raton, Florida).

CLARK, P. F. & B. GALIL, 1993. A revision of the xanthid genus *Pilodius* Dana, 1851 (Crustacea: Brachyura: Xanthoidea). Journal of Natural History, London, **27**: 1119-1206.

EDMONDSON, C. H., 1935. New and rare Polynesian Crustacea. Occasional papers of Bernice P. Bishop Museum, **10** (24): 1-40.

— —, 1962. Xanthidae of Hawaii. Occasional papers of Bernice P. Bishop Museum, **22** (13): 215-309.

FELDER, D. L., F. ÁLVAREZ, J. W. GOY & R. LEMAITRE, 2009. Decapoda (Crustacea) of the Gulf of Mexico, with comments on the Amphionidacea. In: D. L. FELDER & D. K. CAMP (eds.), Gulf of Mexico origin, waters, and biota, **1**, Biodiversity: 1019-1104. (Texas A&M University Press, College Station).

FELDER, D. L. & R. ROBLES, 2009. Molecular phylogeny of the family Callianassidae based on a preliminary analyses of two mitochondrial genes. In: J. W. MARTIN, K. A. CRANDALL & D. L. FELDER (eds.), Decapod crustacean phylogenetics. Crustacean Issues, **18**: 327-342. (Taylor & Francis/CRC Press, Boca Raton, Florida).

FOREST, J. & D. GUINOT, 1961. Crustacés Décapodes Brachyoures de Tahiti et des Tuamotu. Expédition Français sur les Récifs Coralliens de la Nouvelle-Calédonie: 1-195. (A. Lahure, Paris).

GUINOT, D., 1964. Crustacés décapodes brachyoures (Xanthidae) des campagnes de la Calypso en mer Rouge (1952), dans le golfe Persique et a l'île Aldabra (1954). Mémoires du Muséum national d'Histoire naturelle, (Nouvelle Série) (A, Zoologie) **32** (1): 1-108.

— —, 1969. Sur divers Xanthidae, notamment sur *Actaea* De Haan et *Paractaea* gen. nov. (Crustacea Decapoda Brachyura). Cahiers du Pacifique, **13**: 223-267.

— —, 1971. Recherches préliminaries sur les groupements naturels chez les Crustacés Décapodes Brachyoures VIII. Synthèse et bibliographie. Bulletin du Muséum national d'Histoire naturelle, Paris, (2) **42** (5): 1063-1090.

GUINOT, D. & E. MACPHERSON, 1987. Révision du genre *Pilumnoides* Lucas, 1844, avec description de quatre espèces nouvelles et création de Pilumnoidinae subfam. nov. (Crustacea Decapoda Brachyura). Bulletin du Muséum national d'Histoire naturelle, (4) **9** (A, 1): 211-247.

HUELSENBECK, J. P. & F. RONQUIST, 2001. MRBAYES: Bayesian inference phylogeny. Bioinformatics, **17**: 754-755.

JOYCE, E. A. & J. WILLIAMS, 1969. Rationale and pertinent data. Memoires of the Hourglass Cruises, **1**: 1-50.

KATOH, K., K. MISAWA, K. KUMA & T. MIYATA, 2002. MAFFT: a novel method for rapid multiple sequence alignment based on fast Fourier transform. Nucleic Acid Research, **30**: 3059-3066.

MANTELATTO, F. L., R. ROBLES & D. L. FELDER, 2007. Molecular phylogeny of the crab genus *Portunus* (Crustacea, Portunidae) from the western Atlantic. Zoological Journal of the Linnean Society, London, **150**: 211-220.

MANTELATTO, F. L., R. ROBLES, C. D. SCHUBART & D. L. FELDER, 2009. Molecular phylogeny of the genus *Cronius* Stimpson, 1860, with reassignment of *C. tumidulus* and several American species of *Portunus* to the genus *Achelous* De Haan, 1833 (Brachyura: Portunidae). In: J. W. MARTIN, K. A. CRANDALL & D. L. FELDER (eds.), Decapod crustacean phylogenetics. Crustacean Issues, **18**: 567-579. (Taylor and Francis/CRC Press, Boca Raton, Florida).

MARTIN, J. W. & G. E. DAVIS, 2001. An updated classification of the Recent Crustacea. Natural History Museum of Los Angeles County, Science Series, **39**: 1-124.

MILNE-EDWARDS, A., 1873. Descriptions des quelques Crustacés nouveaux ou peu connus provenant du Musée de M. C. Godeffroy. Journal des Museum Godeffroy, **1**: 77-88.

NG, P. K. L., D. GUINOT & P. J. F. DAVIE, 2008. Systema Brachyurorum: Part I. An annotated checklist of extant brachyuran crabs of the world. The Raffles Bulletin of Zoology, (Supplement) **17**: 1-286.

NYLANDER, J. A. A., 2004. MrModeltest 2.3 [program distributed by author]. (Department of Systematic Zoology, EBC, Uppsala University, Sweden).

PALACIOS-THEIL, E., J. A. CUESTA, E. CAMPOS & D. L. FELDER, 2009. Molecular-based re-examination of subfamily relationships and polyphyly in the family Pinnotheridae (Crustacea: Decapoda). In: J. W. MARTIN, K. A. CRANDALL & D. L. FELDER (eds.), Decapod crustacean phylogenetics. Crustacean Issues, **18**: 457-574. (Taylor and Francis/ CRC Press, Boca Raton, Florida).

ROBLES, R., C. D. SCHUBART, J. CONDE, C. CARMONA-SUÁREZ, F. ALVAREZ, J. VILLALOBOS & D. L. FELDER, 2007. Molecular phylogeny of the American *Callinectes* Stimpson, 1860 (Brachyura: Portunidae), based on two partial mitochondrial genes. Marine Biology, Berlin, **150**: 1265-1274.

ROBLES, R., C. C. TUDGE, P. C. DWORSCHAK, G. C. B. POORE & D. L. FELDER, 2009. Molecular phylogeny of the Thalassinidea based on nuclear and mitochondrial genes. In: J. W. MARTIN, K. A. CRANDALL & D. L. FELDER (eds.), Decapod crustacean phylogenetics. Crustacean Issues, **18**: 309-326. (Taylor and Francis/CRC Press, Boca Raton, Florida).

SERÈNE, R., 1980. Notes sur quelques Crustacés Brachyoures provenant principalement de l'île Maurice et conservés au Muséum d'Histoire naturelle de Genéve. Description d'*Etisus zehntneri* sp. nov. Revue Suisse de Zoologie, **87**: 711-722.

——, 1984. Crustacés Décapodes Brachyoures de l'océan Indien occidental et de la mer Rouge, Xanthoidea: Xanthidae et Trapeziidae. Avec un addendum par Crosnier, A.: Carpiliidae et Menippidae. Faune Tropicale, **24**: 1-349.

SVENSON, G. J. & M. F. WHITING, 2004. Phylogeny of Mantodea based on molecular data: evolution of a charismatic predator. Systematic Entomology, **29**: 342-352.

TAKEDA, M. & H. S. KIM, 1977. A new *Actumnus* (Crustacea; Decapoda; Brachyura) from Jeju Island, Korea. Korean Journal of Zoology, **20**: 135-139.

THOMA, B. P., C. D. SCHUBART & D. L. FELDER, 2009. Molecular phylogeny of the western Atlantic representatives of the genus *Hexapanopeus* (Decapoda: Brachyura: Panopeidae). In: J. W. MARTIN, K. A. CRANDALL & D. L. FELDER (eds.), Decapod crustacean phylogenetics. Crustacean Issues, **18**: 551-565. (Taylor and Francis/CRC Press, Boca Raton, Florida).

TSHUDY, D., R. ROBLES, T.-Y. CHAN, K. C. HO, K. H. CHU, S. T. AHYONG & D. L. FELDER, 2009. Phylogeny of marine clawed lobsters (families Nephropidae Dana, 1852 and Taumastochelidae Bate, 1888) based on mitochondrial genes. In: J. W. MARTIN, K. A. CRANDALL & D. L. FELDER (eds.), Decapod crustacean phylogenetics. Crustacean Issues, **18**: 357-368. (Taylor and Francis/CRC Press, Boca Raton, Florida).

First received 6 July 2009.
Final version accepted 11 August 2009.

ON A COLLECTION OF CALAPPOID AND LEUCOSIOID CRABS (DECAPODA, BRACHYURA) FROM VANUATU, WITH DESCRIPTION OF A NEW SPECIES OF LEUCOSIIDAE

BY

BELLA S. GALIL[1,3]) and PETER K. L. NG[2,4])

[1]) National Institute of Oceanography, Israel Oceanographic and Limnological Research, P.O. Box 8030, Haifa 31080, Israel

[2]) Tropical Marine Science Institute and Department of Biological Sciences, National University of Singapore, Lower Kent Ridge Road, Singapore 119260, Republic of Singapore

ABSTRACT

Thirty-eight species of crabs of Calappidae, Matutidae, Iphiculidae, and Leucosiidae were identified from material collected during the SANTO 2006 Expedition to Espiritu Santo Island, Vanuatu. Of these, 32 (14 calappids, one matutid, four iphiculids, and 13 leucosiids) are reported for the first time from Vanuatu. The leucosiid genus *Euclosia* Galil, 2003, is a junior homonym of *Euclosia* Mulsant, Verreaux & Verreaux, 1866 (Aves), and a new name, *Euclosiana*, is proposed. One new species, *Euclosiana guinotae* n. sp., is described as new. It differs from its closest congener, *E. rotundifrons* (Chopra, 1934), in its smooth rather than closely punctate dorsum, and the lack of a diagonal band of granules on the third thoracic sternite.

RÉSUMÉ

Trente huit espèces de crabes Calappidae, Matutidae, Iphiculidae et Leucosiidae ont été identifiées dans le matériel récolté lors de l'expédition SANTO 2006 à l'île Espiritu Santo, Vanuatu. Le genre leucosien *Euclosia* Galil, 2003, est un homonyme plus récent d'*Euclosia* Mulsant, Verreaux & Verreaux, 1866 (Aves), et un nouveau nom, *Euclosiana*, est donc proposé. Parmi les trente deux espèces signalées pour la première fois de ces îles : 14 callapides, un matutide, quatre iphiculides et 13 leucosides. L'une des espèces, *Euclosiana guinotae* n. sp. est décrite; elle diffère de l'espèce proche *E. rotundifrons* (Chopra, 1934) par sa surface lisse au lieu de finement ponctuée et l'absence de la bande diagonale de granules sur le troisième sternite thoracique.

[3]) Corresponding author; e-mail: bella@ocean.org.il

[4]) e-mail: dbsngkl@nus.edu.sg

© Koninklijke Brill NV, Leiden, 2010

INTRODUCTION

An expediton to the island of Espiritu Santo, Vanuatu (SANTO 2006) conducted in 2006 by the Muséum national d'Histoire naturelle, Paris, and the Institut de Recherche pour le Développement (IRD) obtained 38 species of calappid, matutid, iphiculid, and leucosiid crabs. Thirty-two species (14 calappids, one matutid, four iphiculids, 13 leucosiids) are reported herein for the first time from Vanuatu. One species of leucosiid is recognized as new, *Euclosiana guinotae* n. sp.

The leucosiid genus *Euclosia* Galil, 2003, is a junior homonym of *Euclosia* Mulsant, Verreaux & Verreaux, 1866 (type species by original designation *Trochilus flavicaudatus* Fraser, 1840), a hummingbird genus of the family Trochilidae. *Trochilus flavicaudatus* Fraser, 1840, however, is a junior subjective synonym of *Trochilus lafresnayi* Boissoneau, 1840, which is the type species of *Lafresnaya* Bonaparte, 1850, by monotypy. This makes *Euclosia* Mulsant, Verreaux & Verreaux, 1866, a junior synonym of *Lafresnaya* Bonaparte, 1850. Although *Euclosia* Mulsant, Verreaux & Verreaux, 1866, does not appear to have been used as a valid genus name for the bird since 1899, one of the requirements of Article 23.9.1.2 of the Code (which deals with reversal of precedence) nevertheless requires the junior name *Euclosia* Galil, 2003, to have been used for a period of not less than 10 years. As the genus was only named in 2003, Article 23.9 thus cannot be applied, and a replacement name needs to be established for *Euclosia* Galil, 2003. To this effect, we propose the new name *Euclosiana*. The gender is feminine, and the type species is *Leucosia obtusifrons* De Haan, 1841.

Specimens examined are deposited in the Muséum national d'Histoire naturelle (MNHN), Paris; and the Zoological Reference Collection (ZRC), Raffles Museum of Biodiversity Research, National University of Singapore. The abbreviation CL is used for the carapace length, measured along the vertical median line; while G1 and G2 are used for the male first and second gonopods, respectively. All localities listed are in Vanuatu.

TAXONOMY

Family CALAPPIDAE

Calappa bicornis Miers, 1884

Material examined. — One ♂ (CL 48.1 mm) (ZRC 2009.0426), Segond Channel, stn EP1, 15°32.5′S 167°09.0′E, 46-47 m, 10 Sept. 2006.

Distribution. — Madagascar to Japan (Ng, 2002), Vanuatu (new record).

Calappa capellonis Laurie, 1906

Material examined. — One ♂ (CL 36.3 mm) (ZRC 2009.0427), NW Malo Is., stn EP22, 15°37.3′S 167°05.8′E, 78-91 m, 21 Sep. 2006.

Distribution. — Madagascar to New Caledonia (Galil, 1997), Vanuatu (new record).

Calappa clypeata Borradaile, 1903

Material examined. — One ♂ (CL 16.9 mm), 1 ♀ (CL 16.1 mm) (ZRC 2009.0430), Turtle Bay, stn AT38, 15°21.4′S 167°12.8′E, 29-58 m, 27 Sep. 2006.

Distribution. — East Africa to New Caledonia (Galil, 1997), Vanuatu (new record).

Calappa hepatica (Linnaeus, 1758)

Material examined. — One ♂ (CL 43.4 mm) (ZRC 2009.0432), Santo Is., Rose Point, stn VM2, 15°34.9′S 167°02.4′E, intertidal, 10 Sep. 2006.

Distribution. — Red Sea to Hawaii (Balkis & Ceviker, 2003), Vanuatu (new record).

Calappa japonica Ortmann, 1892

Material examined. — One ♀ (CL 103.7 mm) (ZRC 2009.0436), W Malo Is., stn AT64, 15°39.6′S 167°01.9′E, 249-252 m, 4 Oct. 2006.

Distribution. — Red Sea to New Caledonia (Lai & Ng, 2006), Vanuatu (new record).

Calappa liaoi Ng, 2002

Material examined. — One ♀ (CL 21.0 mm) (ZRC 2009.0437), Segond Channel, stn EP1, 15°32.5′S 167°09.0′E, 46-47 m, 10 Sep. 2006.

Distribution. — Philippines (Galil & Ng, 2009), Vanuatu (new record).

Calappa lophos (Herbst, 1782)

Material examined. — One juv. (CL 8.7 mm) (MNHN), Segond Channel, stn ED5, 15°31.7′S 167°09.4′E, 9-13 m, 12 Sep. 2006.

Distribution. — Indian Ocean to Australia (Lai et al., 2006), Vanuatu (new record).

Calappa ocularia Ng, 2002

Material examined. — One ♂ (CL 13.4 mm) (ZRC 2009.0441), NW Malo Is., Bruat Channel, stn AT5, 15°39.9′S 167°03.8′E, 114-132 m, 15 Sep. 2006.

Distribution. — Philippines (Ng, 2002), Vanuatu (new record).

Calappa philargius (Linnaeus, 1758)

Material examined. — One juv. (CL 10.7 mm) (ZRC 2009.0446), Segond Channel, stn ED5, 15°31.7′S 167°09.4′E, 9-13 m, 12 Sep. 2006.

Distribution. — Japan to Samoa (Ng et al., 2002), Vanuatu (new record).

Calappa pustulosa Alcock, 1896

Material examined. — One ♀ (CL 13.1 mm) (ZRC 2009.0449), E Urélapa Is., stn AT30, 15°36.7′S 167°02.6′E, 83-120 m, 25 Sep. 2006.

Distribution. — Indian Ocean to Solomon Is. (Galil, 1997, 2007), Vanuatu (new record).

Calappa undulata Dai & Yang, 1991

Material examined. — One ♂ (CL 35.0 mm), 1 ♀ (CL 45.1 mm) (ZRC 2009.0450), NW Malo Is., stn EP19, 15°37.5′S 167°05.1′E, 80-94 m, 19 Sep. 2006.

Distribution. — Andaman Sea to South China Sea (Ng et al., 1999), Vanuatu (new record).

Calappa gallus (Herbst, 1803)

Material examined. — One ovigerous ♀ (CL 45.9 mm) (ZRC 2009.0561), E Aoré Is., off Aimbué Bay, stn EP34, 15°33.2/33.3′S 167°12.8/12.9′E, 40-60 m, 14 Oct. 2006.

Distribution. — Madagascar to Hawaii (Ng, 2000), Vanuatu (new record).

Mursia diwata Galil & Takeda, 2004

Material examined. — One ♂ (CL 14.2 mm) (ZRC 2009.0453), W Malo Is., stn AT8, 15°40.5′S 167°01.5′E, 366-389 m, 17 Sep. 2006.

Distribution. — Previously known only from the Philippines (Galil & Takeda, 2004), Vanuatu (new record).

Paracyclois milneedwardsii Miers, 1886

Material examined. – Three juvs. (CL 12.2-13.1 mm) (ZRC 2009.0458), W Malo Is., stn AT18, 15°41.3′S 167°02.6′E, 321-336 m, 21 Sep. 2006; 1 ♀ (CL 47.8 mm) (ZRC 2009.0459), Big Bay, stn AT98, 15°06.6′S 166°50.8′E, 347-394 m, 14 Oct. 2006; 1 juv. (MNHN), W Malo Is., stn AT121, 15°38.7′S 167°01.2′E, 275-290 m, 19 Oct. 2006.

Distribution. — Japan to Solomon Is. (Galil & Ng, 2009), Vanuatu (new record).

Family MATUTIDAE

Izanami curtispina (Sakai, 1961)

Material examined. — One ♂ (CL 18.2 mm), 4 ♀♀ (CL 12.1-20.1 mm) (ZRC 2009.0460), 1 ♂ (CL 17.7 mm), 3 ♀♀ (MNHN), NW Malo Is., stn AT5, 15°39.9′S 167°03.8′E, 114-132 m, 15 Sep. 2006.

Distribution. — Madagascar to Solomon Is. (Galil, 2007), Vanuatu (new record).

Family IPHICULIDAE

Iphiculus convexus Ihle, 1918

Material examined. — One ♀ (CL 25.4 mm) (ZRC 2009.0461), E Aoré Is., stn EP10, 15°34.5′S 167°05.1′E, 45-101 m, 15 Sep. 2006; 1 ♂ (CL 22.5 mm) (ZRC 2009.0462), 1 ♂ (CL 19.0 mm) (MNHN), S Urélapa Is., stn EN20, 15°37.2′S 167°01.9′E, 85 m, 19 Sep. 2006; 1 ♀ (CL 24.7 mm) (ZRC 2009.0463), N Malo Is., Bruat Channel, stn FP50, 15°36.8′S 167°08.7′E, 25 m, 4 Oct. 2006; 1 ♀ (CL 15.0 mm) (ZRC 2009.0464), NE Aoré Is., AT85, 15°32.6′S 167°15.7′E, 114-196 m, 12 Oct. 2006.

Distribution. — Philippines, Indonesia (Galil & Ng, 2009), Vanuatu (new record).

Iphiculus spongiosus Adams & White, 1848

Material examined. — Two ♂♂ (CL 11.5 mm, 11.9 mm), 2 juvs. (ZRC 2009.0465), SE Aésé Is., stn AT24, 15°27.0′S 167°16.1′E, 190-210 m, 23 Sep. 2006.

Distribution. — Red Sea to Solomon Is. (Galil, 2007), Vanuatu (new record).

Pariphiculus agariciferus Ihle, 1918

Material examined. — One ♂ (CL 9.2 mm), 1 ♀ (CL 12.9 mm) (ZRC 2009.0470), S Aésé Is., Palikulu Bay, stn AT13, 15°27.8′S 167°15.7′E, 146-153 m, 19 Sep. 2006.

Distribution. — Japan to Indonesia (Galil & Ng, 2007), Vanuatu (new record).

Pariphiculus mariannae (Herklots, 1852)

Material examined. — One juv. (MNHN), E Urélapa Is., stn AT41, 15°36.7′S 167°02.7′E, 88-118 m, 28 Sep. 2006.

Distribution. — Arabian Sea to Solomon Is. (Galil, 2007), Vanuatu (new record).

Family LEUCOSIIDAE

Alox rugosum (Stimpson, 1858)

Material examined. — One ♀ (CL 12.6 mm) (ZRC 2009.0476), Malparavu Is., stn LD6, 15°21.9′S 167°11.5′E, 2-5 m, 27 Sep. 2006.

Distribution. — Madagascar to New Caledonia (Tan, 1996), Vanuatu (new record).

Arcania cornuta (MacGilchrist, 1905)

Material examined. — One ♂ (CL 30.3 mm) (ZRC 2009.0477), Tutuba Is., stn EP6, 15°34.2′S 167°15.6′E, 52 m, 13 Sep. 2006.

Distribution. — Persian Gulf to Fiji (Galil, 2001a, 2007).

Arcania elongata Yokoya, 1933

Material examined. — One ♂ (CL 17.8 mm), 1 ♀ (CL 18.8 mm), 1 juv. (ZRC 2009.0480), S Tutuba Is., stn AT29, 15°35.9′S 167°01.3′E, 83-90 m, 25 Sep. 2006.

Distribution. — Japan to New Caledonia (Galil & Ng, 2007), Vanuatu (new record).

Arcania gracilis (Henderson, 1893)

Material examined. — One juv. (ZRC 2009.0483), NW Tutuba Is., stn AT2, 15°32.5′S 167°16.1′E, 160-175 m, 14 Sep. 2006.

Distribution. — Red Sea to Vanuatu (Galil, 2001a, 2007).

Arcania septemspinosa (Fabricius, 1787)

Material examined. — One ♀ (CL 20.9 mm) (ZRC 2009.0562), Segond Channel, stn AT53, 15°31.8′S 167°13.6′E, 62-71 m, 2 Oct. 2006.

Distribution. — Across Indo-West Pacific region (Galil & Ng, 2007), Vanuatu (new record).

Euclosiana guinotae n. sp. (figs. 1-3)

Type material. — Holotype: ♂ (CL 28.5 mm) (MNHN), Tutuba Is., stn EP6, 15°34.2′S 167°15.6′E, 52 m, 13 Sep. 2006; paratypes: 1 ♂ (CL 29.4 mm) (ZRC 2009.0552), Scorff Passage, Santo Is., stn EP12, 15°31.9′S 167°15.1′E, 97 m, 16 Sep. 2006; 2 ♂♂ (CL 27.9 mm, 28.1 mm), 1 ovigerous ♀ (CL 29.9 mm) (ZRC 2009.0553), 1 ♂, 1 ovigerous ♀ (CL 29.8 mm) (MNHN), W Urélapa Is., stn EP24, 15°36.5′S 167°00.9′E, 108-121 m, 25 Sep. 2006.

Etymology. — Named in honour of Prof. Danièle Guinot, for her seminal work on brachyuran crab systematics.

Description. — Carapace subpentagonal, globose; regions of carapace indistinct. Dorsal surface of carapace glabrous, shiny, smooth, sparsely punctate. Frontal region produced, upcurved, laterally concave. Frontal margin medially dentate. Antennular fossa sealed by basal antennular segment. Antennae short, inserted between antennular fossa, orbit. Orbit small, rounded, outer orbital margin unsutured, anterior margin of efferent branchial channel forms part of lower orbital margin. Eyes retractable. External maxillipeds concealing trapezoid buccal opening; endopod of merus triangular, as long as ischium; female endopod with setose, lengthwise fringe. Anterolateral margin of carapace weakly convex along hepatic region. Lateral angle of carapace rounded, overhanging thoracic sinus, margin prominently beaded. Thoracic sternites smooth. Margin of thoracic loop barely visible in dorsal view. Pterygostomian region smooth. Sinuous anterior margin of thoracic sinus bearing coalescent granules; row of perliform granules above cheliped basis, median granules largest. Posterolateral margins of carapace rounded, granular, granules smaller, further apart posteriorly. Posterior margin sinuous in male, rounded in female. Chelipeds subequal, robust, long; longer in adult male than in female. Cheliped merus tubular, bearing prominent perliform tubercles on anterior, posterior margins, upper surface bearing coalesced cluster of 6-8 granules proximally, followed by 2 diverging rows of tubercles, distally smooth; lower surface coarsely tuberculate anteriorly, pitted proximally. Carpus inflated, bearing row of granules on inner margin. Propodus smooth except granular lower surface, beaded line along inner lower margin; upper margin carinate. Fingers as long as propodus, inner margins serrate. Pereiopods slender, short. Pereiopodal meri with obsolescent rows of granules on dorsal, ventral surfaces; last pereiopod propodi keeled dorsally, ventrally. Dactyli longer than propodi, lanceolate. Male abdominal sulcus deep, elongated, nearly reaching buccal cavity; lateral margin with distinct ridge fitting into suture between abdominal somites. Male abdomen with horizontally narrow first somite, second somite spindle-shaped, minute; third to fifth somites fused, vertically furrowed proximally; sixth somite with rounded longitudinal ridge bearing median denticle; telson triangular, one-third as long as sixth somite. Female abdomen with first and second somites narrow, arched; third to fifth somites fused, greatly enlarged, shield-like, telson laciniated. G1 elongated, shaft stout, sinuous, tightly coiled 3 times on itself, bearing ruff of preapical setae, cornute apical process, slender, arrow-shaped, bent at angle to rest of shaft. G2 short, curved, scoop-like apex.

Colour in life. — Dorsum pale pinkish brown, paler posteriorly; gastric region with three pale circles, median circle largest; large bright orange spots on either side of the intestinal region; three orange-coloured marks on lateral margins (fig. 1). Carapace ivory coloured ventrally. Chelipeds pale pinkish brown, paler distally. Perliform tubercles on upper surface of cheliped merus pale, ringed with orange, tubercles on lower surface ivory coloured; fingers proximally pale orange, distally white. Pereiopods white, meri subdistally orange; carpi, propodi, dactyli proximally ringed with orange.

Remarks. — *Euclosiana guinotae* n. sp. shares with *E. rotundifrons* (Chopra, 1934), its closest congener, a slender, arrow-shaped apical process of the G1 (fig. 3; Galil, 2003b, figs. 4B, 6F), but *E. guinotae* is readily distinguished by its smooth, rather than closely punctuate dorsum (figs. 1, 2A) and the lack of a diagonal band of granules on the third thoracic sternite (fig. 2B). The species is so far known only from Vanuatu.

Heterolithadia fallax (Henderson, 1893)

Material examined. — One ♂ (CL 9.4 mm) (ZRC 2009.0511), Mavéa Is., stn EP37, 15°23.4′S 167°13.1′E, 50-61 m, 16 Oct. 2006.

Distribution. — India to Philippines (Tan, 1996), Vanuatu (new record).

Myra curtimana Galil, 2001

Material examined. — One ♀ (CL 14.7 mm), 3 juvs. (MNHN), NW Malo Is., Bruat Channel, stn AT5, 15°39.9′S 167°03.8′E, 114-132 m, 15 Sep. 2006.

Distribution. — Philippines to Fiji (Galil & Ng, 2009), Vanuatu (new record).

Myra tumidospina Galil, 2001

Material examined. — Three ♂♂ (CL 29.6-33.7 mm), 2 ♀♀ (CL 33.6 mm, 33.7 mm) (ZRC 2009.0515), Tutuba Is., stn EP6, 15°34.2′S 167°15.6′E, 52 m, 13 Sep. 2006.

Distribution. — Philippines to Fiji (Galil & Ng, 2007).

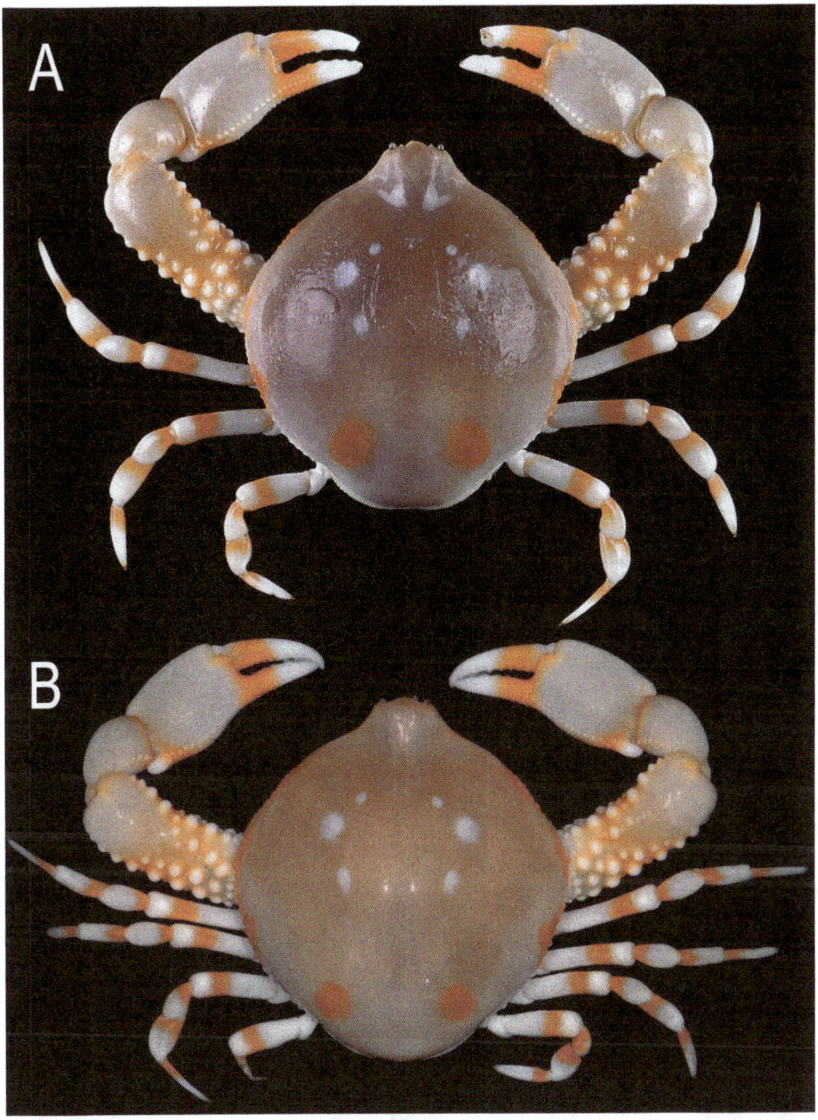

Fig. 1. *Euclosiana guinotae* n. sp. Colour in life. A, holotype ♂ (CL 28.5 mm) (MNHN), Tutuba Is.; B, paratype ♂ (CL 29.4 mm) (ZRC 2009.0553), Scorff Passage, Santo Is.

Nursilia dentata Bell, 1855

Material examined. — One ♂ (CL 4.5 mm), 1 ♀ (CL 5.5 mm) (ZRC 2009.0535), NW Malo Is., stn AT76, 15°38.7′S 167°03.6′E, 105-135 m, 10 Oct. 2006.

Distribution. — Red Sea to Fiji (Tan, 1996), Vanuatu (new record).

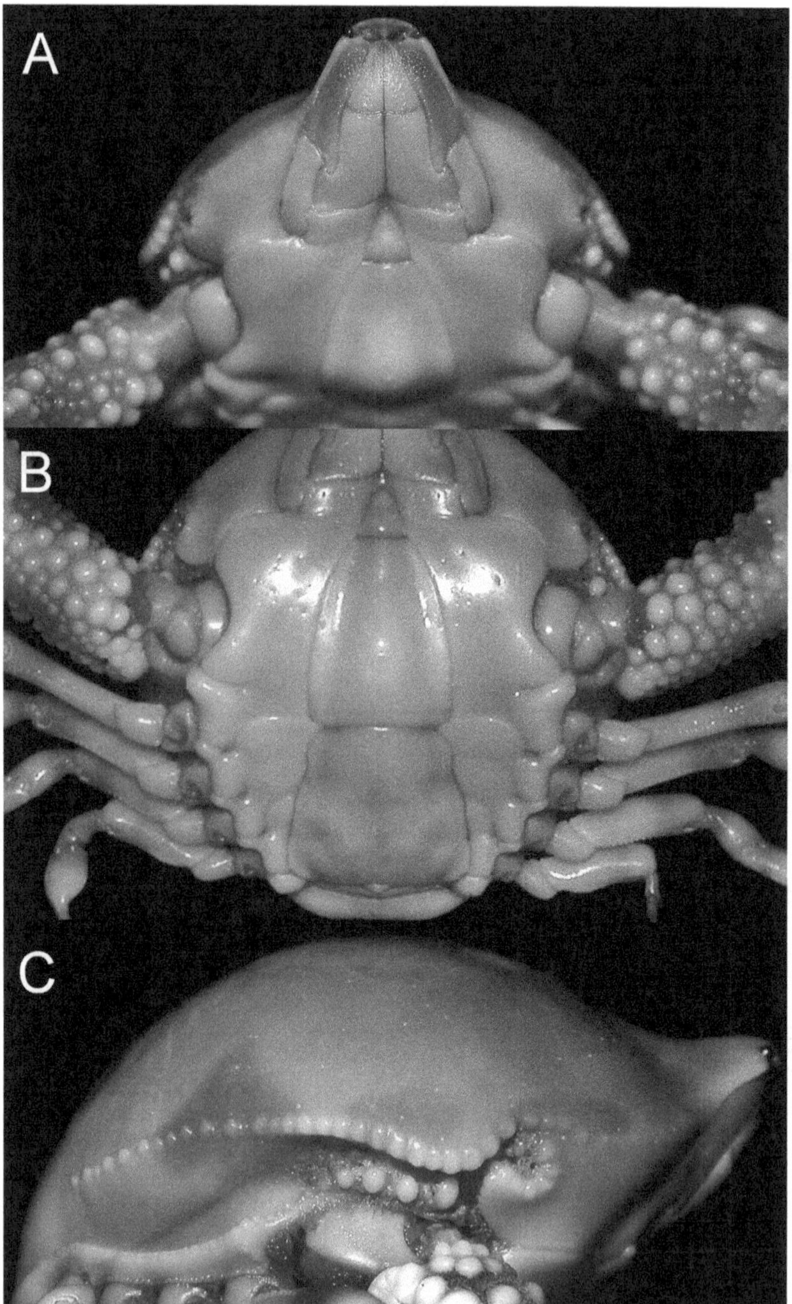

Fig. 2. *Euclosiana guinotae* n. sp. Paratype ♂ (CL 29.4 mm) (ZRC 2009.0553), Scorff Passage, Santo Is. A, frontal view showing third maxillipeds; B, ventral view showing thoracic sternum and abdomen; C, lateral view showing thoracic sinus.

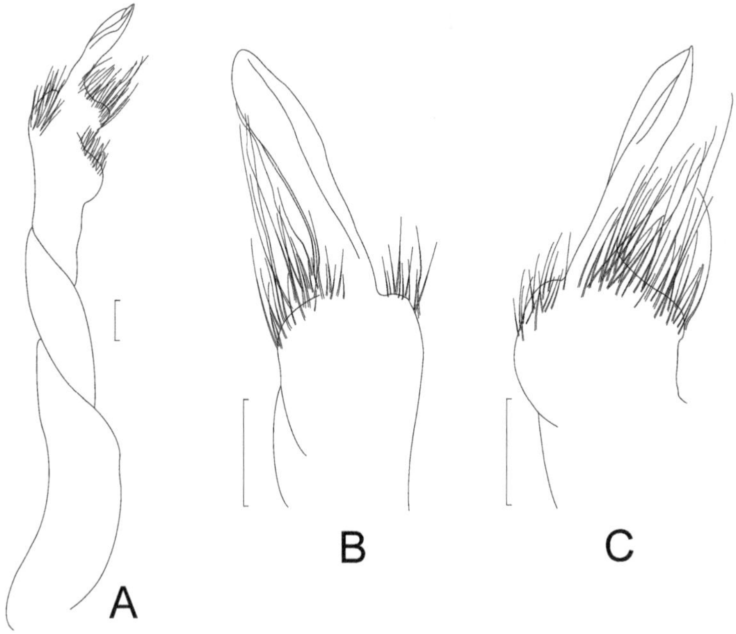

Fig. 3. *Euclosiana guinotae* n. sp., holotype ♂ (CL 28.5 mm) (MNHN), Tutuba Is. A, G1; B, C, distal part of G1. Scale bars: 1.0 mm.

Oreophorus reticulatus Adams & White, 1849

Material examined. — One ♀ (CL 14.7 mm) (ZRC 2009.0536), Bruat Channel, stn AT74, 15°37.0′S 167°09.7′E, 50-64 m, 10 Oct. 2006.

Distribution. — Persian Gulf to Australia (Hewitt, 2004), Vanuatu (new record).

Raylilia coniculifera Galil, 2001

Material examined. — One ♂ (CL 9.2 mm) (ZRC 2009.0537), Vaucluse Passage, stn AG12, 15°22.5′S 167°12.3′E, 74 m, 18 Sep. 2006.

Distribution. — Indonesia, New Caledonia (Galil, 2001b), Vanuatu (new record).

Raylilia uenoi (Takeda, 1995)

Material examined. — One ♂ (CL 10.2 mm) (ZRC 2009.0538), Segond Channel, NW Aoré Is., stn ZR4, 15°33.1′S 167°09.6′E, 0-45 m, 17 Sep. 2006.

Distribution. — Madagascar to New Caledonia (Galil, 2001b), Vanuatu (new record).

Soceulia alainia Galil, 2006

Material examined. — One ♂ (CL 35.2 mm) (ZRC 2009.0554), Aoré Is., Aimbuel Bay, stn EP23, 15°32.8′S 167°12.1′E, 74-88 m, 22 Sep. 2006.

Distribution. — Indonesia, New Caledonia (Galil, 2006), Vanuatu (new record).

Tokoyo cirrata Galil, 2003

Material examined. — One juv. (ZRC 2009.0540), NW Tutuba Is., stn AT2, 15°32.5′S 167°16.1′E, 160-175 m, 14 Sep. 2006.

Distribution. — Australia, Solomon Is., Vanuatu (Galil, 2007).

Toru pilus (Tan, 1996)

Material examined. — Two ♂♂ (CL 10.3 mm, 10.9 mm) (MNHN), W Malo Is., stn AT8, 15°40.5′S 167°01.5′E, 366-389 m, 17 Sep. 2006.

Distribution. — Philippines to Fiji (Galil, 2007).

Toru septimus Galil, 2003

Material examined. — Two juvs. (ZRC 2009.0548), NW Tutuba Is., stn AT1, 15°32.4′S 167°16.4′E, 167-367 m, 14 Sep. 2006.

Distribution. — Solomon Is. to Tonga (Galil, 2007).

Urnalana insularis (Takeda & Kurata, 1976)

Material examined. — One ♂ (CL 8.8 mm) (ZRC 2009.0558), Segond Channel, N Aoré Is., stn NB12, 15°33.1′S 167°09.6′E, 20 m, 19 Sep. 2006.

Distribution. — Japan to Fiji (Galil, 2005), Vanuatu (new record).

Urnalana pulchella (Bell, 1855)

Material examined. — One ovigerous ♀ (CL 9.0 mm) (ZRC 2009.0559), NNW Mavéa Is., stn AT39, 15°22.4′S 167°12.6′E, 57-81 m, 27 Sep. 2006.

Distribution. — Red Sea to Fiji (Galil & Ng, 2007), Vanuatu (new record).

ACKNOWLEDGEMENTS

The SANTO 2006 Expedition was organized by MNHN, Pro Natura International, and Institut de Recherche pour le Développement, under a permit issued to Philippe Bouchet by the Environment Unit of the Government of Vanuatu. The "Marine Biodiversity" component is a part of the Census of Marine Life's CReefs programme, which was funded by grants from the Total and Sloan Foundations. We also thank B. Richer de Forges who collected many of the specimens on board the "Alis". S. H. Tan, J. C. Mendoza, and M. Manuel-Santos helped with the photography and fieldwork, with T.-Y. Chan in charge of photography. The first author's study in Singapore was partially funded by a research fellowship from the Raffles Museum.

REFERENCES

BALKIS, H. & D. CEVIKER, 2003. A new exotic crab species *Calappa hepatica* (Linnaeus, 1758) for the Mediterranean fauna. Israel Journal of Zoology, **49** (4): 320-321.

CHOPRA, B., 1934. Further notes on Crustacea Decapoda in the Indian Museum. III. On the decapod Crustacea collected by the Bengal Pilot Service off the mouth of the river Hughli. Dromiacea and Oxystomata. Records of the Indian Museum, **35**: 25-52.

GALIL, B. S., 1997. Crustacea Decapoda: a revision of the Indo-Pacific species of genus *Calappa* Weber, 1795 (Calappidae). In: A. CROSNIER (ed.), Résultats des Campagnes MUSORSTOM, **18**. Mémoires du Muséum national d'Histoire naturelle, Paris, **176**: 271-335.

— —, 2001a. A revision of the genus *Arcania* Leach, 1817 (Crustacea: Decapoda: Leucosioidea). Zoologische Mededelingen, Leiden, **75** (11): 169-206.

— —, 2001b. A new genus and species of leucosiid crab (Crustacea, Decapoda Brachyura) from the Indo-Pacific Ocean. Zoosystema, **23** (1): 65-75.

— —, 2003a. Four new genera of leucosiid crabs (Crustacea: Brachyura: Leucosiidae) for three new species and nine species previously in the genus *Randallia* Stimpson, 1857, with a redescription of the type species, *R. ornata* (Randall, 1939). Proceedings of the Biological Society of Washington, **116** (2): 395-422.

— —, 2003b. Contribution to the knowledge of Leucosiidae. II. *Euclosia* gen. nov. (Crustacea: Brachyura). Zoologische Mededelingen, Leiden, **77** (20): 331-348.

— —, 2005. Contribution to the knowledge of Leucosiidae III. *Urnalana* gen. nov. (Crustacea: Brachyura). Zoologische Mededelingen, Leiden, **79** (2): 9-40.

— —, 2006. Contributions to the knowledge of Leucosiidae VI. *Soceulia* gen. nov. (Crustacea: Brachyura). Zoologische Mededelingen, Leiden, **80** (6): 71-79.

— —, 2007. The deep-water Calappidae, Matutidae and Leucosiidae of the Solomon Islands, with a description of a new species of *Euclosia* Galil, 2003 (Crustacea, Decapoda, Brachyura). Zoosystema, **29** (3): 555-563.

GALIL, B. S. & P. K. L. NG, 2007. Leucosiid crabs from Panglao, Philippines, with description of three new species (Crustacea: Decapoda: Brachyura). Raffles Bulletin of Zoology, (Supplement) **16**: 83-98.

— — & — —, 2009. Calappoidea and Leucosioidea (Crustacea: Decapoda: Brachyura) from Luzon, Philippines, with descriptions of two new species of *Mursia*. Zootaxa, **2085**: 45-60.

GALIL, B. & M. TAKEDA, 2004. On a collection of *Mursia* (Crustacea, Decapoda, Brachyura, Calappidae) from Balicasag Island, Philippines. Bulletin of the National Science Museum, Tokyo, (A) **30** (1): 23-35.

HEWITT, M. A., 2004. Crustacea (excluding Cirripedia) of the Dampier Archipelago, Western Australia. Records of the Western Australian Museum, (Supplement) **66**: 169-219.

LAI, J. C. Y. & P. K. L. NG, 2006. A new species of *Calappa* Weber, 1795 (Crustacea: Decapoda: Calappidae) from East and South Africa. Zootaxa, **1358**: 39-48.

MULSANT, M. É., J. VERREAUX & É. VERREAUX, 1866. Essai d'une classification méthodique des Trochilidés ou Oiseaux-mouche, contenant le catalogue de toutes les espèces de ces oiseaux. Mémoires de la Société impériale des sciences naturelles de Cherbourg, **12** [Series 2, Tome II]: 207 [19 pp.].

NG, P. K. L., 2000. *Calappa pokipoki*, a new species of box crab (Crustacea: Decapoda: Brachyura: Calappidae), from Hawaii. Proceedings of the Biological Society of Washington, **113** (4): 945-955.

— —, 2002. New species and new records of box crabs (*Calappa*) (Crustacea: Decapoda: Brachyura: Calappidae) from the Philippines. Journal of the National Taiwan Museum, **55** (1): 41-60.

NG, P. K. L., H. L. CHEN & T.-Y. CHAN, 1999. Taxonomic notes on three Indo-West Pacific species of *Calappa* (Decapoda: Brachyura: Calappidae). Raffles Bulletin of Zoology, **47** (2): 607-616.

NG, P. K. L., J. C. Y. LAI & C. AUNGTONYA, 2002. The box and moon crabs of Thailand, with description of a new species of *Calappa* (Crustacea: Brachyura: Calappidae, Matutidae). Phuket Marine Biological Center Special Publication, **23** (2): 341-360.

TAN, C. G. S., 1996. Leucosiidae of the *Albatross* expedition to the Philippines, 1907-1910 (Crustacea: Brachyura: Decapoda). Journal of Natural History, London, **30**: 1021-1058.

First received 14 May 2009.
Final version accepted 13 July 2009.

CHASTITY BELTS AND PLANKTOTROPHIC LARVAE: CONSTRAINTS ON GECARCINID REPRODUCTIVE BEHAVIOUR

BY

RICHARD G. HARTNOLL[1,2])

[1]) School of Biological Sciences, University of Liverpool and the Marine Biological Association of the U.K., Plymouth

ABSTRACT

Land crabs of the family Gecarcinidae all have planktotrophic larval development in the sea, with five or six zoeae and one megalopa. This imposes two constraints on the crabs. Firstly, females must migrate to the sea, sometimes over considerable distances, to release the larvae. This stresses them and limits reproductive investment and larval output. Secondly, at the end of their larval development the crabs must recolonize land, often onto small and remote islands. Limited larval output, and the vagaries of currents, climate and planktonic circulation, potentially restrict recruitment. In addition, female gecarcinids are available for mating only during the brief interval of decalcification of the genital operculum (the 'chastity belt'). This may occur before, during, and/or after migration to the sea, depending on species. This diversity of mating opportunity constrains male migration patterns to optimize mating opportunity.

RÉSUMÉ

Les crabes terrestres de la famille des Gecarcinidae ont tous des développements larvaires en mer, comprenant cinq stades zoé et un stade mégalope. Cela impose deux contraintes aux crabes. Premièrement, les femelles doivent migrer vers la mer, souvent sur de longues distances, pour délivrer leurs larves. Cela les perturbe et limite leur investissement reproductif dans la production des larves. Deuxièmement, à la fin du développement larvaire les crabes doivent regagner la terre, qui consiste souvent en petites îles éloignées. Une émission larvaire limitée et les aléas de la circulation planctonique limitent potentiellement le recrutement. Les femelles Gecarcinidae sont disponibles pour l'accouplement seulement pendant la brève période de décalcification de l'opercule génital (la « ceinture de chasteté »). Cela se produit, pendant, et/ou après la migration vers la mer, selon les espèces. Cette diversité d'opportunités d'accouplements contraint la stratégie de migration des mâles pour optimiser ces possibilités d'accouplement.

[2]) e-mail: hartnoll@advsys.co.uk

© Koninklijke Brill NV, Leiden, 2010

INTRODUCTION

Crabs of the family Gecarcinidae are generally referred to as 'land crabs', but they are not the only land-living crabs, and in some respects not the best adapted to a terrestrial life. They are generally medium to large crabs of robust build, with six known genera and 21 recognized species worldwide (Ng & Guinot, 2001; Ng et al., 2008). A number of gecarcinid species exceed 100 mm carapace width (the little-studied genus *Epigrapsus* is rather smaller at <40 mm, see Liu & Jeng, 2005): *Cardisoma guanhumi* Latreille, 1828, is the largest, reaching 130 mm carapace width (Türkay, 1970). Adult gecarcinids show a variable degree of dependence on water. *Gecarcinus*, *Gecarcoidea*, and *Johngarthia* (and probably some *Epigrapsus* species) often live far from permanent water bodies (fig. 1A), and can be found a considerable distance inland and at substantial altitudes: in Jamaica, *Gecarcinus ruricola* (Linnaeus, 1758) reaches altitudes in excess of 1000 m (Britton et al., 1982). In contrast, *Cardisoma* and *Discoplax*, whilst sometimes living well away from the sea, are more dependent on water (see Hartnoll, 1988, for references). They either live close to permanent water, or inhabit burrows which penetrate to the water table. Nevertheless gecarcinids are all poorly adapted to desiccation stress in comparison to insects, and are predominantly active by night, or in humid conditions during and after rain. All gecarcinids concur in that their larval development shows no concession to a terrestrial life style. The larvae must hatch into the sea, where they pass through an extended series of planktotrophic stages before returning to land. This requirement imposes the first two constraints on reproductive behaviour: the necessity for the female to migrate to the sea for breeding and the need for the juvenile stages to re-colonize the land.

Gecarcinids are not alone in having this combination of adult terrestrial adaptation together with a lack of larval modification. The same pattern is seen in various hermit crabs: many *Calcinus* species, and the Coenobitidae, comprising *Coenobita* and the coconut or robber crab *Birgus latro* (Linnaeus, 1767). In the Grapsidae, *Geograpsus crinipes* (Dana, 1851) and *G. grayi* (H. Milne Edwards, 1853) are similar, as are a number of Sesarmidae species such as *Armases ricordi* (H. Milne Edwards, 1853), *Chiromantes eulimene* (De Man, 1895) and *C. ortmanni* (Crosnier, 1965). In the Ocypodidae, several species of *Ocypode* regularly burrow well above high tide, including *O. africana* (De Man, 1881), *O. cordimanus* Latreille, 1818, *O. pallidula* Jacquinot, 1846 and *O. saratan* (Forskål, 1775). However, none of the above species are as terrestrial as the better adapted gecarcinids. Further details of the extent of terrestrial adaptation of these and other species are reviewed in Hartnoll (1988).

However, there are other land crabs where the larval development has been modified to avoid the necessity for females to migrate to the sea ('landlocked' crabs: many largely aquatic, but others terrestrial to varied degrees). Some of these crabs have normal planktotrophic larvae, but the zoeas are hatched into flowing freshwater (which they can tolerate), which carries them to the sea where they complete their development. Examples are the sesarmids *Chiromantes haematocheir* (De Haan, 1833) (cf. Hashimoto, 1965; Saigusa, 1981; as a *Sesarma*), and *Armases roberti* (H. Milne Edwards, 1853) (cf. Diesel & Schuh, 1998). In *A. roberti*, the larvae can be released up to 5 km from the sea. However, the juveniles of all these crabs must still migrate from the sea back to the adult habitat.

Greater modifications take place in those land-locked crabs which have abbreviated larvae or direct development, where this occurs within the habitat of the adult female. In such cases there is no need for the females to migrate to the sea, or for a return migration by the juveniles. Two taxonomic groups dominate this category. One is the Sesarmidae, where abbreviated larval development occurs in three clusters. In Jamaica there is the bromeliad crab *Metopaulias depressus* Rathbun, 1896 (cf. Hartnoll, 1964; Diesel, 1989) and seven (Reimer et al., 1998) or perhaps eight (Schubart et al., 1998) endemic species of *Sesarma*. Secondly, there are the 48 species of *Geosesarma* (see Ng et al., 2008). Soh (1969, as a *Sesarma*) describes the abbreviated development of *G. perracae* (Nobili, 1903), while Ng & Tan (1995) record complete development in *G. notophorum* Ng & C. G. S. Tan, 1995. Thirdly, in Southeast Asia, there are a number of cavernicolous species of *Karstarma* (see Ng et al., 2008 for list of species and references), including *K. cerberus* (Holthuis, 1964) (cf. Holthuis, 1964, as a *Sesarma*) and *K. jacobsoni* (Ihle, 1912) (cf. Ihle, 1912, as a *Sesarma*) which can occur some distance from the sea. However, from what is known they have relatively small eggs suggesting that either the adults or larvae may need to return to the sea. The other group is the 'freshwater' or 'river' crabs, which have direct development with the juvenile crabs brooding under the female abdomen. The group comprises four very species-rich (or synonym-rich, depending upon viewpoint) superfamilies: the Gecarcinucoidea (~350 species), the Potamoidea (~600 species), the Pseudothelphusoidea (~250 species), and the Trichodactyloidea (~50 species). Despite the name, many 'freshwater crabs' are quite terrestrial (Yeo et al., 2008; Cumberlidge & Ng, 2009). Hartnoll (1988) provided a review of some of these terrestrial species, with the Australian *Austrothelphusa transversa* (Von Martens, 1868) providing a particularly striking example (Taylor & Greenaway, 1979, as a *Holthuisana*). To this list should also be added several further candidates. The American ocypodid *Uca subcylindrica* Stimpson, 1859, can live over 30 km from the shore, with abbreviated lecithotrophic larvae developing in ephemeral pools (Rabalais & Cameron, 1985). Species of the Southeast Asian dotillid genus *Potamocypoda* Tweedie, 1938, live in fresh water (Ng, 2004). The goneplacoid *Trogloplax joliveti* Guinot, 1986 (Chasmocarcinidae) is found in caves 70 km inland in New Britain (Guinot, 1986). Finally there is a diversity of the Hymenosomatidae which live in fresh water, have large eggs, and direct development (e.g., Ng, 1991; Ng & Chuang, 1996; Naruse & Ng, 2007; Naruse et al., 2008). It is among the most terrestrial of these various 'landlocked' crabs that the species most fully adapted to life on land occur, rather than in the 'land crabs' of the Gecarcinidae.

Two constraints on gecarcinid behaviour have been mentioned above, resulting from these terrestrial crabs retaining a marine planktotrophic larval phase: females must migrate to the sea to release the larvae and the young crabs must recolonize the land at the end of the larval phase. A third constraint arises from the limited opportunities available for mating, the 'chastity belt' of the title. The timing of this availability, in relation to the migration of the females, constrains the migratory behaviour of the males. These three constraints, affecting different components of the gecarcinid life cycle, will be examined below.

The taxonomy of the Brachyura has undergone a number of revisions at all levels in recent years: some of these are now in regular usage, others are not. In this study I have followed the nomenclature in Ng et al. (2008).

EGGS, HATCHING AND LARVAL DEVELOPMENT

As stated above, gecarcinids undergo an extended planktotrophic larval phase in the sea. Not all species have been comprehensively studied, but there is substantial supporting evidence from a variety of studies.

The eggs are small and numerous, characteristics of crabs with a long planktotrophic development. They are 0.42 mm diameter in *Gecarcinus ruricola* and 0.50 mm in *Johngarthia lagostoma* (H. Milne Edwards, 1837) (unpubl. data), whereas they are typically >1.5 mm in crabs with abbreviated development (Hartnoll, 1964). Egg numbers range from 20 000 to >600 000 in *Cardisoma guanhumi* (cf. Henning, 1975), 12 000 to 58 000 in *Epigrapsus notatus* (Heller, 1865) (cf. Liu & Jeng, 2005), 20 000 to 110 000 in *Gecarcinus lateralis*

(Fréminville, 1835) (cf. Klaassen, 1975), 18 000 to 213 000 in *G. ruricola* (cf. Hartnoll et al., 2007), and 36 000 to 109 000 in *J. lagostoma* (cf. Hartnoll et al., 2010). These egg numbers are comparable to (though somewhat lower than) those for marine crabs of similar size with planktotrophic development (Hines, 1982). There are a number of observations indicating that the first stage zoeas hatch directly into the sea, as in *Epigrapsus notatus* (cf. Liu & Jeng, 2005), *G. ruricola* (cf. Hartnoll et al., 2007), *Gecarcoidea natalis* (Pocock, 1888) (cf. Hicks et al., 1990) and *J. lagostoma* (cf. Hartnoll et al., 2010). Hatching is typically nocturnal and a very rapid process. The females enter shallow water, and release the zoeae in a matter of seconds by rapid fanning of the abdomen, often during immersion by a single wave. In some cases the females do not even enter the water, but release the egg mass from vertical rock faces to drop into the sea below (Klaassen, 1975; Bliss et al., 1978; Hicks, 1985; Hicks et al., 1990; Liu & Jeng, 2005; Hartnoll et al., 2010).

Complete larval development has been described for the following species: five zoeal stages plus megalopa in *Cardisoma carnifex* (Herbst, 1796) (cf. Kannupandi et al., 1980), *C. guanhumi* (cf. Costlow & Bookhout, 1968), *Discoplax hirtipes* (Dana, 1852) (cf. Shokita & Shikatani, 1990, as a *Cardisoma*) and *Johngarthia planatus* (Stimpson, 1860) (cf. Cuesta et al., 2007); five or six zoeal stages plus megalopa in *Gecarcinus lateralis* (cf. Willems, 1982, as *G. lateralis lateralis*); and six zoeal stages plus megalopa in *C. armatum* Herklots, 1851 (cf. Cuesta & Anger, 2005). The studies concur that development occurs in normal salinity (34-35‰) water, and that all larval stages will feed on *Artemia* nauplii during rearing.

All of this evidence, from egg size, egg number, site of release, and number and habit of larval stages, concurs in indicating a 'normal' (ancestral) planktotrophic development. There is no indication that any gecarcinid species diverges from this pattern. A clear consequence of this development strategy is that gecarcinids, in contrast to 'landlocked' land crabs, show low species diversity, little endemism, and extensive distributions for most species. There are only six genera and 21 gecarcinid species worldwide, despite the fact that the group is abundant and clearly successful in many areas. This species-paucity contrasts with at least eight endemic sesarmids in Jamaica, 48 species of *Geosesarma* in the Indo-West Pacific, and some 1200 species of potamids (freshwater crabs) worldwide (references above, in Introduction). However, whilst this wide dispersal and genetic flow has advantages, it comes at a price.

FEMALE BREEDING MIGRATION

Females must migrate to the sea in order for the larvae to be released directly into the sea. Such migrations are widely documented: in *Discoplax hirtipes* Dana, 1851 (cf. Gibson-Hill, 1947; Shokita, 1971, as a *Cardisoma*), *C. guanhumi* Latreille, 1825 (cf. Gifford, 1962; Henning, 1975), *Epigrapsus notatus* (cf. Liu & Jeng, 2005), *Gecarcinus lateralis* (cf. Klaassen, 1975; Bliss et al., 1978; Wolcott & Wolcott, 1982), *G. ruricola* (cf. Hartnoll et al., 2007), *Gecarcoidea lalandii* H. Milne Edwards, 1837 (cf. Liu & Jeng, 2007), *G. natalis* (cf. Gibson-Hill, 1947; Hicks, 1985; Adamczewska & Morris, 2001), and *Johngarthia lagostoma* (cf. Hartnoll et al., 2010). These migrations can involve large concentrations of crabs, which may even disrupt traffic flow on coastal roads (fig. 1B). On Christmas Island it pays not to leave the front door of your house open because migrating red crabs may be found in the lounge or often emerging from the back door. The same crabs are a novel hazard on the golf course requiring special local course rules (Hicks et al., 1990).

Migrations can follow specific pathways, particularly where island populations are concerned. Breeding may be restricted to specific shores, not necessarily those closest to the normal residential areas for adults. *Gecarcoidea natalis* on Christmas Island (Adamczewska & Morris, 2001), *Gecarcinus ruricola* in the San Andres Archipelago (Hartnoll et al., 2007) and *J. lagostoma* on Ascension Island (Hartnoll et al., 2010) all migrate to restricted areas of the respective islands. Tracking and marking experiments showed that *Gecarcoidea natalis* resident in an area tended to migrate to the same shore, and that crabs used the same migration paths in successive years (Adamczewska & Morris, 2001). It is not clear whether migration is to the shores where the crabs initially landed, or to those shores which will maximize the likelihood of larval return. Both factors will potentially favour the same shore areas, since the crabs will have landed in areas favouring larval return! Selective pressures will favour the choice of such shores (Adamczewska & Morris, 2001).

Gecarcinids all have delimited breeding seasons (during which migration, mating and incubation occur), but these can be quite extensive: for example, seven months in *Discoplax hirtipes* (cf. Shokita & Shikatani, 1990, as a *Cardisoma*); six months in *C. guanhumi* (cf. Gifford, 1962); five months in *Gecarcinus ruricola* (cf. Hartnoll et al., 2007) and *Gecarcoidea lalandii* (cf. Liu & Jeng, 2007). However, in any given year breeding may not extend throughout this period, and migratory activity is often stimulated by rainfall (Adamczewska & Morris, 2001). In other species the seasons are shorter:

Fig. 1. A, *Johngarthia lagostoma* (H. Milne Edwards, 1837) at an altitude of 800 m on Ascension Island, an island with no permanent bodies of water; B, road sign on the island of Cozumel, Mexico, warning traffic of migrating *Cardisoma guanhumi* Latreille, 1828.

three months in *Gecarcoidea natalis* (cf. Hicks, 1985), and two months in *Epigrapsus notatus* (cf. Liu & Jeng, 2005). It has been suggested that the duration of the breeding season is related to the duration of availability of water (Liu & Jeng, 2005, 2007). However, species in arid habitats may also have extended seasons, related to the unpredictability of rainfall, as in *J. lagostoma* (cf. Hartnoll et al., 2010). In such species peak breeding may occur at different times each year.

The timing of migration is constrained by several factors. Overall it tends to correspond with the local rainy season (see references in Hartnoll et al., 2007, 2010), which is of benefit in limiting desiccation stress. Variations in annual rainfall account for much of the fluctuation between years in timing and intensity of migration. There are also variable degrees of lunar entrainment (see references in Hartnoll et al., 2007, 2010), with different phase preferences between species. Mangrove crabs, such as *Cardisoma*, may benefit from hatching at the highest tides of new or full moon, to facilitate immersion and maximize larval export. Conversely, species hatching on exposed rocky shores may benefit from the lesser wave action associated with neap tides in the first and last quarters, as in *Gecarcoidea* and *Johngarthia* (cf. Hartnoll et al., 2010). Ideally such studies should be based upon data on larval release, rather than on migration shorewards: the study on *G. lalandii* (cf. Liu & Jeng, 2007) is an excellent example of the former, and shows a particularly clear-cut entrainment. In most species migration occurs at night (see references in Hartnoll et al., 2007), the only exception being *G. natalis*, which is active by day (Hicks, 1985; Green, 1997). This is in marked contrast to its congener *G. lalandii* which shows the usual nocturnal activity (Liu & Jeng, 2007).

The annual breeding migration will clearly stress the females taking part, potentially exposing them to desiccation and predation. On Ascension Island, the migration involves crossing several km of barren arid volcanic rock, and numbers of dead female *J. lagostoma*, many bearing eggs, were found near the shore (Hartnoll et al., 2010). In contrast, in the less stressful terrain of Taiwan, migrating *G. lalandii* showed "no weak or dead females as a result of desiccation" (Liu & Jeng, 2007). Hicks (1985) records predation on migrating *G. natalis* by various predators, both on land and in water. In many areas, human activity, particularly traffic, poses a major threat (Hicks, 1985). Overall, female migration, whilst essential, comes at a cost. Whilst the males of a number of species also migrate (see below), the cost will be less. In a number of gecarcinid species males reach a larger size than females (Hartnoll et al., 2009): the higher mortality of females, associated with migration, may be a contributory factor.

It has been noted above that reproduction is seasonal, but that the season may extend over a number of months, involving a sequence of migration episodes. An important question is whether individual females breed only once during the annual season, or more than once, a problem that is not easy to resolve. From the lengthy duration of ovarian maturation and the lack of ovarian re-ripening in females with mature eggs, it has been concluded that some species breed only once a year, as in *Gecarcinus ruricola* (cf. Hartnoll et al., 2007) and *J. lagostoma* (cf. Hartnoll et al., 2010). Marking experiments, and the short breeding season, indicate that *Epigrapsus notatus* also breeds only once (Liu & Jeng, 2005). Seasonal iteroparity is nevertheless suggested for *Cardisoma guanhumi* (cf. Gifford, 1962) and *Gecarcinus lateralis* (cf. Klaassen, 1975) based on ovarian condition and captive material, respectively. Marking programmes can provide unambiguous evidence of iteroparity: in *Gecarcoidea lalandii* this demonstrated that at least some females bred twice, but it was not clear whether this was a widespread pattern (Liu & Jeng, 2007). Even in these iteroparous species, however, the demands of migration result in an annual reproductive investment that is low compared to tropical marine species such as the grapsid *Grapsus adscensionis* (Osbeck, 1765) (cf. Hartnoll, 2009). This will reduce annual larval output, contributing to the problems of recruitment.

RETURN MIGRATION OF THE JUVENILES

The planktonic larval phase of gecarcinids lasts some three weeks (Costlow & Bookhout, 1968; Willems, 1982; Hicks, 1985), and at the end of this period the megalopa or first crab stage (depending on species, see Hartnoll & Clark, 2006) must return to land or perish. This poses a problem for populations of those species which inhabit the margins of continents, such as *Cardisoma* and *Discoplax* species and *Gecarcinus lateralis*: they may be swept offshore and fail to reach land in time. The problem is even greater, however, for those populations on islands, which are often small and remote, because their return relies on the vagaries of winds and currents. This group includes the species of *Epigrapsus*, *Gecarcoidea*, and *Johngarthia*, as well as *Gecarcinus ruricola*. Here, the chances of a timely landfall are much reduced, and the problem is compounded by the remoteness from alternative sources of recruitment. The situation of *Johngarthia lagostoma* on Ascension Island is a prime example (Hartnoll et al., 2006a).

There are not many records of the return of juvenile gecarcinids, which could be for two reasons: either it is not noticed or it is infrequent. There is probably an element of truth in both options. The records which exist are mostly of mass returns, in numbers so great that they cannot be missed, and are consequently reported even if monitoring is not in place. Such records exist for *Gecarcoidea natalis* (cf. Hicks et al., 1990), *Gecarcinus ruricola* (cf. Hartnoll & Clark, 2006) (fig. 2A), *J. lagostoma* (cf. Hartnoll et al., 2009), and *J. planatus* (cf. Niaussat & Ehrhardt, 1971, as a *Gecarcinus*). For the first two species, mass returns are recorded every five years or so, and records are reliable. In *J. lagostoma* returns have been recorded only twice in the last forty years, but the most probable return sites are remote and little visited, and episodes could well have been missed. But in all species these spasmodic mass returns cannot be the sole recruitment episodes, since size structure analysis does not display the modal peaks which would result if they were (Hartnoll et al., 2006b, 2009). Small specimens occur in years when no mass recruitment has been recorded, and so there must be some recruitment in most, possibly all, years. Since the breeding season extends over several months, there is scope for extended 'trickle' recruitment which would go unrecorded in the absence of specific monitoring. Detailed study of *Gecarcoidea natalis* over several years did detect at least some annual recruitment (Hicks, 1985). Mass returns will result only when optimal conditions for larval survival, and favourable current and wind patterns, coincide with major peaks of larval release. Thus the mass return of *Gecarcinus ruricola* on Old Providence Island, Colombia, in June 2004 followed the biggest spawning episode of the year three weeks earlier in May (Hartnoll & Clark, 2006). It is in any case problematic as to whether these mass returns can function as effective recruitment agents to the populations. In other crab species, e.g., the majoid *Chionoecetes opilio* (Fabricius, 1788), it has been shown that 'saturated' populations of adults can cannibalize and exclude small recruits (Lovrich & Sainte-Marie, 1997, and references therein). The survival rate of mass returns has not been followed up, but a slow and steady recruitment may constitute a more effective process.

Whilst it appears that recruitment is perhaps not always such a dire problem as the dearth of observations might suggest, there are cases where it does seem restricted. The remoteness of Ascension Island has been highlighted, and the population of *J. lagostoma* there is extremely skewed. Of more than 4000 crabs examined, only 25 were below the size of maturity (Hartnoll et al., 2009), which clearly indicates problems with recruitment. This contrasts with continental margin species such as *Gecarcinus lateralis*, where small crabs are

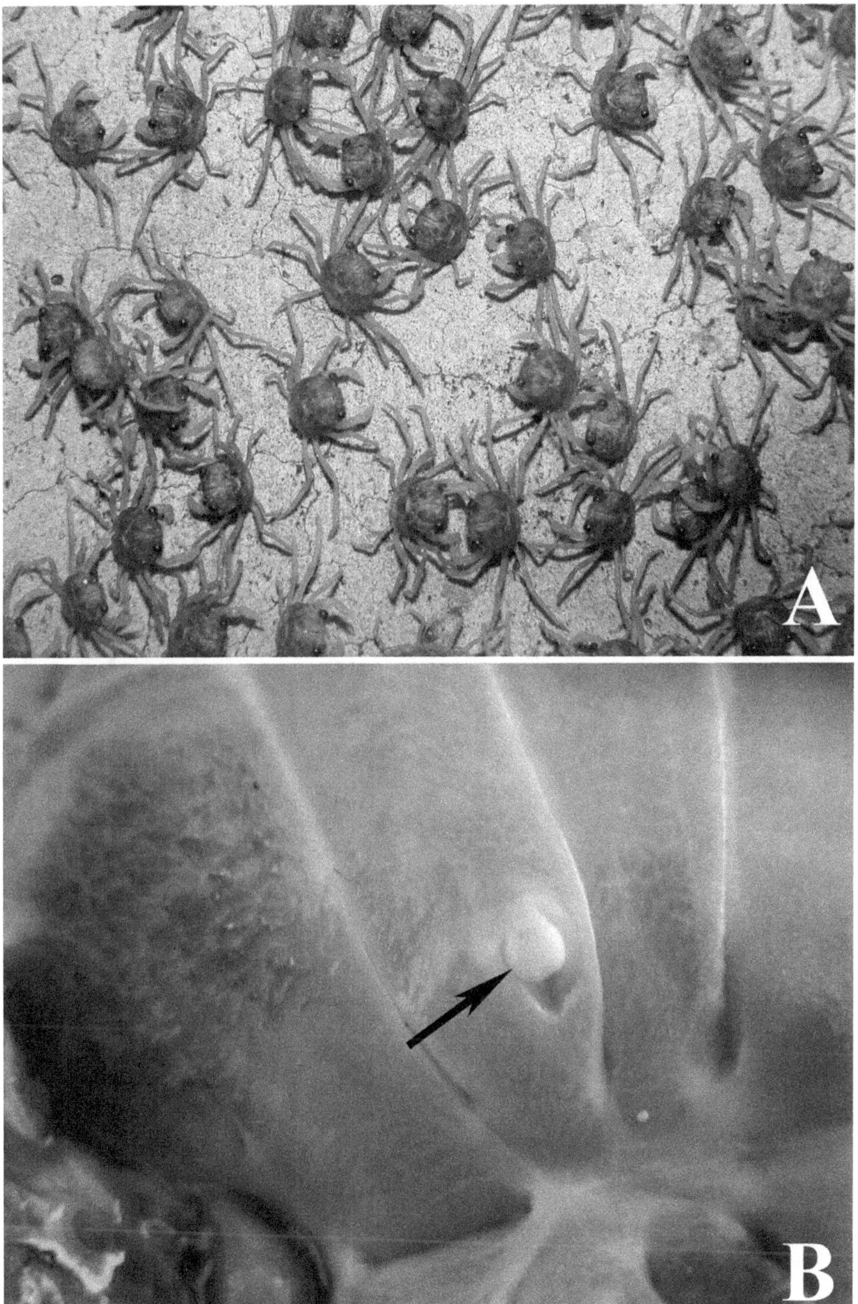

Fig. 2. A, *Gecarcinus ruricola* (Linnaeus, 1758): first crabs crossing the coastal road in Old Providence Island, Colombia, during the mass return in June 2004; B, *Johngarthia lagostoma* (H. Milne Edwards, 1837): female sternum showing genital operculum (arrow).

regularly found in considerable abundance (Bliss et al., 1978), suggesting that recruitment is both consistent and ample.

THE TIMING OF MATING — THE CHASTITY BELT EFFECT

A general feature of reproduction in crabs is that mating occurs only during limited periods in the life cycle of females. In some species it occurs only when the females are soft-shelled after moulting, as in such well-studied species as the portunids *Callinectes sapidus* Rathbun, 1896 (cf. Chidchester, 1911) and *Carcinus maenas* (Linnaeus, 1758) (cf. Broekhuysen, 1936), and the cancrid *Cancer pagurus* Linnaeus, 1758 (cf. Edwards, 1966). Soft-shell mating is a general feature in the Cancridae and Portunidae (Hartnoll, 1969). In other groups, females mate only in the hard shelled condition, a general phenomenon in xanthid, grapsid, and ocypodid crabs (Hartnoll, 1969). For crabs with a terminal moult, such as spider crabs, mating may occur in both soft and hard-shelled states (Jones & Hartnoll, 1997). In most of the above the temporal limitation of mating is the result of periodic chemical attractiveness of the females (Eales, 1974; Gleeson, 1980), and/or receptive behaviour by the females (Christy, 1987). There is generally no physical impediment to mating at other times. Egg laying does not necessarily occur soon after mating, and can take place up to two years later (Paul, 1984). In gecarcinids, all observations agree that the females mate only in the hard condition (e.g., Henning, 1975; Klaassen, 1975; Bliss et al., 1978; Hicks, 1985; Hartnoll et al., 2006a).

In some crabs, however, there is a physical constraint to mating in the form of calcified opercula occluding the female genital openings — the 'chastity belt' of the title. These opercula are normally rigidly shut, and cannot be moved without fracturing the integument (Hartnoll, 1965) so that neither mating nor egg-laying are possible as long as this condition prevails. The structure of the operculum, and the basis of its immobility, is described for the grapsid *Pachygrapsus marmoratus* (Fabricius, 1887) in Hartnoll (1968a). Immobile opercula occur in at least the following families: Corystidae, Dorippidae, Macrophthalmidae, Ocypodidae, Grapsidae, Sesarmidae and Gecarcinidae (cf. Hartnoll, 1968a): the list becomes longer following the increased family diversity recorded in Ng et al. (2008) (e.g., in some Goneplacidae as well, Castro, 2007). The opercula are flexible for a short time after moulting, before the integument has fully recalcified. It was once wrongly assumed that mating must occur within this interval (Hartnoll, 1964, 1965). However, all subsequent observations have shown that mating, followed shortly by laying, occurs instead within a short period of decalcification during the intermoult period when the integument is otherwise fully hardened (see references in Hartnoll, 1969; Brockerhof & McLay, 2005). This decalcification lasts only a matter of days: 12-20 days in *Corystes cassivelaunus* (Pennant, 1777) (cf. Hartnoll, 1968b); ~4-5 days in *Medorippe lanata* (Linnaeus, 1767) (cf. Hartnoll, 1968a, as a *Dorippe*); 7-14 days in *Hemigrapsus sexdentatus* (H. Milne Edwards, 1837) (cf. Brockerhoff & McLay, 2005); 6-15 days in *Ilyoplax pusilla* (De Haan, 1835) (cf. Henmi & Murai, 1999). There are further records of species being receptive for short periods related to opercular decalcification: *Gaetice depressus* (De Haan, 1833) (cf. Fukui, 1993); *Macrophthalmus hirtipes* (Jacquinot, in Hombron & Jacquinot, 1846) (cf. Jennings et al., 2000); and *Uca pugnax* (Smith, 1870) (cf. Greenspan, 1982). These observations concur that the physical availability of females, when the operculum decalcifies, coincides with them becoming "attractive" to males, and receptive to mating.

The presence of such immobile genital opercula appears a general feature of gecarcinids. It occurs in *Gecarcinus ruricola* (cf. Hartnoll, 1968a), species of *Cardisoma*, *Discoplax*, *Gecarcoidea* and *Johngarthia* (cf. Hartnoll et al., 2007), and presumably in the rest of the family. The genital operculum in *J. lagostoma* is shown in fig. 2B. There are no direct observations on the duration of opercular decalcification in gecarcinids, but by analogy with the species described above, it is presumably only a matter of days. Mating always occurs within a few days before laying in *Gecarcinus lateralis* (cf. Bliss et al., 1978; Klaassen, 1975). From three days before laying, the female responds to male courtship, and mates; after laying she ceases to be receptive (Klaassen, 1975). In *J. lagostoma* a number of females were examined, some with ripe ovaries, some with newly laid eggs but only one of the latter had flexible opercula (Hartnoll et al., 2007), confirming the brevity of decalcification. Thus there is a brief window of opportunity for mating in gecarcinids, and the timing of this window dictates the migratory pattern (or lack of migration) of males.

MALE BREEDING MIGRATIONS

The timing and location of laying by the female, which occurs within the short time span of opercular decalcification, flags the opportunity for mating. Such limited opportunity constrains the male migration pattern. Whereas mating is generally a brief and rarely recorded event (though see *Gecarcinus lateralis*, above), becoming ovigerous is an overt and well documented phenomenon, and a valuable index of the 'window of opportunity'. A problem is that during incubation, females are highly cryptic in their behaviour (see references in Hartnoll et al., 2010), tending to become surface-active only as the eggs approach hatching. Nevertheless, they must emerge at some stage and complete their migration to the sea, their emergence revealing the site of laying, which at the same time indicates the site of mating. Differences in the site where mating can occur constrains the patterns of male migration.

At one end of the spectrum male migration is absent, although for species which live very close to the sea, the question of whether or not males migrate becomes to a degree academic. In such species, females mate and lay their eggs in or close to the residential zones, and later make the often short trip to the sea for hatching. The males have no need to migrate. This is case of *Gecarcinus lateralis*, where the females normally migrate <100 m (Klaassen, 1975; Bliss et al., 1978). Klaassen (1975) does state that males move closer to the sea at the start of the breeding season and establish mating territories, but only by <50 m. A similar behaviour pattern, where males do not migrate, occurs in the cavernicolous *Discoplax longipes* A. Milne-Edwards, 1867. Here eggs are laid and incubated in the caves where the adults live, restricting mating opportunities to this location. The females with ripe eggs then migrate directly to the sea, a distance of 100-200 m: no males migrate, and it would not benefit them to do so (Ng & Guinot, 2001). *Cardisoma guanhumi* inhabits areas where its burrows can reach the water table, but these areas may be several kilometres from the shore. Laying of eggs (and of necessity therefore, mating) occurs in these areas. Males do not migrate, but when the eggs are ripe females migrate to the shore over distances up to 5-6 km (Gifford, 1962). On Christmas Island in the Indian Ocean, *Discoplax hirtipes* also shows a similar lack of male migration (Hicks et al., 1990).

Other species have an intermediate pattern of male migration, whereby males migrate variable distances in different proportions: some do not migrate at all, others do so all the way to the sea. *Johngarthia lagostoma* on Ascension Island is of this type. Some females lay in the residential uplands, a number along the migration routes, and the rest after reaching the shore. This offers a diversity of mating opportunities, and consequently males migrate to variable degrees: some not at all, some partway to the shore, and some all the way (a distance up to 4-5 km). Marking studies showed that some males spend considerable time in locations on the migration route, or at the shore, in each case waiting for 'available' females (Hartnoll et al., 2010). During the breeding season, as a result, males predominate in the residential uplands, the sexes are equal along the migration routes, and females predominate at the shore. *Gecarcinus ruricola*, in the San Andrés Archipelago, Colombia, shows a similar pattern of intermediate male migration, again associated with females laying at various stages of the seaward migration. The crabs reaching the shore are 80% female (Hartnoll et al., 2007). Here the maximum migration distance is also several kilometres.

The other end of the spectrum is where all males migrate to the shore. This is so in *Gecarcoidea natalis*, where males and females migrate to the sea in similar numbers, over distances of several kilometres: only after both indulge in 'dipping' in the sea, does courtship and mating occur, followed by egg laying. The males then return inland, whilst females remain at the shore until the eggs are ready to hatch (Hicks, 1985; Hicks et al., 1990). There is no mating opportunity until the shore is reached, so all males must migrate in order to breed. The reproductive pattern in *G. lalandii* is not so clear from the one published account (Liu & Jeng, 2007). However, the authors do state that ovigerous females are not found in the residential zone, and that males are found in the surf zone accompanying females: possibly the basic pattern mirrors that of *G. natalis*.

CONCLUSION

It has been shown how basic aspects of gecarcinid behaviour have been constrained by two aspects of their biology. One is the retention in these terrestrial crabs of marine planktotrophic development. This imposes a requirement on the females to migrate to the sea to release larvae, and on the juveniles to recolonize land at the end of the larval phase. Both involve risk, and could potentially limit the success of the group. The other aspect is the short window of opportunity for mating. The interaction of this with the female migration varies between species, and determines the migration patterns of the males.

ACKNOWLEDGEMENTS

This paper is dedicated to Danièle Guinot, whose work has been an example and inspiration to me over many years. I am grateful for the advice and help she has always given so freely.

REFERENCES

ADAMCZEWSKA, A. M. & S. MORRIS, 2001. Ecology and behaviour of *Gecarcoidea natalis*, the Christmas Island red crab, during the annual breeding migration. Biological Bulletin, Woods Hole, **200**: 305-320.

BLISS, D. E., J. VAN MONTFRANS, M. VAN MONTFRANS & J. R. BOYER, 1978. Behaviour and growth of the land crab *Gecarcinus lateralis* (Fréminville) in southern Florida. Bulletin of the American Museum of Natural History, **160**: 111-152.

BRITTON, J. C., G. C. KROH & C. GOLIGHTLY, 1982. Biometric and ecological relationships in two sympatric Caribbean Gecarcinidae (Crustacea: Decapoda). Journal of Crustacean Biology, **2**: 207-222.

BROCKERHOFF, A. M. & C. L. MCLAY, 2005. Factors influencing the onset and duration of receptivity of female purple rock crabs, *Hemigrapsus sexdentatus* (H. Milne Edwards, 1837) (Brachyura: Grapsidae). Journal of Experimental Marine Biology and Ecology, **314**: 123-135.

BROEKHUYSEN, G. J., 1936. On development, growth and distribution of *Carcinides maenas* (L.). Archives Néerlandaises de Zoologie, **2**: 257-399.

CASTRO, P., 2007. A reappraisal of the family Goneplacidae MacLeay, 1838 (Crustacea, Decapoda, Brachyura) and revision of the subfamily Goneplacinae, with the description of ten new genera and eighteen new species. Zoosystema, **29** (4): 609-773.

CHIDCHESTER, F. E., 1911. The mating habits of four species of the Brachyura, Biological Bulletin, Woods Hole, **21**: 235-248.

CHRISTY, J. H., 1987. Competitive mating, mate choice and mating associations of brachyuran crabs. Bulletin of Marine Science, **41**: 177-191.

COSTLOW, J. D. & C. G. BOOKHOUT, 1968. The complete larval development of the land crab, *Cardisoma guanhumi* Latreille in the laboratory (Brachyura, Gecarcinidae). Crustaceana, (Supplement) **2**: 259-270.

CUESTA, J. & K. ANGER, 2005. Larval morphology and salinity tolerance of a land crab from West Africa, *Cardisoma armatum* (Brachyura: Grapsoidea: Gecarcinidae). Journal of Crustacean Biology, **25**: 640-654.

CUESTA, J. A., M. U. GARCÍA-GUERRERO & M. E. HENDRICKX, 2007. The complete larval development of *Johngarthia planatus* (Brachyura: Grapsoidea: Gecarcinidae) described from laboratory reared material, with notes on the affinity of *Gecarcinus* and *Johngarthia*. Journal of Crustacean Biology, **27**: 263-277.

CUMBERLIDGE, N. & P. K. L. NG, 2009. Systematics, evolution and biogeography of freshwater crabs. In: J. W. MARTIN, K. A. CRANDALL & D. L. FIELDER (eds.), Decapod crustacean phylogenetics. Crustacean Issues, **18**: 491-508. (Taylor and Francis/CRC Press, Boca Raton, Florida).

DIESEL, R., 1989. Parental care in an unusual environment: *Metopaulias depressus* (Decapoda: Grapsidae), a crab that lives in epiphytic bromeliads. Animal Behaviour, **38**: 561-575.

DIESEL, R. & M. SCHUH, 1998. Effects of salinity and starvation on larval development of the crabs *Armases ricordi* and *A. roberti* (Decapoda: Grapsidae) from Jamaica, with notes on the biology and ecology of adults. Journal of Crustacean Biology, **18**: 423-436.

EALES, A. J., 1974. Sex pheromone in the shore crab, *Carcinus maenas*, and the site of its release from females. Marine Behaviour and Physiology, **2**: 345-355.

EDWARDS, E., 1966. Mating behaviour in the European edible crab (*Cancer pagurus* L.). Crustaceana, **16**: 161-181.

FUKUI, Y., 1993. Timing of copulation in the moulting and reproductive cycles in a grapsid crab, *Gaetice depressus* (Crustacea: Brachyura). Marine Biology, Berlin, **117**: 221-226.

GIBSON-HILL, C. A., 1947. Field notes on the terrestrial crabs. Bulletin of the Raffles Museum, **18**: 43-52.

GIFFORD, C. A., 1962. Some observations on the general biology of the land crab, *Cardisoma guanhumi* (Latreille) in South Florida. Biological Bulletin, Woods Hole, **123**: 207-223.

GLEESON, R. A., 1980. Pheromone communication in the reproductive behaviour of the blue crab *Callinectes sapidus*. Marine Bchaviour and Physiology, **7**: 119-134.

GREEN, P. T., 1997. Red crabs in rain forest on Christmas Island, Indian Ocean: activity patterns, density and biomass. Journal of Tropical Ecology, **13**: 17-38.

GREENSPAN, B. N., 1982. Semi-monthly reproductive cycles in male and female fiddler crabs, *Uca pugnax*. Animal Behaviour, **30**: 1084-1092.

GUINOT, D., 1986. Description d'une crabe cavernicole aveugle de Nouvelle-Bretagne (Papouasia Nouvelle-Guinée), *Trogloplax joliveti* gen. nov. sp. nov., et établissment d'une sous-famille nouvelle, Trogloplacinae subfam. nov. Compte Rendu de l'Academie des Sciences, Paris, (3) **303**: 307-312.

HARTNOLL, R. G., 1964. The freshwater grapsid crabs of Jamaica. Proceedings of the Linnean Society of London, **175**: 145-169.

— —, 1965. Notes on the marine grapsid crabs of Jamaica. Proceedings of the Linnean Society of London, **176**: 113-147.

— —, 1968a. Morphology of the genital ducts in female crabs. Journal of the Linnean Society, (Zoology) **47**: 279-300.

— —, 1968b. Reproduction in the burrowing crab *Corystes cassivelaunus* (Pennant, 1777) (Decapoda, Brachyura). Crustaceana, **15**: 165-170.

— —, 1969. Mating in the Brachyura. Crustaceana, **16**: 161-181.

— —, 1988. Evolution, systematics and geographical distribution. In: W. W. BURGGREN & B. R. MCMAHON (eds.), Biology of the land crabs: 6-54. (Cambridge University Press).

— —, 2009. Sexual maturity and reproductive strategy in the rock crab *Grapsus adscensionis* (Osbeck, 1765) (Brachyura, Grapsidae) on Ascension Island. Crustaceana, **82**: 275-291.

HARTNOLL, R. G., M. S. P. BAINE, A. BRITTON, Y. GRANDAS, J. JAMES, A. VELASCO & M. G. RICHMOND, 2007. Reproduction of the black land crab, *Gecarcinus ruricola*, in the San Andres Archipelago, western Caribbean. Journal of Crustacean Biology, **27**: 425-436.

HARTNOLL, R. G., M. S. P. BAINE, Y. GRANDAS, J. JAMES & H. ATKIN, 2006b. Population biology of the black land crab, *Gecarcinus ruricola*, in the San Andres Archipelago, western Caribbean. Journal of Crustacean Biology, **26**: 316-325.

HARTNOLL, R. G., A. C. BRODERICK, B. J. GODLEY, S. MUSICK, M. PEARSON, S. A. STROUD & K. E. SAUNDERS, 2010. Reproduction in the land crab *Johngarthia lagostoma* on Ascension Island. Journal of Crustacean Biology, **30**: 83-92.

HARTNOLL, R. G., A. C. BRODERICK, B. J. GODLEY & K. E. SAUNDERS, 2009. Population structure of the land crab *Johngarthia lagostoma* on Ascension Island. Journal of Crustacean Biology, **29**: 57-61.

HARTNOLL, R. G. & P. CLARK, 2006. A mass recruitment event in the land crab *Gecarcinus ruricola* (Linnaeus, 1758) [Brachyura: Grapsoidea: Gecarcinidae], and a description of the megalop. Zoological Journal of the Linnaean Society, **146**: 149-164.

HARTNOLL, R. G., T. MACKINTOSH & T. J. PELEMBE, 2006a. *Johngarthia lagostoma* (H. Milne Edwards, 1837) on Ascension Island: a very isolated land crab population. Crustaceana, **79**: 197-215.

HASHIMOTO, H., 1965. The spawning of the terrestrial grapsid crab, *Sesarma haematocheir* (De Haan). Zoological Magazine, Tokyo Zoological Society, **74**: 82-87.

HENMI, Y. & M. MURAI, 1999. Decalcification of vulvar operculum and mating in the ocypodid crab *Ilyoplax pusilla*. Journal of Zoology, London, **247**: 133-137.

HENNING, H. G., 1975. Okologische, ethologische und sinnesphysiologische Untersuchungen an der Landkrabbe *Cardisoma guanhumi* Latreille (Decapoda, Brachyura) in Nordcolumbien. Forma et Functio, **8**: 253-304.

HICKS, J. W., 1985. The breeding behaviour and migrations of the terrestrial crab *Gecarcoidea natalis* (Decapoda: Brachyura). Australian Journal of Zoology, **33**: 127-142.

HICKS, J., H. RUMPFF & H. YORKSTON, 1990. Christmas crabs (2nd ed.): 1-81. (Christmas Island Natural History Association, Christmas Island, Australia).

HINES, A. S., 1982. Allometric constraints and variables of reproductive effort in brachyuran crabs. Marine Biology, Berlin, **69**: 309-320.

HOLTHUIS, L. B., 1964. *Sesarma* (*Sesarma*) *cerberus*, a new cavernicolous crab from Amboina. Zoologische Mededelingen, Leiden, **40**: 65-72.

IHLE, J. E. W., 1912. Ueber eine klein Brachyuren-Sammlung aus unterirdischen Flussen von Java. Notes from the Leyden Museum, **34**: 177-182.

JENNINGS, A. C., C. L. MCLAY & A. M. BROCKERHOFF, 2000. Mating behaviour of *Macrophthalmus hirtipes* (Brachyura: Ocypodidae). Marine Biology, Berlin, **137**: 267-276.

JONES, D. R. & R. G. HARTNOLL, 1997. Mate selection and mating behaviour in spider crabs. Estuarine and Coastal Marine Science, **44**: 185-193.

KANNUPANDI, T., A. KHAN, M. THOMASS, S. SUNDARAMOORTHY & R. NATARAJAN, 1980. Larval development of the land crab *Cardisoma carnifex* Herbst (Brachyura Gecarcinidae) reared in the laboratory. Indian Journal of Marine Science, **9**: 271-277.

KLAASSEN, F., 1975. Okologische und ethologische Untersuchungen zur Fortpflanzungsbiologie von *Gecarcinus lateralis* (Decapoda, Brachyura). Forma et Functio, **8**: 101-174.

LIU, H.-C. & M.-S. JENG, 2005. Reproduction of *Epigrapsus notatus* (Brachyura: Gecarcinidae) in Taiwan. Journal of Crustacean Biology, **25**: 135-140.

— — & — —, 2007. Some reproductive aspects of *Gecarcoidea lalandii* (Brachyura: Gecarcinidae) in Taiwan. Zoological Studies, **46**: 347-354.

LOVRICH, G. A. & B. SAINTE-MARIE, 1997. Cannibalism in the snow crab, *Chionoecetes opilio* (O. Fabricius) (Brachyura: Majidae), and its potential importance to recruitment. Journal of Experimental Marine Biology and Ecology, **211**: 225-245.

NARUSE, T. & P. K. L. NG, 2007. On the taxonomy of the genus *Hymenicoides* Kemp, 1917 (Crustacea: Decapoda: Brachyura: Hymenosomatidae), with resurrection of *Limnopilos* Chuang & Ng, 1991, and descriptions of two new species. Zootaxa, **1621**: 17-31.

NARUSE, T., P. K. L. NG & D. GUINOT, 2008. Two new genera and two new species of troglobitic false spider crabs (Crustacea: Decapoda: Brachyura: Hymenosomatidae) from Indonesia, with notes on *Cancrocaeca* Ng, 1991. Zootaxa, **1739**: 21-40.

NG, P. K. L., 1991. *Cancrocaeca xenomorpha*, new genus and species, a blind troglobitic freshwater hymenosomatid (Crustacea: Decapoda: Brachyura) from Sulawesi, Indonesia. Raffles Bulletin of Zoology, **39**: 59-63.

— —, 2004. Crustacea: Decapoda, Brachyura. In: C. YULE & H. S. YOUNG (eds.), Freshwater invertebrates of the Malaysian region: 311-336. (Malaysian Academy of Sciences).

NG, P. K. L. & C. T. N. CHUANG, 1996. The Hymenosomatidae (Crustacea: Decapoda: Brachyura) of Southeast Asia, with notes on other species. Raffles Bulletin of Zoology, (Supplement) **3**: 1-82.

NG, P. K. L. & D. GUINOT, 2001. On the land crabs of the genus *Discoplax* A. Milne. Edwards, 1867 (Crustacea: Decapoda: Brachyura: Gecarcinidae), with description of a new cavernicolous species from the Philippines. Raffles Bulletin of Zoology, **49**: 311-338.

NG, P. K. L., D. GUINOT & P. J. F. DAVIE, 2008. Systema Brachyurorum: Part I. An annotated checklist of extant brachyuran crabs of the world. Raffles Bulletin of Zoology, (Supplement) **17**: 1-286.

NG, P. K. L. & C. G. S. TAN, 1995. *Geosesarma notophorum* sp. nov. (Decapoda, Brachyura, Grapsidae, Sesarminae), a terrestrial crab from Sumatra, with novel brooding behaviour. Crustaceana, **68**: 390-395.

NIAUSSAT, P. & J. P. EHRHARDT, 1971. Quelques aspects de l'éthologie du crabe terrestre *Gecarcinus planatus* Stimpson de l'atoll de Clipperton. Vie et Milieu, (Supplement) **22**: 167-191.

PAUL, A. J., 1984. Mating frequency and viability of stored sperm in the tanner crab *Chionoecetes bairdi* (Decapoda, Majidae). Journal of Crustacean Biology, **4**: 375-381.

RABELAIS, N. N. & J. N. CAMERON, 1985. Physiological and morphological adaptations of adult *Uca subcylindrica* to semi-arid environments. Biological Bulletin, Woods Hole, **168**: 135-146.

REIMER, J., C. D. SCHUBART & R. DIESEL, 1998. Description of a new freshwater crab of the genus *Sesarma* Say, 1817 (Brachyura, Grapsidae, Sesarminae) from western Jamaica. Crustaceana, **71**: 185-196.

SAIGUSA, M., 1981. Adaptive significance of a semilunar rhythm in the terrestrial crab *Sesarma*. Biological Bulletin, Woods Hole, **160**: 311-321.

SCHUBART, C. D., R. DIESEL & S. B. HEDGES, 1998. Rapid evolution to terrestrial life in Jamaican crabs. Nature, London, **393**: 363-365.

SHOKITA, S., 1971. On the spawning habits of the land crab *Cardisoma hirtipes* Dana from Ishigaki Island, in the Ryukyu Islands. Biological Magazine, Okinawa, **7**: 27-32.

SHOKITA, S. & S. SHIKATANI, 1990. Complete larval development of the land crab *Cardisoma hirtipes* Dana (Brachyura Gecarcinidae) reared in the laboratory. Researches on Crustacea, **18**: 1-14.

SOH, C. L., 1969. Abbreviated development in a non-marine crab, *Sesarma* (*Geosesarma*) *perracae* (Brachyura; Grapsidae), from Singapore. Journal of Zoology, London, **158**: 357-370.

TAYLOR, H. H. & P. GREENAWAY, 1979. The structure of the gills and lungs of the arid-zone crab, *Holthuisana* (*Austrothelphusa*) *transversa* (Martens) (Sundathelphusidae: Brachyura) including observations on arterial vessels within the gills. Journal of Experimental Biology, **111**: 103-122.

TÜRKAY, M., 1970. Die Gecarcinidae Amerikas. Mit einem Anhang uber *Ucides* Rathbun (Crustacea: Decapoda). Senckenbergiana Biologica, **51**: 333-354.

WILLEMS, K. A., 1982. Larval development of the land crab *Gecarcinus lateralis lateralis* (Fréminville, 1835) (Brachyura: Gecarcinidae) reared in the laboratory. Journal of Crustacean Biology, **2**: 180-201.

WOLCOTT, T. G. & D. L. WOLCOTT, 1982. Larval loss and spawning behaviour in the land crab *Gecarcinus lateralis* (Freminville). Journal of Crustacean Biology, **2**: 477-485.

YEO, D. C. J., P. K. L. NG, N. CUMBERLIDGE, C. MAGALHÃES, S. R. DANIELS & M. R. CAMPOS, 2008. Global diversity of crabs (Crustacea: Decapoda: Brachyura) living in freshwater. In: E. V. BALIAN, C. LÉVÊQUE, H. SEGERS & K. MARTENS (eds.), Freshwater animal diversity assessment. Hydrobiologia, **575**: 275-286.

First received 9 June 2009.
Final version accepted 31 July 2009.

NECROCARCINUS ORNATISSIMUS FORIR, 1887, AND *PREHEPATUS WERNERI* FRAAYE & COLLINS, 1987 (UPPER MAASTRICHTIAN, THE NETHERLANDS) REVISITED, WITH NOTES ON OTHER CRETACEOUS DYNOMENID CRABS (DECAPODA, BRACHYURA)

BY

JOHN W. M. JAGT[1,5]), RENÉ H. B. FRAAIJE[2,6]), BARRY W. M. VAN BAKEL[2,3,7]) and PEDRO ARTAL[4,8])

[1]) Natuurhistorisch Museum Maastricht, De Bosquetplein 6-7, NL-6211 KJ Maastricht, Netherlands

[2]) Oertijdmuseum De Groene Poort, Bosscheweg 80, NL-5283 WB Boxtel, Netherlands

[3]) Nationaal Natuurhistorisch Museum (Naturalis), P.O. Box 9517, NL-2600 RA Leiden, Netherlands

[4]) Museo Geológico del Seminario de Barcelona, Diputación 231, E-08007 Barcelona, Spain

ABSTRACT

The recently traced type specimens of decapod crustaceans erected by H. Forir from the Upper Cretaceous (upper Maastrichtian) of southern Limburg and Liège (southeast Netherlands, northeast Belgium) include *Necrocarcinus ornatissimus*, a species based on a single left claw. This species is shown to be conspecific with another claw-based taxon from the same area and stratigraphic level, *Prehepatus werneri* Fraaye & Collins, 1987. *Necrocarcinus ornatissimus* is thought to be a dynomenid.

RÉSUMÉ

Les spécimens types de crustacés décapodes récemment découverts et décrits par H. Forir du Crétacé supérieur (Maastrichtien supérieur) du sud du Limbourg et de Liège (sud-est Pays-Bas, nord-est Belgique), comprennent *Necrocarcinus ornatissimus*, une espèce décrite à partir d'une seule pince gauche. Il est montré que cette espèce est conspécifique de *Prehepatus werneri* Fraaye & Collins, 1987, taxon du même niveau stratigraphique, décrit également d'après la pince. Il est proposé de rattacher *N. ornatissimus* aux Dynomenidae.

[5]) Corresponding author; e-mail: john.jagt@maastricht.nl

[6]) e-mail: info@oertijdmuseum.nl

[7]) e-mail: barryvanbakel@gmail.com

[8]) e-mail: partal@optimus.es

© Koninklijke Brill NV, Leiden, 2010

INTRODUCTION

During the last two decades, the upper portion of the Maastricht Formation (Emael, Nekum, and Meerssen members; latest Maastrichtian, *Belemnitella junior* and *Belemnella* (*Neobelemnella*) *kazimiroviensis* cephalopod zones) in the type area of that unit, has yielded rich and diverse decapod crustacean faunas (Collins et al., 1995; Fraaye, 1996a, b, 1997; Fraaye & Van Bakel, 1998; Jagt et al., 2000; Swen et al., 2001; Fraaije, 2002, 2003; Van Bakel et al., 2003; Fraaije et al., 2008). With the exception of some raninid taxa, most of this material comprises isolated remains of carapaces and dissociated chelae, some of which may actually represent moults. In general, complete specimens are exceedingly rare. The consistent association of claws with certain types of carapaces may be used to infer conspecificity of some of these, in particular necrocarcinids (paranecrocarcinines). In a few cases, subsequent finds of well-preserved material have substantiated such interpretations.

Forir (1887b: 169, 172 [51, 54], pl. 7 fig. 7a-d) described an isolated left claw from 'la couche à bryozoaires de l'étage maestrichtien supérieur (danien) de la Montagne St-Pierre à Maestricht' as a new species, *Necrocarcinus ornatissimus*. In current lithostratigraphic terminology, this would correspond to one of the bryozoan-rich intervals of the lower/middle Meerssen Member (Maastricht Formation). Despite several attempts during recent years (P. J. Felder, pers. comm.), Forir's original material could not be traced at the Université d'État (Liège) where it was stated to have been deposited. In November 2003, the first author came across two collection trays in the basement of the Institut royal des Sciences naturelles de Belgique (Brussels), containing most of Forir's labelled originals (fig. 1A-F). It appears that the late Victor Van Straelen had borrowed this material, but failed to return it. Additional specimens in Forir's Collection, e.g., a xanthoid and additional dromioids (Forir, 1887a, c, 1889), have been described in detail elsewhere (Schweitzer et al., 2007a) or will be revised shortly (Jagt et al., work in progress).

A left claw similar to *N. ornatissimus* was recorded one hundred years later under the name of *Prehepatus werneri* by Fraaye & Collins (1987) from the same stratigraphic unit, but at a different locality (the former Blom quarry, Berg en Terblijt), a few kilometres east of Maastricht. However, those authors failed to mention Forir's (1887b) paper. Subsequently, Collins et al. (1995: 169) stated that, 'The last-named species [= *Necrocarcinus ornatissimus*] has a superficial resemblance to *Prehepatus werneri* Fraaye & Collins, 1987'. Now that the type of *N. ornatissimus* has been traced, it can be demonstrated beyond doubt that *P. werneri* is a junior synonym.

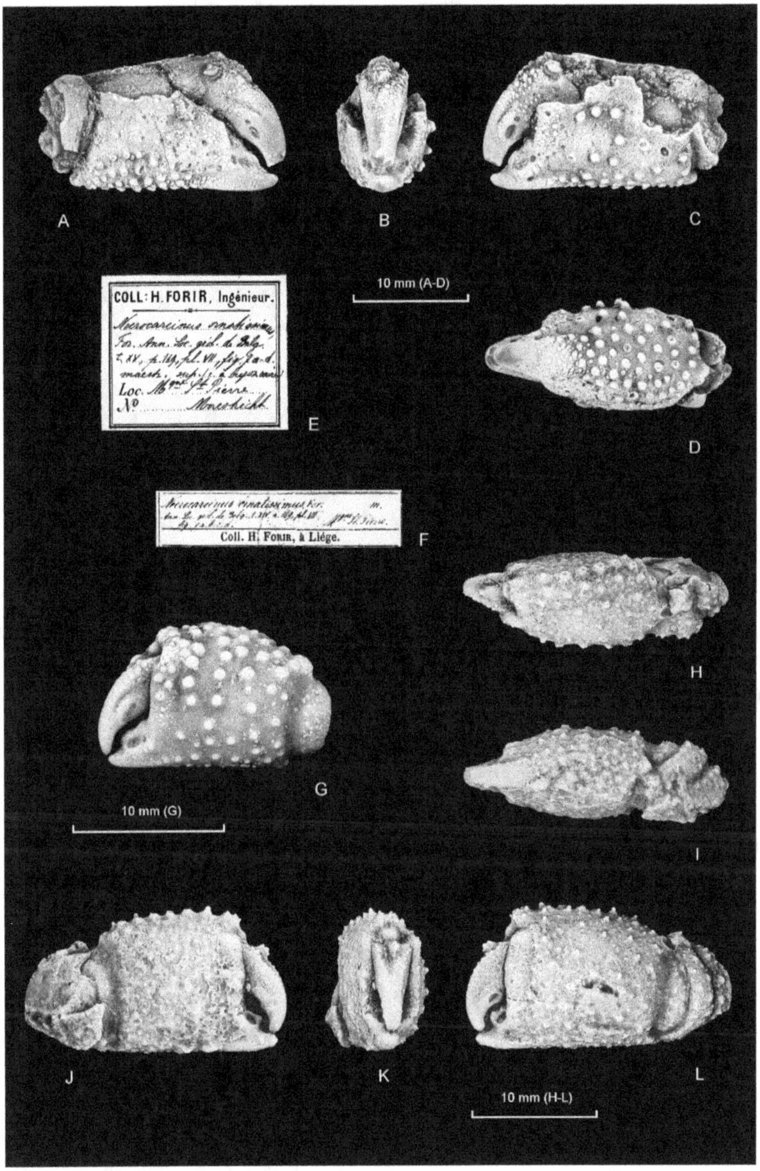

Fig. 1. Cretaceous dynomenid claws, assignable to the 'form genus' *Roemerus* Bishop, 1983; A-F, *Roemerus ornatissimus* (Forir, 1887b), holotype (IRScNB MI 11005) of *Necrocarcinus ornatissimus*, with original labels in Forir's handwriting; lower/middle Meerssen Member (Maastricht Formation), St Pietersberg, south of Maastricht (Netherlands); G, *Roemerus ornatissimus* (Forir, 1887b), holotype (MAB k. 1007) of *Prehepatus werneri* Fraaye & Collins, 1987; lower/middle Meerssen Member (Maastricht Formation), former Blom quarry, Berg en Terblijt (Netherlands); H-L, *Roemerus* sp. (MGSB 68.581), lower Aptian, Morella (Castellón, Spain).

Numerous carapaces and isolated chelae of necrocarcinids (paranecrocarcinines) have been collected from the upper Maastricht Formation (Nekum and Meerssen members, in particular) as to cast serious doubts over either a necrocarcinid or calappid nature of *N. ornatissimus* (see fig. 2). Moreover, a literature search and examination of newly acquired fossil material from northern Spain, made us reflect on the possibility that this, and other, claw-based species represented dynomenids rather than necrocarcinids. Previous authors (e.g., Förster, 1968; Wright & Collins, 1972) already hinted at this. The fact that 'necrocarcinid' claws appear consistently associated with dynomenid carapaces throughout the Lower and Upper Cretaceous can thus no longer be ignored. Some of these isolated claws have unfortunately been formally named, as in the case of associated carapaces. So long as specimens preserving chelae are not forthcoming, it is proposed here to use parataxonomy for such cases, by using the 'form genus' *Roemerus* Bishop, 1983 (type species, by original designation, *R. robustus* Bishop, 1983). This taxon, from the lower Albian (Lower Cretaceous) Glen Rose Limestone of central Texas, originally regarded as a paguroid claw, is here interpreted as a dynomenid, which may be conspecific with the co-occurring, carapace-based species, *Trachynotocarcinus naglei* (Bishop, 1983), originally referred to the genus *Palaeodromites* A. Milne-Edwards, 1865 (see Collins & Donovan, 2007: 62). Vega et al. (2006: 331, fig. 6.7) subsequently recorded *R. robustus* from the Sierra Madre Formation, a lower Albian plattenkalk facies, in Chiapas (southeast Mexico). Given that the specimen preserved similar-sized left and right chelae, these authors considered it to be a diogenid hermit crab.

Additional evidence that claws of a type similar to *Roemerus robustus* are in fact dynomenid rather than paguroid (diogenid) or necrocarcinid, comes from still unpublished early Aptian and Maastrichtian material from Castellón and Navarra, Spain (*Roemerus* sp. and *Graptocarcinus texanus* Roemer, 1887; see figs. 1H-L, 3), from collections briefly exhibited by Àlex Ossó during the Third Symposium on Mesozoic and Cenozoic Decapod Crustaceans at Milan (May, 2007) and from specimens from the Cenomanian of northern France (*Graptocarcinus* sp.; fig. 4E-H), contained in the L. de Putter Collection, Axel, Netherlands (casts are MAB k. 2545 and k. 2546). Previous records of so-called necrocarcinid claws (i.e., assigned to *Necrocarcinus*) are invariably associated with dynomenid carapaces of the genera *Palaeodromites* (= *Cyphonotus* [sensu Bell, 1863]; *Distefania* Checcia-Rispoli, 1914), *Graptocarcinus* Roemer, 1887, and *Trachynotocarcinus* Wright & Collins, 1972.

It appears that on such features as size, ornament, development of fixed and movable fingers plus the presence or absence of ovate depressions in

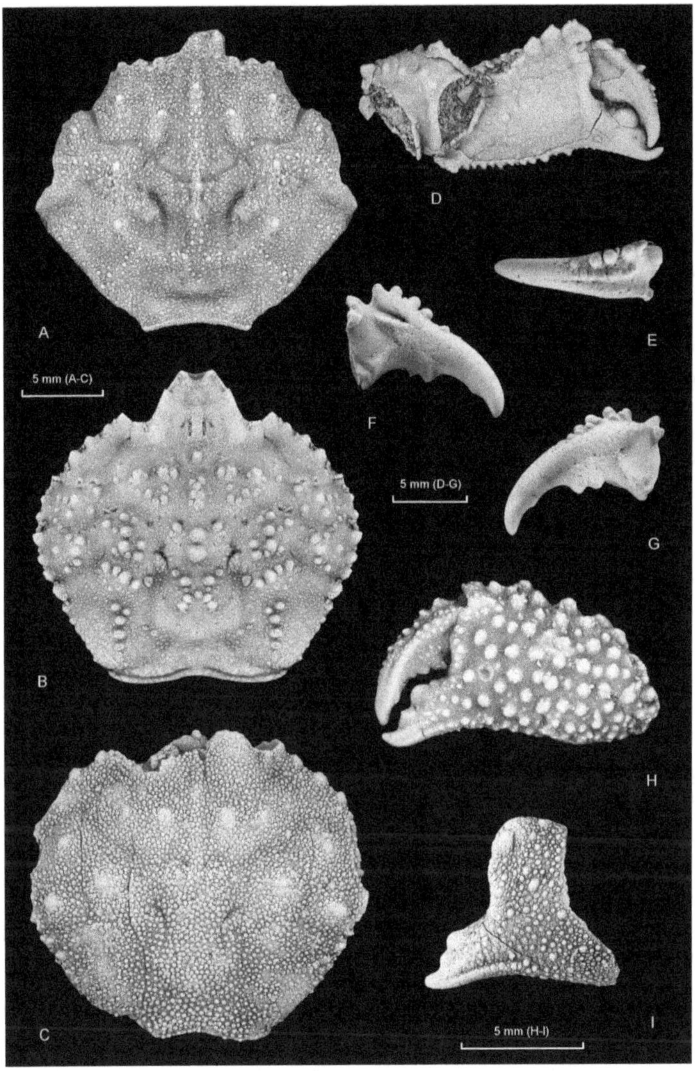

Fig. 2. Necrocarcinids (paranecrocarcinines) and orithopsid from the Maastricht Formation (Emael, Nekum, and Meerssen members; see also Collins et al., 1995; Fraaije, 2002) in the Maastrichtian type area; A, *Orithopsis angelicus* (Fraaije, 2002), carapace (MAB k. 2548, composite photograph); B, D, H, *Paranecrocarcinus quadriscissus* (Noetling, 1881) (=*Dromiopsis ubaghsi* Forir, 1887c), carapace MAB k. 2549 and two left chelae (NHMM JJ 6737 and MAB k. 0720, respectively), in inner and outer respect, respectively; C, I, *Paranecrocarcinus vanbirgeleni* Fraaije, 2002, holotype (MAB k. 2278) carapace, and referred fragmentary left chela (MAB k. 2550), respectively; E-G, '*Pseudomicippe granulosa* (Bosquet MS) Pelseneer, 1889', isolated left dactylus (holotype, IRScNB MI 10670), here held to be conspecific with *Paranecrocarcinus quadriscissus* (see fig. 3D, H). All specimens coated with ammonium chloride prior to photography.

TABLE I

Dynomenid claws, previously assigned to the genera *Necrocarcinus*, *Prehepatus*, and *Roemerus*, linked to associated dynomenid carapaces in the same faunal assemblages (data from Bosquet, 1854; Forir, 1887b; Fritsch & Kafka, 1887; Gründel, 1974; Bishop, 1983; Fraaye & Collins, 1987; Collins et al., 1995; Collins & Dieni, 1996; Collins & Donovan, 2007); *Forir (1887b: 165 [47], pl. 7 fig. 5a, b) assumed claws associated with a partial carapace which he named *Dromiopsis gigas* to be conspecific

Claw-based taxa	Carapace-based taxa
Necrocarcinus avicularis Fritsch in Fritsch & Kafka, 1887	(*Cancer*?) *modestus* (?= *Graptocarcinus texanus*; see Wright & Collins, 1972)
Roemerus ornatissimus (Forir, 1887b) (= *Prehepatus werneri* Fraaye & Collins, 1987)	*Stephanometopon granulatum* Bosquet, 1854
Necrocarcinus woodwardi [sensu Gründel, 1974]	*Palaeodromites incertus* (Bell, 1863)
Necrocarcinus woodwardi [sensu Bell, 1863]	*Palaeodromites incertus* (Bell, 1863) *Trachynotocarcinus sulcatus* (Bell, 1863)
Roemerus robustus Bishop, 1983	*Trachynotocarcinus naglei* (Bishop, 1983)
Pagurinae gen. indet.	*Graptocarcinus bellonii* Collins & Dieni, 1996
Dromiopsis gigas Forir, 1887b*	*Dromiopsis gigas* Forir, 1887b

both dactylus and fixed finger (see below), species might be based on isolated dynomenid chelae here assigned to the 'form genus' *Roemerus*, but we have refrained from introducing new taxa (for available names, see table I) and hope other authors will follow our example. It has previously been noted (e.g., Collins, 1999) that isolated crab claws are notoriously difficult to identify, but that 'educated guesses' regarding association with carapaces have at times proved correct. We can only hope for finds of more or less complete specimens with claws preserved in every example referred to below. Until that time, we prefer to use parataxonomy for those taxa that have already been described on the basis of isolated cheliped material. In those cases where chelae are found attached to carapaces, we propose carapace-based names take priority over those erected on the basis of isolated chelae.

SYSTEMATIC PART

Abbreviations used to denote the repositories of specimens referred to in the text are as follows: ANSP, Academy of Natural Sciences, Philadelphia, Pennsylvania, U.S.A.; GSCM, Georgia Southern College Museum, Statesboro, Georgia, U.S.A.; MAB, Oertijdmuseum De Groene Poort, Boxtel, Netherlands (formerly Museum de Ammonietenhoeve, Boxtel); MGSB, Museo Geológico del Seminario de Barcelona, Barcelona, Spain; MNB, Museum für Naturkunde, Berlin, Germany; MNH-UABCS, Museo de Historia Natural, Universidad Autonóma de Baja California Sur, La Paz, Mexico; NHMM, Natuurhistorisch Museum Maastricht, Maastricht, Netherlands (with collection letter codes: JJ — J. W. M. Jagt; RD — R. W. Dortangs); SDSNH, San Diego Natural History Museum, San Diego, California, U.S.A.; UT, University of Texas, Austin, Texas, U.S.A.

Superfamily DROMIOIDEA De Haan, 1833

Family DYNOMENIDAE Ortmann, 1892

'Form genus' **Roemerus** Bishop, 1983

Type species. — *Roemerus robustus* Bishop, 1983: 42, pl. 3 figs. 20-31, text-fig. 8G, by original designation; lower Albian Glen Rose Limestone, Blanco and Kendall counties, Texas. The holotype, left claw (UT 45704), and one paratype, left propodus (SDSNH 23662), have been described by Bishop (1983).

Diagnosis. — 'Chelae elongate, similar, with rectangular palm, outer face transversely convex, inner flat. Carpal and dactyl articulations perpendicular to lower margin, which is straight except for convexity below base of fixed finger. Fingers short and turned slightly inward. Fixed finger has at least 1 proximal tooth. Tip of movable finger overlaps fixed finger in smaller claws and closes onto outer edge of tip in large claws. Surface sparsely covered with large, low granules which become numerous and prominent on lower edge. The upper surface surmounted by a low, oblique ridge accentuated on the proximal inner face by a few low granules and becoming less conspicuous as it runs toward the top center of the distal margin of the claw' (Bishop, 1983: 42).

Remarks. — As here understood, the 'form genus' *Roemerus* comprises only those forms that have oval depressions (for setae; see below) in both dactylus and fixed finger (fig. 1). It appears that this 'form genus' can be linked to carapaces of *Palaeodromites*, while forms without such oval depressions belong to *Graptocarcinus* or some closely allied dynomenid.

Roemerus ornatissimus (Forir, 1887b) (fig. 1A-G)

Necrocarcinus ornatissimus Forir, 1887b: 169 [51], pl. 7 fig. 7a-d.
Prehepatus werneri Fraaye & Collins, 1987: 549, figs. 1, 2; Collins et al., 1995: 198, fig. 11c-g.

Types. — The holotype, by monotypy, of *N. ornatissimus*, left claw (IRScNB MI 11005) (fig. 1A-D; original labels, see fig. 1E, F). The holotype of *P. werneri*, by original designation, left claw (MAB k. 0017) (fig. 1G); paratypes, fragmentary right fixed finger (MAB k. 0018) and fragmentary right movable finger (MAB k. 0019).

Material examined. — In addition to the types of both taxa, several isolated left claws and movable fingers (MAB k. 1037-k. 1038, NHMM JJ 12653, NHMM JJ 13264, NHMM RD unregistered), as well as a left cheliped (MAB k. 2547, leg. J. W. M. Jagt) are available.

Description. — Left manus subrectangular in outline; upper margin slightly depressed proximally, outer surface being moderately convex longitudinally, transversely, passing sharply into broad, gently rounded upper surface; broad furrow separating finely granular carpal articulation from remainder of manus; straight distal margin bordered on upper surface by row of fine granules, which become larger, sparser on outer surface where bounded by short, vertical groove and anteriormost tubercle of 3 rows of 4 large tubercles, each forming strongly curved line; above these, less regularly arranged tubercles give way to crowded granules on upper surface; below, row of 6 tubercles, followed by straight row of 9 or 10 decreasing in size distally, terminating at margin halfway along fixed finger, joined at an angle by median one of 3 rows of fine granules on fixed finger, uppermost lining base of occlusal teeth on opposing margin; several other tubercles scattered among transverse rows, some tending to form longitudinal rows; lower margin weakly concave to straight, more or less in line with the fixed finger, with row of 7 or 8 tubercles decreasing in size distally. Fixed finger about one-third of manus length; deep depression occurring below second of 2 large cusps on opposing margin; moderately curved moveable finger with narrow, granular tubercle proximally; of 3 granular ridges, the lowermost, strongest issuing from articulating facet, continuing almost to tip; shallow ovate depression, ridged above, coincident with depression below proximal cusp on fixed finger; both depressions found on inner faces as well, all with perforated floor.

Discussion. — Fraaye & Collins (1987) originally referred this claw type to the otherwise exclusively North American, claw-based genus *Prehepatus* Rathbun, 1935 (p. 47; type species, by original designation, *P. cretaceus* Rathbun, 1935), as the first and only European representative. Rathbun (1935: 47) interpreted *P. cretaceus* as a calappid, differing from others, 'in [...] the palmar portion bent horizontally to form an upper surface'. In the absence of carapaces preserving such claws, it is difficult to assign them correctly, but Schweitzer et al. (2007b: 32) discussed this matter in detail, and favoured placement within the Hepatidae Stimpson, 1871, at least for the time being. To date, six species have been assigned to *Prehepatus*, all from North and Central America, as follows:

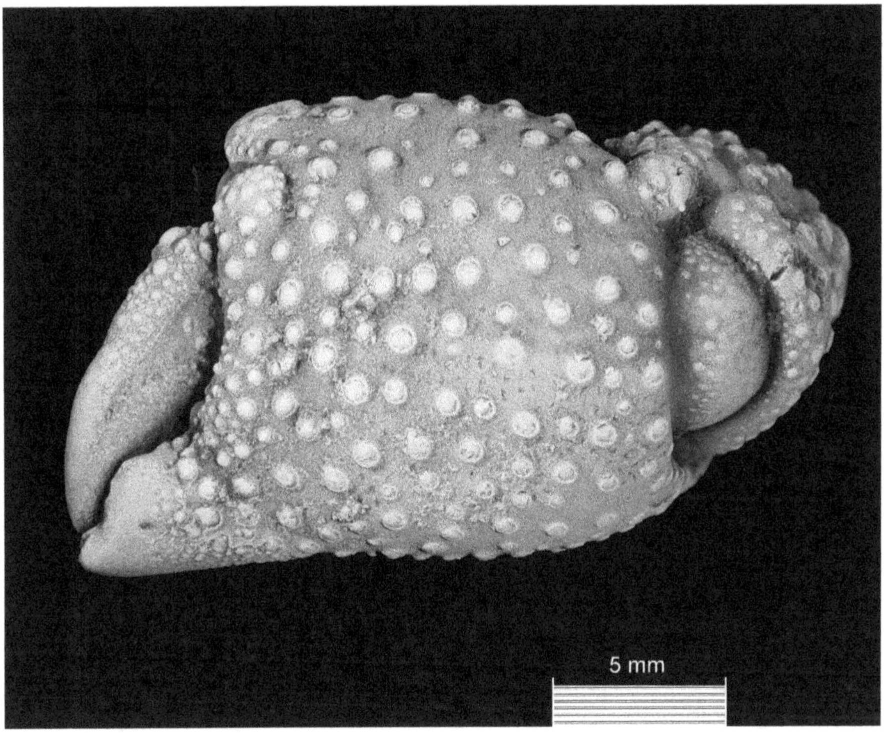

Fig. 3. *Graptocarcinus texanus* (Roemer, 1887), Maastrichtian, Puerto de Urbasa (Navarra, Spain), MGSB 28.148b.

1. *Prehepatus cretaceus* Rathbun, 1935 (p. 47, pl. 11 figs. 29, 30), from the Duck Creek Limestone (upper Albian) of Grayson County, Texas (holotype UT 211). Rathbun (1935: 47) opined that, 'The dactylus, "*Pseudomicippe*" *granulosa* Pelseneer may belong to the same genus as the species above described' [=*P. cretaceus*]. We disagree, and rather consider *Pseudomicippe granulosa* to be conspecific with *Paranecrocarcinus quadriscissus* (see below, and figs. 2D-H, 3).
2. *Prehepatus pawpawensis* Rathbun, 1935 (p. 48, pl. 11 figs. 26-28), from the Pawpaw Shale (upper Albian; see Kennedy, 2004) of Tarrant County, Texas (holotype is an unregistered specimen in UT collections).
3. *Prehepatus dilksi* Roberts, 1962 (p. 183, pl. 89 fig. 2), from the Merchantville Formation (latest Santonian to mid-Campanian; see Miller et al., 2004) of Burlington County, New Jersey (holotype ANSP 19728).
4. *Prehepatus hodgesi* Bishop, 1983 (p. 50, pl. 3 figs. 1-19; text-figs. 8F, 11, 12), from the Glen Rose Limestone (lower Albian) of Blanco and Kendall counties, Texas (holotype SDSNH 23655).

5. *Prehepatus harrisi* Bishop, 1985 (p. 1029, figs. 1.1-1.5, 2.1-2.9), from the Coon Creek and Prairie Bluff formations (lower-middle Maastrichtian) of Union County, Mississippi (holotype GSCM 1686). This species has subsequently been recorded from the lower Maastrichtian of northeast Mexico by Vega et al. (1995: 247, figs. 3.1, 3.2).
6. *Prehepatus mexicanus* Schweitzer et al., 2007b (p. 33, figs. 2.10, 2.11), the sole Cenozoic member, from the Eocene of Baja California Sur (northwest Mexico) (holotype MNH-UABCS/Ba10-8).

Bishop (1985: 1031; see also Bishop, 1986) recognized two lineages in Mid- and Late Cretaceous *Prehepatus*, one containing species with triangular claws, such as *P. hodgesi*, *P. harrisi* and *P. pawpawensis*, confined to the Gulf Coastal Plain of the United States, the other with more quadrate claws (i.e., *P. cretaceus*, *P. dilksi*) occurring in the Gulf Coastal and northern Atlantic Coastal plains.

All these forms differ considerably from the claw type represented by *R. ornatissimus* in that they are more triangular, of smaller size, with differently shaped fixed and movable fingers (notably lacking elongate-oval pits or depressions on opposing faces, and a tendency to become more or less spatulate), and with an oblique, rather than (near-)straight manus/carpal articulation. For that reason, hepatids can be ruled out for *Roemerus*.

Records of carapaces preserving appendages and/or venters are extremely sparse (compare Schweitzer et al., 2003; Larghi, 2004) amongst necrocarcinids. However, of two species of the Paranecrocarcininae Fraaije, Van Bakel, Jagt & Artal, 2008 from the Maastrichtian type area, associations of carapaces and chelae are so consistent that the latter may safely be taken to belong here (fig. 2; see also Fraaije, 2002). Of the third species, originally referred to *Necrocarcinus* (see Fraaije, 2002), but now placed in the Orithopsidae Schweitzer, Feldmann, Fam, Hessin, Hetrick, Nyborg & Ross, 2003 (see Guinot, Vega & Van Bakel, 2008: 27), only carapaces are so far known. These taxa are (see fig. 2A-I):

1. *Paranecrocarcinus quadriscissus* (Noetling, 1881) (p. 368, pl. 20 fig. 4a, b; holotype MNB A.239), inferred to be from the Meerssen Member (Maastricht Formation) of Valkenburg aan de Geul, southern Limburg (the Netherlands).
2. *Paranecrocarcinus vanbirgeleni* Fraaije, 2002 (p. 916, figs. 1.3, 3.1, 3.2; holotype MAB k. 2278), from the lower Meerssen Member (Maastricht Formation) of the ENCI-HeidelbergCementGroup quarry (Maastricht).

3. *Orithopsis angelicus* (Fraaije, 2002) (p. 914, figs. 1.1, 2, as *Necrocarcinus angelicus*; holotype MAB k. 1008), from the middle Meerssen Member (Maastricht Formation) of the same quarry.

Although still not found connected with carapaces, claws of the first two species differ considerably from *R. ornatissimus* (compare Collins et al., 1995, fig. 11A, B; Fraaije, 2002, figs. 3.8, 3.9). The 'species' *Pseudomicippe granulosa* (Bosquet MS) Pelseneer, 1886 (p. 170, fig. 6; holotype IRScNB 10670), as based on a single left movable finger from the upper Maastrichtian of the St Pietersberg area, south of Maastricht (compare fig. 2E-G), is here considered conspecific with *Paranecrocarcinus quadriscissus*, and thus unrelated to the majine genus *Pseudomicippe* Heller, 1861 (see Ng et al., 2008).

With such differences in claw morphology, paranecrocarcinines can also be ruled out, as can necrocarcinines, for that matter, demonstrated by a comparison with a late Albian specimen of *Necrocarcinus labeschii* which preserves a near-complete left claw (fig. 4A-D). Outline, ornament, merus/carpal articulation and size and shape of fixed and movable fingers are different to *R. ornatissimus*, and can easily be distinguished from claws of *Cenomanocarcinus* as well (see Breton & Collins, 2007, fig. 4).

PREVIOUS RECORDS OF CRETACEOUS 'NECROCARCINID' CHELAE

Several examples of comparable 'necrocarcinid' claws in the literature refer to dynomenids. Bell (1863: 21, pl. 5 fig. 4) illustrated a claw from the lower Cenomanian Greensand of Wiltshire (southwest England), which he assigned to a new species, *Necrocarcinus woodwardii*, noting the claw of '*N. inflatus*' (= *Cenomanocarcinus inflatus*) had 'the same general aspect'. That this is in error is clearly seen when comparing Bell's illustration with that in Breton & Collins (2007, fig. 4). It is instead much closer to dynomenid claws here referred to the 'form genus' *Roemerus*. Moreover, on the basis of carapaces, the dynomenids *Palaeodromites incertus* and *Trachynotocarcinus sulcatus* have been recorded from these levels (see Wright & Collins, 1972: 9, 13).

Schlüter (in Von der Marck & Schlüter, 1868: 298) referred with a question mark to an isolated claw from the lower Turonian of the Bochum area (northern Germany) as '*Necrocarcinus woodwardi* Bell?', noting that it resembled the claw illustrated by Bell (1863, pl. 5 fig. 4). Schlüter (1879: 609, pl. 17 fig. 3) referred to this specimen as *Necrocarcinus* sp. n. This is also best referred to *Roemerus* sp. until such time that it can be positively assigned to a dynomenid carapace.

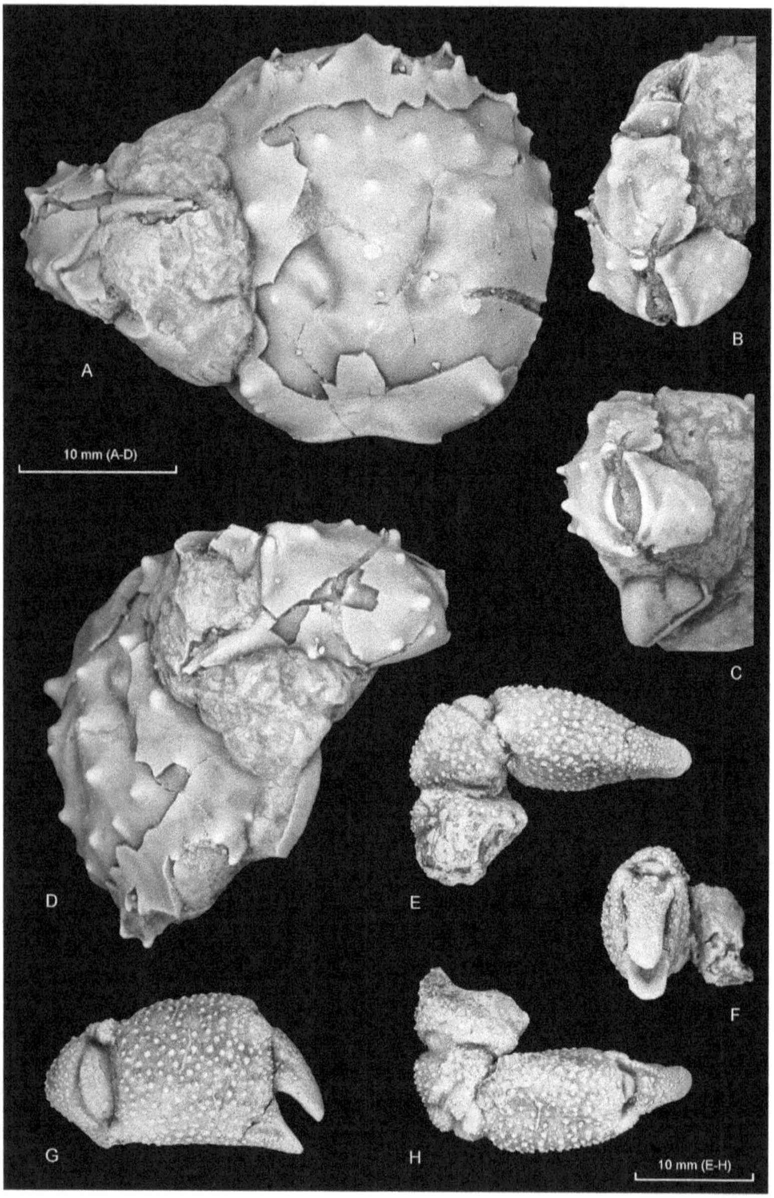

Fig. 4. Mid-Cretaceous necrocarcinid and dynomenid; A-D, *Necrocarcinus labeschii* (J. A. [Eudes-] Deslongchamps, 1835), upper Albian, Escalles, northern France (MAB k. 2549 [cast]), carapace with associated left claw, of a type similar to that of *Cenomanocarcinus inflatus* Van Straelen, 1936 (see Breton & Collins, 2007, fig. 4); E-H, *Graptocarcinus* sp., lower Cenomanian, Escalles, northern France (MAB k. 2546); note the lack of pits on opposing faces in the fixed and movable fingers.

Fritsch (in Fritsch & Kafka, 1887) assigned two 'species', both based on isolated claws, to *Necrocarcinus* Bell, 1863. *Necrocarcinus avicularis* Fritsch (in Fritsch & Kafka, 1887: 47, pl. 10 figs. 2a-f, 10-12), of Cenomanian age from the localities Zbyslav and Kamajk (Bohemia, Czech Republic), was distinguished from taxa known at the time on account of a more roundish shape and finer ornament. *Necrocarcinus perlatus* Fritsch (in Fritsch & Kafka, 1887: 47, pl. 10 fig. 14), from the younger (?Turonian) Priesener Schichten at Černodol, near Laun (Bohemia), was based on a claw with very short fingers and relatively large tubercles. For the time being, the first of these 'species' would also be best assigned to the 'form genus' *Roemerus*, especially so since it co-occurred with a dynomenid carapace, (*Cancer*?) *modestus* Fritsch (in Fritsch & Kafka, 1887: 49, pl. 10 fig. 12), which Wright & Collins (1972: 54) questionably synonymized with *Graptocarcinus texanus* Roemer, 1887. The status of the second 'species' is not certain, and without having seen the original specimen, we cannot discuss this matter further (see below).

Mertin (1941: 156, 240, pl. 8 fig. 15) recorded a single claw as *Necrocarcinus* sp. from the basal 'Untersenon', possibly lower Santonian of Heudeber, east of Halberstadt (eastern Germany), which could also represent a dynomenid claw, but the photograph leaves much to be desired. The fine granular ornament appears distinctive, but without having seen the original nothing more can be said.

Förster (1968: 183-185, pl. 13 figs. 5, 6) discussed and illustrated a number of what he referred to as 'Necrocarcininae-Scheren'. He noted that finds of still-connected claws and carapaces of necrocarcinids were extremely rare, and that he only knew of such instances for *N. labeschii*, *N. woodwardii* and *Cenomanocarcinus vanstraeleni*. The last-named species is now referred to its own family, Cenomanocarcinidae Guinot, Vega & Van Bakel, 2008. Despite this drawback, previous authors named several species on the basis of isolated claws, mostly with reference to *N. labeschii* from the 'Gault' (Albian) of England, which revealed such a typical claw shape. Förster (1968) rightly pointed out that assignment to genera was problematic, especially when the very different claws of *C. vanstraeleni* were considered. In addition to taxa referred to above, Förster (1968) listed *N. woodwardi* [sensu Schlüter, 1879: 608, pl. 17 fig. 2], from the Cenomanian of northern Germany, *Necrocarcinus* sp. [sensu Schlüter, 1879: 609, pl. 17 fig. 3], from the lower Turonian of northern Germany, *N. franconicus* Lehner, 1937 (p. 214, pl. 19 figs 21, 22), from the lower Turonian of southern Germany, *N.* cf. *senoniensis* [sensu Mertin, 1941: 239, pl. 8 fig. 14], from the upper Campanian of northern

Germany. These records, as well as earlier ones from the Lower Cretaceous of France and Switzerland, as noted by Förster (1968) (e.g., *Gebia controversa* de Tribolet, 1874: 657, pl. 15 figs. 1-5, Valanginian, western Switzerland; *Gebia muensteri* Robineau-Desvoidy, 1849: 137, pl. 5 fig. 17, Hauterivian, central France) should be revised in order to determine if they can be linked to co-occurring dynomenid carapaces. Förster (1968: 185) also suggested that *N. perlatus* was not a necrocarcinid claw, but that in shape and ornament there was a close match with *Caloxanthus formosus* (A. Milne-Edwards, 1864; Cenomanian, France). Of especial note is Förster's (1968: 185) observation that, '[. . .] denn eine Reihe kretazischer Dynomeniden dürfte ähnlich gebaute Scheren besessen haben. Da bisher jedoch nur bei *Necrocarcinus* Funde im Zusammenhang mit dem Carapax vorliegen, sollen diese charakteristischen Scheren vorerst weiterhin als *Necrocarcinus*-Scheren geführt werden'. This translates as, '[. . .] because a number of Cretaceous dynomenids are likely to have had similarly built chelae. Given that so far only finds in association with carapaces are known for *Necrocarcinus*, such characteristic chelae could still be referred to as *Necrocarcinus* chelae in the future']. We agree with Förster's observation.

Gründel (1974: 90, pl. 4 figs. 1, 2) recorded a claw assigned to *Necrocarcinus woodwardi* Bell, 1862 [sic] from the *plenus* Zone (uppermost Cenomanian; 'Klippenfazies' near Plauen, eastern Germany) and added the dynomenid, *Palaeodromites incertus* (Bell, 1863) (see Gründel, 1974, pl. 3 fig. 3), from the same stratigraphic level and area.

Collins & Dieni (1996: 70, figs. 2/1, 2, 4) recorded a new dynomenid species, *Graptocarcinus bellonii*, from Cenomanian rudist limestones in northeast Italy. An associated claw, interpreted by these authors as 'Pagurinae genus indet'. (Collins & Dieni, 1996, figs. 2/3, 5, 6) is remarkably close to *Roemerus* and other dynomenid claws illustrated in the literature. It is here assumed that these forms are identical to each other.

Necrocarcinus ornatissimus and *P. werneri*, as well as material described in the literature and on recently acquired, mostly unpublished specimens, are also associated to dynomenids, and to *Palaeodromites* and *Graptocarcinus* in particular (see figs. 1H-L, 3; tables I, II). The use of the 'form genus' *Roemerus* is proposed here for such claws, so long as finds of carapaces preserving claws are lacking for *Palaeodromites* in particular.

McLay (1999) briefly commented on fossil dynomenid genera, while Guinot (2008: 20, 21) was of the opinion that the genus *Diaulax* Bell, 1863 was a dynomenid, and that the family Diaulacidae Wright & Collins, 1972 (see also

TABLE II

Cretaceous dynomenid crabs (from Wright & Collins, 1972; Förster, 1975; Via Boada, 1981; Collins et al., 1995; Collins & Dieni, 1996; Breton & Collins, 2007; Jagt et al., 2007; Guinot, 2008; Jagt et al., work in progress); *originally recorded by Collins et al. (1995) from the Maastrichtian type area (Maastricht, Netherlands); subsequent studies have shown that this material actually came from the upper Maastrichtian (Saint-Symphorien Calcarenite Formation) of the Mons Basin, southern Belgium

Genus and species	Stratigraphic and geographic range
Diaulax feliceps Wright & Collins, 1972	middle-upper Albian; England
Diaulax millerae Bishop, 1992	lower Campanian; Delaware
Diaulax oweni (Bell, 1850) (= *carteriana* Bell, 1863)	lower Albian-middle Cenomanian; England
Diaulax roddai Bishop, 1983	lower Albian; Texas
Diaulax yokoii Collins, Kanie & Karasawa, 1993	Santonian; Japan
Dromiopsis mosae Collins, Fraaye & Jagt, 1995*	upper Maastrichtian; Belgium
Dromiopsis praelaevior Collins, Fraaye & Jagt, 1995	upper Maastrichtian; Netherlands
Dromiopsis cf. *rugosa* (Von Schlotheim, 1820)	upper Campanian; Germany
Dromiopsis sp.	upper Campanian; Belgium
Graptocarcinus bellonii Collins & Dieni, 1996	Cenomanian; Italy
Graptocarcinus maastrichtensis Fraaye, 1996b	upper Maastrichtian; Netherlands
Graptocarcinus muiri Stenzel, 1944	Aptian; Mexico
Graptocarcinus texanus Roemer, 1887	?Albian, lower-middle Cenomanian; Texas, England, Spain, Czech Republic, Mexico
Palaeodromites autissiodorensis (Van Straelen, 1936)	Hauterivian-Aptian; France
Palaeodromites centrosus (Van Straelen, 1940)	Cenomanian; Spain
Palaeodromites crypticus Jagt, van Bakel & Fraaije, 2007	lower-middle Cenomanian; Belgium
Palaeodromites himerensis (Checchia-Rispoli, 1914)	Cenomanian; Sicily
Palaeodromites incertus (Bell, 1863)	lowermost-upper Cenomanian; England, France, Germany, Austria
Palaeodromites octodentatus A. Milne-Edwards, 1865	Hauterivian; France
Palaeodromites sinuososulcatus Wright & Collins, 1972	lower Albian; England
Palaeodromites transiens Wright & Collins, 1972	upper Albian-lower Cenomanian; England, Spain
Stephanometopon granulatum Bosquet, 1854	upper Maastrichtian; Netherlands, Belgium
Trachynotocarcinus naglei (Bishop, 1983)	lower Albian; Texas
Trachynotocarcinus sulcatus (Bell, 1863)	lower-upper Cenomanian; England, France

Collins et al., 1993) was perhaps best treated as a subfamily, close to, or possibly synonymous with, the Metadynomeninae. *Diaulax* currently comprises five Mid- to Late Cretaceous species, two from England, one from Japan and two from the United States. Finds of carapace-associated chelae are known in four of these (table II), showing them to be of a general dynomenid aspect, but to differ quite considerably from those here assigned to *Roemerus*. *Diaulax carteriana* Bell, 1863, the type species by monotypy, is a junior subjective synonym of *Platypodia oweni* Bell, 1850, as discussed by Wright & Collins (1972: 56). It is known from the lower Albian to middle Cenomanian. Wright & Collins, 1972 (p. 58, pl. 10 fig. 1a, b) described the chelae of *D. oweni* '[. . .] equal and a little shorter than the carapace. The hand has a tumid outer surface and a flat palm, slightly excavate at the rear end. The fingers are about a third of the length of the hand; the fixed finger is reflexed; a groove lined with pits for hairs runs along the upper margin of the movable finger. The carpus is half as long as the hand, the outer surface very convex with a shallow longitudinal depression. The convex outer surfaces of hand and wrist have, in large Cenomanian specimens, many small distinct well-spaced tubercles tending to a linear arrangement but they are absent or only weakly developed in Cambridge Greensand specimens. The merus is large, with slightly convex outer and flat inner surface; there are a few large granules on the posterior margin, but the upper surface has minute close flat granules like those on the cephalothorax'. The chelae of *D. feliceps* (Wright & Collins, 1972: 59, pl. 10 figs. 2-6), from the middle and upper Albian of Kent, England, were characterized as, 'The surface ornament is as in *Diaulax oweni* and the same fine ornament covers the inflated untuberculate chelae'. Collins et al. (1993: 297) added that, 'The median furrow on the carpus is common to both *D. oweni* and *D. feliceps*, but only *D. feliceps* has a groove on the propodus, that of *D. oweni* being ornamented by longitudinal rows of granules continued onto the carpus'.

Claws of *D. millerae* (see Bishop, 1992: 559, fig. 1A), from the lower Campanian of the United States Atlantic Coastal Plain, were described as, '[. . .] approximately equal; merus stout; carpus pentagonal, moderately sized; propodus short and high, strongly granulate (at least on inner face); dactylus long, narrow, and strongly curved downward [. . .]'. Those of *D. yokoii* Collins, Kanie & Karasawa, 1993 from the Santonian of Japan (see also Karasawa et al., 2006), were typified as follows (Collins et al., 1993: 297), 'The propodus of the cheliped is trapezoidal in lateral view, with the basal margin about one-third longer than the dorsal and interdigital margins. There is a wide convexity in the basal margin at the base of the stout, moderately deflexed, slightly inturned fixed finger. The interdigital margin is nearly straight and bounded by a shallow groove and a wide, deep groove bounds the carpal articulation which is rather more oblique in the left chela, the right and left propodus have the finely granulated lateral surface smoothly rounded and have an incipient furrow extending parallel to the dorsal margin, but it is closer to the midline in the right chela. The slender dactylus is about as long as the fixed finger. The carpus is a little shorter than the propodus, rounded triangular in section and the strongly convex lateral surface is smooth with a median furrow'. In short, claws of *Diaulax* differ from those here referred to *Roemerus* in details or ornament and in structure of fixed and movable fingers, which both lack ovate depressions.

Stratigraphic range. — *Roemerus ornatissimus* appears to be confined to the upper Nekum and Meerssen members (Maastricht Formation) in southern Limburg and contiguous Belgian territory, and the range of the dynomenid *Stephanometopon granulatum* Bosquet, 1854 (p. 137, pl. 10 fig. 12a, b, non 12c) is much the same. The holotype carapace fragment (IRScNB 10652), since damaged, was illustrated by Bosquet (1854) together with a claw. This claw (Bosquet, 1854, pl. 10 fig. 12c) undoubtedly belongs to a different form, perhaps a species of *Dromiopsis* Reuss, 1859, which is known from the same levels (*D. praelaevior* Collins, Fraaye & Jagt, 1995) (see table II).

MORPHOLOGY OF CHELAE

The claws of the 'form genus' *Roemerus* have a distinctive morphology, which can be linked to their mode of feeding. The genera *Graptocarcinus* and *Palaeodromites* are considered dynomenids on the basis of carapace outline and areolation, groove system as well as construction of lateral margins and orbitofrontal area (see also McLay, 1999: 434). Their remains are found in reefal or near-reefal deposits, like extant descendants. Dynomenids are interpreted as one of the most ancient families, with an impressive record from the Upper Jurassic onwards (Guinot, 2008: 3). The wide range of cheliped morphology undoubtedly reflects different feeding habits (see McLay, 1999). Dynomenid chelipeds generally differ from those of dromiids, as a result of their different ecology. Dromiids camouflage their bodies, hence their common name 'sponge crabs' (see, e.g., McLay, 1993; Guinot et al., 1995; Guinot & Tavares, 2003; Van Bakel et al., 2009). Normally dromiids are larger than dynomenids; the latter often hide and live in coral crevices. In many dynomenids, the propodus/carpus and carpus/merus articulations show a bulge. There is no opening between these articles when bent. This modification is likely to serve as protection for the arm when moving through coral colonies or during feeding.

The cheliped fingers of dynomenids show particular setal modifications. Dynomenids are filter feeders and their diet apparently consists of food obtained by sieving fine sediment, or perhaps coral mucus (McLay, 1999: 428). They have clumps of stiff setae at the inner and outer margins of both fixed and movable fingers, which collectively close the space between the fingers and form a sieving device. McLay (1999: 449) observed that, '… *D. praedator* show that these setae help to sift out food particles which are passed to the third maxillipeds'. This 'sieving screen' was observed by McLay (1999: 476, 484, 490) in several extant species (e.g., *Dynomene hispida* (Latreille, in Milbert, 1812), *D. praedator* A. Milne-Edwards, 1879 and *D. filholi* Bouvier, 1894). The modified articulation of the propodus and carpus, and the clumps of setae on both sides of the fingers, can be considered characteristic of the Dynomenidae (see fig. 1A-F, H-L). These distinct characters have not been observed in fossil dromiid or necrocarcinid crabs, but are seen in all 'species' here assigned to the 'form genus' *Roemerus*.

ACKNOWLEDGEMENTS

We thank Danièle Guinot for her kind help with examining extant dromioid crabs in MNHN, and for fruitful discussions on comparative morphology. For assistance in various ways, we thank Joe S. H. Collins, Yvonne Coole, Sjeuf Felder (deceased), Rodney M. Feldmann, Rafael Lemaitre, Pat McLaughlin, Àlex Ossó, Louis de Putter, and Carrie E. Schweitzer. One of us (JWMJ) acknowledges the support of the European Community-Access to Research Infrastructure action of the Improving Human Reseach Potential Programme (ABC grant), which allowed him to visit the IRScNB in November-December 2003, where Isabella Van de Velde, Annie V. Dhondt (deceased), Wilfried Miseur, and Dirk Anne were extremely helpful.

REFERENCES

BAKEL, B. W. M. VAN, P. ARTAL, R. H. B. FRAAIJE & J. W. M. JAGT, 2009. A new Oligocene crab (Decapoda, Brachyura, Dromiacea) from northwest Belgium, with comments on its palaeobiology. Geologica Belgica, **12**: 45-57.

BAKEL, B. W. M. VAN, J. W. M. JAGT, R. H. B. FRAAIJE & Y. COOLE, 2003. New data on the crab *Binkhorstia ubaghsii* (Late Maastrichtian, NE Belgium, SE Netherlands). In: J. W. M. JAGT, B. W. M. VAN BAKEL & R. H. B. FRAAIJE (eds.), Second Symposium on Mesozoic and Cenozoic Decapod Crustaceans, Boxtel/Maastricht, the Netherlands, September 3-6, 2003. Contributions to Zoology, **72**: 85-89.

BELL, T., 1850. In: F. DIXON, Thc gcology and fossils of the Tertiary and Cretaceous formations of Sussex: i-xvi, 1-422. (Longman, London).

— —, 1863. A monograph of the fossil malacostracous Crustacea of Great Britain. Part II. Crustacea of the Gault and Greensand. Monograph of the Palaeontographical Society London, **1863**: i-viii, 1-40.

BISHOP, G. A., 1983. Fossil decapod crustaceans from the Lower Cretaceous, Glen Rose Limestone of central Texas. Transactions of the San Diego Society of Natural History, **20**: 27-55.

— —, 1985. A new crab, *Prehepatus harrisi* (Crustacea, Decapoda) from the Coon Creek and Prairie Bluff formations, Union County, Mississippi. Journal of Paleontology, **59**: 1028-1032.

— —, 1986. Occurrence, preservation, and biogeography of the Cretaceous crabs of North America. In: R. H. GORE & K. L. HECK (eds.), Biogeography of the Crustacea. Crustacean Issues, **4**: 111-142. (A. A. Balkema, Rotterdam/Boston).

— —, 1992. A new Cretaceous crab, *Diaulax millerae* (Crustacea: Decapoda), from the northern Atlantic Coastal Plain, U.S.A. Proceedings of the Biological Society of Washington, **105**: 555-561.

BOSQUET, J., 1854. Les Crustacés fossiles du Terrain Crétacé du Limbourg. Verhandelingen uitgegeven door de Commissie belast met het vervaardigen eener geologische Beschrijving en Kaart van Nederland, **2**: 1-127 [10-137]. (A. C. Kruseman, Haarlem).

BRETON, G. & J. S. H. COLLINS, 2007. Decapod fauna from the Cenomanian stratotype. In: A. GARASSINO, R. M. FELDMANN & G. TERUZZI (eds.), 3rd Symposium on Mesozoic and Cenozoic Decapod Crustaceans, Museo di Storia Naturale di Milano, May 23-25, 2007. Memorie della Società Italiana di Scienze Naturali e Museo Civico di Storia Naturale di Milano, **35**: 17-20.

CHECCHIA-RISPOLI, G., 1914. « *Distefania* », nuovo genere di Brachiuri del Cenomaniano della Sicilia. Bollettino della Società Zoologica Italiana, (3) **3**: 173-186.

COLLINS, J. S. H., 1999. Fossils explained 25: Crab claws. Geology Today, May/June 1999: 114-117.

COLLINS, J. S. H. & I. DIENI, 1996. New decapod crustaceans from the Cenomanian rudist limestones of NE Italy. Bulletin of the Mizunami Fossil Museum, **22** [for 1995]: 67-72.

COLLINS, J. S. H. & S. K. DONOVAN, 2007. New decapod crustaceans from the Palaeogene of Jamaica. Bulletin of the Mizunami Fossil Museum, **33** [for 2006]: 59-65.

COLLINS, J. S. H., R. H. B. FRAAYE & J. W. M. JAGT, 1995. Late Cretaceous anomurans and brachyurans from the Maastrichtian type area. Acta Palaeontologica Polonica, **40**: 165-210.

COLLINS, J. S. H., Y. KANIE & H. KARASAWA, 1993. Late Cretaceous crabs from Japan. Transactions and Proceedings of the Palaeontological Society of Japan, (new series) **172**: 292-310.

DESLONGCHAMPS, E. [EUDES-], J.-A., 1835. Mémoire pour servir à l'histoire naturelle des Crustacés fossiles. Mémoires de la Société linnéenne de Normandie, **5**: 37-41.

FORIR, H., 1887a. Contributions à l'étude du système crétacé de la Belgique. I. Sur quelques poissons et crustacés nouveaux ou peu connus. Annales de la Société Géologique de Belgique, **14**: 25-56.

— —, 1887b. Contributions à l'étude du système crétacé de la Belgique. II. Études complémentaires sur les crustacés. Annales de la Société Géologique de Belgique, **14**: i-ii, 155-175.

— —, 1887c. Contributions à l'étude du système crétacé de la Belgique. III. Bibliographie et tableau des thoracostracés crétacés décrits jusqu'à ce jour. Annales de la Société Géologique de Belgique, **14**: 176-195.

— —, 1889. Contributions à l'étude du système crétacé de la Belgique. IV. Troisième note sur des poissons et crustacés nouveaux ou peu connus. Annales de la Société Géologique de Belgique, **16**: 445-460.

FÖRSTER, R., 1968. *Paranecrocarcinus libanoticus* n. sp. (Decapoda) und die Entwicklung der Calappidae in der Kreide. Mitteilungen aus der Bayerischen Staatssammlung für Paläontologie und Historische Geologie, **8**: 167-195.

— —, 1975. Ein Krebs aus dem oberen Campan von Misburg. Berichte der Naturhistorischen Gesellschaft zu Hannover, **119**: 285-294.

FRAAYE, R. H. B., 1996a. A new Tethyan migrant: *Cretachlorodius enciensis* n. gen., n. sp. (Crustacea, Decapoda), from the Maastrichtian type area. Journal of Paleontology, **70**: 293-296.

— —, 1996b. Two new crabs, *Graptocarcinus maastrichtensis*, and *Caloxanthus kuypersi* (Crustacea, Decapoda), from the type Maastrichtian of the Netherlands. Journal of Paleontology, **70**: 463-465.

— —, 1997. Late Cretaceous swimming crabs: radiation, migration, competition, and extinction. Acta Geologica Polonica, **46** [for 1996]: 269-278.

— —, 2002. New calappid crabs (Crustacea, Decapoda, Brachyura) from the Late Maastrichtian of the Netherlands. Journal of Paleontology, **76**: 913-917.

— —, 2003. The first record of albuneid crabs (Crustacea, Decapoda) from the Cretaceous. Bulletin of the Mizunami Fossil Museum, **29** [for 2002]: 69-72.

FRAAYE, R. H. B. & B. W. M. VAN BAKEL, 1998. New raninid crabs (Crustacea, Decapoda, Brachyura) from the Late Maastrichtian of the Netherlands. Geologie en Mijnbouw, **76**: 293-299.

FRAAIJE, R. H. B., B. W. M. VAN BAKEL, J. W. M. JAGT & P. ARTAL, 2008. New decapod crustaceans (Anomura, Brachyura) from mid-Cretaceous reefal deposits at Monte Orobe (Navarra, northern Spain), and comments on related type-Maastrichtian material. In: E. STEURBAUT, J. W. M. JAGT & E. A. JAGT-YAZYKOVA (eds.), Annie V. Dhondt Memorial Volume. Bulletin de l'Institut royal des Sciences naturelles de Belgique, (Sciences de la Terre) **78**: 193-208.

FRAAYE, R. H. B. & J. S. H. COLLINS, 1987. A new crab, *Prehepatus werneri* (Crustacea, Decapoda) from the Maastrichtian of the Netherlands. Journal of Paleontology, **61**: 549-551.

FRITSCH, A. & J. KAFKA, 1887. Die Crustaceen der böhmischen Kreideformation: i-iii, 1-53, i. (Fr. Řivnáč, Prague).

GRÜNDEL, J., 1974. Die Decapoda (Crustacea) der sächsischen Kreide. Abhandlungen aus dem Staatlichen Museum für Mineralogie und Geologie zu Dresden, **21**: 79-100.

GUINOT, D., 2008. A re-evaluation of the Dynomenidae Ortmann, 1892 (Crustacea, Decapoda, Brachyura, Podotremata), with the recognition of four subfamilies. Zootaxa, **1850**: 1-26.

GUINOT, D., D. DOUMENC & C. C. CHINTIROGLOU, 1995. A review of the carrying behaviour in brachyuran crabs, with additional information on the symbioses with sea anemones. The Raffles Bulletin of Zoology, **43**: 377-416.

GUINOT, D. & M. TAVARES, 2003. A new subfamilial arrangement for the Dromiidae De Haan, 1833, with diagnoses and description of new genera and species. Zoosystema, **25**: 43-129.

GUINOT, D., F. J. VEGA & B. W. M. VAN BAKEL, 2008. Cenomanocarcinidae n. fam., a new Cretaceous podotreme family (Crustacea, Decapoda, Brachyura, Raninoidia), with comments on related families. Geodiversitas, **30**: 5-43.

HAAN, W. DE, 1833-1850. Crustacea. In: P. F. VON SIEBOLD, Fauna japonica sive descriptio animalium, quae itinere per Japoniam jussu et auspiciis superiorum, qui summum in India Batavia Imperium tenent, suscepto, annis 1832-1830 cc legit, notis observationibs et adumbrationibus illustravit: ix-xvi, vii-xvii, 1-243. (J. Muller, Amsterdam).

HELLER, C., 1861. Synopsis der im Rothen Meere vorkommenden Crustaceen. Verhandlungen der Zoologisch-Botanischen Gesellschaft zu Wien, **11**: 3-32.

JAGT, J. W. M., B. W. M. VAN BAKEL & R. H. B. FRAAIJE, 2007. *Palaeodromites crypticus*, a new early-middle Cenomanian dynomenid crab (Crustacea, Decapoda) from southern Belgium. Bulletin de l'Institut royal des Sciences naturelles de Belgique, (Sciences de la Terre) **77**: 77-82.

JAGT, J. W. M., R. H. B. FRAAYE & B. W. M. VAN BAKEL, 2000. Late Cretaceous decapod crustacean faunas of northeast Belgium and the southeast Netherlands. In: 1st Workshop on Mesozoic and Tertiary decapod crustaceans, Montecchio Maggiore (Vicenza), 6-8 October 2000. Studi i Ricerche, Associazione di Amici Museo — Museo civico 'G. Zannato', 2000: 37-42.

KARASAWA, H., H. KATO & K. TERABE, 2006. A new member of the family Prosopidae (Crustacea: Decapoda: Brachyura) from the Lower Cretaceous of Japan. Revista Mexicana de Ciencias Geológicas, **23**: 344-349.

KENNEDY, W. J., 2004. Ammonites from the Pawpaw Shale (Upper Albian) in northeast Texas. Cretaceous Research, **25**: 865-905.

LARGHI, C., 2004. Brachyuran decapod Crustacea from the Upper Cretaceous of Lebanon. Journal of Paleontology, **78**: 528-541.

LEHNER, L., 1937. Fauna und Flora der Fränkischen Albüberdeckenden Kreide. II. Fauna, 2. Teil und Flora. Palaeontographica, (A) **87**: 158-230.

MARCK, W. VON DER & C. SCHLÜTER, 1868. Neue Fische und Krebse aus der Kreide von Westphalen. Palaeontographica, **15**: 269-310.

MCLAY, C. L., 1993. Crustacea Decapoda: the sponge crabs (Dromiidae) of New Caledonia and the Philippines with a review of the genera. In: A. CROSNIER (ed.), Résultats des Campagnes MUSORSTOM, **10**. Mémoires du Muséum national d'Histoire naturelle, Paris, **156**: 111-251.

— —, 1999. Crustacea Decapoda: revision of the family Dynomenidae. In: A. CROSNIER (ed.), Résultats des Campagnes MUSORSTOM, **20**. Mémoires du Muséum national d'Histoire naturelle, Paris, **180**: 427-569.

MERTIN, H., 1941. Decapode Krebse aus dem subhercynen und Braunschweiger Emscher und Untersenon, sowie Bemerkungen über einige verwandte Formen in der Oberkreide. Nova Acta Leopoldina, (neue Folge) **10** (68): i-ii, 149-264.

MILLER, K. G., P. J. SUGARMAN, J. V. BROWNING, M. A. KOMINZ, R. K. OLSSON, M. D. FEIGENSON & J. C. HERNÁNDEZ, 2004. Upper Cretaceous sequences and sea-level history, New Jersey Coastal Plain. Geological Society of America Bulletin, **116**: 368-393.

MILNE-EDWARDS, A., 1864. Monographie des crustacés fossiles de la famille des Cancériens, partie 3. Annales des Sciences naturelles, (Zoologie) (5) **1**: 31-88.

— —, 1865. Note sur deux nouveaux crustacés fossiles du terrain néocomien du département de l'Yonne. Bulletin de la Société des Sciences historiques et naturelles de l'Yonne, **19**: 341-347.

NG, P. K. L., D. GUINOT & P. J. F. DAVIE, 2008. Systema Brachyurorum: Part I. An annotated checklist of extant brachyuran crabs of the world. The Raffles Bulletin of Zoology, (Supplement) **17**: 1-286.

NOETLING, F., 1881. Ueber einige Brachyuren aus dem Senon von Mastricht [sic] und dem Tertiär Norddeutschlands. Zeitschrift der Deutschen Geologischen Gesellschaft, **33**: 357-371.

ORTMANN, A. E., 1892. Die Dekapoden-Krebse des Strassburger Museums, mit besonderer Berücksichtigung der von Herrn Dr. Döderlein bei Japan und bei den Liu-Kiu-Inseln gesammelten und zur Zeit im Strassburger Museum aufbewahrten Formen. Zoologische Jahrbücher, (Systematik) **6**: 1-58, 241-326, 532-588.

PELSENEER, P., 1886. Notice sur les crustacés décapodes du Maestrichtien du Limbourg. Bulletin du Musée Royal d'Histoire Naturelle de Belgique, **4**: 161-175.

RATHBUN, M. J., 1935. Fossil Crustacea of the Atlantic and Gulf Coastal Plain. Geological Society of America, Special Papers, **2**: 1-160.

REUSS, A. E., 1859. Zur Kenntniss fossiler Krabben. Denkschrifte der Königlichen Akademie der Wissenschaften zu Wien, **17**: 1-90.

ROBERTS, H. B., 1962. The Upper Cretaceous decapod crustaceans of New Jersey and Delaware. In: The Cretaceous fossils of New Jersey. Bulletin of the New Jersey Bureau of Geology and Topography, **61**: 163-191.

ROBINEAU-DESVOIDY, J. B., 1849. Mémoire sur les crustacés du terrain néocomien de St Sauveur-en-Puisaye. Annales de la Société entomologique de France, (2) **7**: 95-141.

ROEMER, F. A., 1887. *Graptocarcinus texanus*, ein Brachyure aus der oberen Kreide von Texas. Neues Jahrbuch für Mineralogie, Geologie und Paläontologie, **1887**: 173-176.

SCHLÜTER, C., 1879. Neue und weniger gekannte Kreide- und Tertiär-Krebse des nördlichen Deutschlands. Zeitschrift der Deutschen Geologischen Gesellschaft, **31**: 586-615.

SCHWEITZER, C. E., P. ARTAL, B. VAN BAKEL, J. W. M. JAGT & H. KARASAWA, 2007 (cf. a). Revision of the genus *Titanocarcinus* (Decapoda: Brachyura: Xanthoidea) with two new genera and one new species. Journal of Crustacean Biology, **27**: 278-295.

SCHWEITZER, C. E., R. M. FELDMANN, J. FAM, W. A. HESSIN, S. W. HETRICK, T. G. NYBORG & R. L. M. ROSS, 2003. Cretaceous and Eocene decapod crustaceans from southern Vancouver Island, British Columbia, Canada: 1-66. (NRC Research Press, Ottawa).

SCHWEITZER, C. E., R. M. FELDMANN, G. GÓNZALEZ-BARBA & V. ĆOSOVIĆ, 2007 (cf. b). New Decapoda (Anomura, Brachyura) from the Eocene Bateque and Tepetate formations, Baja California Sur, Mexico. Bulletin of the Mizunami Fossil Museum, **33** [for 2006]: 21-45.

SCHWEITZER, C. E., K. J. LACOVARA, J. B. SMITH, M. C. LAMANNA, M. A. LYON & Y. ATTIA, 2003. Mangrove-dwelling crabs (Decapoda: Brachyura: Necrocarcinidae) associated with dinosaurs from the Upper Cretaceous (Cenomanian) of Egypt. Journal of Paleontology, **77**: 888-894.

STENZEL, H. B., 1944. A new Cretaceous crab, *Graptocarcinus muiri*, from Mexico. Journal of Paleontology, **18**: 550-552.

STIMPSON, W., 1870-1881. Brachyura, Part 1. In: L. F. DE POURTALÈS (ed.), Preliminary report on the Crustacea dredged in the Gulf Stream in the Straits of Florida. Bulletin of the Museum of comparative Zoology at Harvard College, **2**: 163-191.

SWEN, K., R. H. B. FRAAIJE & G. J. VAN DER ZWAAN, 2001. Polymorphy and extinction of the Late Cretaceous burrowing shrimp *Protocallianassa faujasi* and first record of the genera *Corallianassa* and *Calliax* (Crustacea, Decapoda, Thalassinoidea) from the Cretaceous. Contributions to Zoology, **70**: 85-98.

TRIBOLET, M. DE, 1874. Sur une nouvelle espèce de crustacé décapode macroure (*Gebia controversa*) des terrains valanginiens et urgonien de Sainte-Croix. Bulletin de la Société vaudoise des Sciences naturelles, **13**: 657-660.

VAN STRAELEN, V., 1936. Crustacés décapodes nouveaux ou peu connus de l'époque crétacique. Bulletin du Musée Royal d'Histoire Naturelle de Belgique, **12** (45): i-ii, 1-50.

— —, 1940. Crustacés décapodes nouveaux du Crétacique de la Navarre. Bulletin du Musée Royal d'Histoire Naturelle de Belgique, **16** (4): 1-5.

VEGA, F. J., R. M. FELDMANN & J. L. VILLALOBOS-HIRIART, 1995. Additions to the crustacean (Decapoda) fauna from the Portrerillos Formation (Late Cretaceous) in northeastern Mexico. Annals of the Carnegie Museum, **64**: 239-249.

VEGA, F. J., P. GARCÍA-BARRERA, M. DEL C. PERRILLIAT, M. A. COUTIÑO & R. MARIÑO-PÉREZ, 2006. El Espinal, a new plattenkalk facies locality from the Lower Cretaceous Sierra Madre Formation, Chiapas, southeastern Mexico. Revista Mexicana de Ciencias Geológicas, **23**: 323-333.

VIA BOADA, L., 1981. Les crustacés décapodes du Cénomanien de Navarra (Espagne): premiers résultats de l'étude des *Galatheidae* [sic]. Géobios, **14**: 247-251.

WRIGHT, C. W. & J. S. H. COLLINS, 1972. British Cretaceous crabs. Monograph of the Palaeontographical Society London, **126** (533): 1-114.

First received 30 May 2009.
Final version accepted 22 June 2009.

A NEW SPECIES OF *ISOLAPOTAMON* BOTT, 1968 (BRACHYURA, POTAMIDAE) FROM MINDANAO, PHILIPPINES

BY

MARIVENE MANUEL-SANTOS[1])
Carcinology Section, Zoology Division, National Museum of the Philippines,
Padre Burgos Avenue, Ermita 1000 Manila, Philippines

ABSTRACT

Isolapotamon danielae is a new species of freshwater crab (Potamidae) described from Mindanao, Philippines and the fourth species of *Isolapotamon* recorded from the region. Only three species are known so far in Mindanao, *I. mindanaoense* (Rathbun, 1904), *I. sinuatifrons* (H. Milne Edwards, 1853), and *I. spatha* Ng & Takeda, 1992. This species is distinguished from allied congeners by its gonopod and carapace.

RÉSUMÉ

Isolapotamon danielae est une nouvelle espèce de crabe d'eau douce (Potamidae) décrite de Mindanao aux Philippines et la quatrième espèce d'*Isolapotamon* trouvée dans cette région. Jusqu'à présent les trois seules espèces connues de Mindanao étaient *I. mindanaoense* (Rathbun, 1904), *I. sinuatifrons* (H. Milne Edwards, 1853), et *I. spatha* Ng & Takeda, 1992. Cette nouvelle espèce se distingue de ses congénères par les caractéristiques de sa carapace et de ses gonopodes.

INTRODUCTION

The potamid genus *Isolapotamon* Bott, 1968, currently contains 17 species (Ng et al., 2008). Most of the species are found in Borneo, but three species have been recorded from Mindanao in the Philippines: *I. mindanaoense* (Rathbun, 1904), *I. sinuatifrons* (H. Milne Edwards, 1853), and *I. spatha* Ng & Takeda, 1992 (see Ng & Takeda, 1992; Ng & Tan, 1998).

[1]) e-mail: vinrmanuel@yahoo.com

© Koninklijke Brill NV, Leiden, 2010

A fourth species from Mindanao, *I. danielae* sp. nov. is described herein. This species is easily distinguished from other species of *Isolapotamon* by the distinctive form of its male first gonopod and several features of the carapace. The terminology follows Ng & Tan (1998). Measurements of the carapace width (cw) and length (cl) are given in millimetres. The abbreviations G1 and G2 are used for the male first and second gonopods, respectively. The material examined is deposited in the National Museum of the Philippines, Carcinology Section (NMCR), Manila, Philippines; and the Zoological Reference Collection (ZRC) of the Raffles Museum of Biodiversity Research, National University of Singapore.

TAXONOMY

Family POTAMIDAE Ortmann, 1896

Isolapotamon Bott, 1968

Isolapotamon danielae sp. nov. (figs. 1, 2)

Type material. — Holotype male, cw 52.7 mm, cl 41.9 mm; Philippines, Mindanao, Agusan del Norte, Maguinca, Sumili, Dagandang Salt Spring; coll. Mindanao State University staff, 14 November 1991 (NMCR 25000); paratype female, cw 39.7 mm, cl 32.0 mm; same data as holotype (NMCR 25036); paratype male, cw 44.7 mm, cl 36.2 mm; Philippines, Mindanao, South Cotabatao, Sebu, Barangay Tupi, Sito Lote Malaic, Bolol; unknown collector, 15 January 1995 (ZRC 2009.0583).

Diagnosis. — Carapace (fig. 1A) broader than long, dorsal surface smooth to rugose, relatively flat, regions well defined; branchial regions moderately swollen, anterolateral margin convex, inflated, cristate, upturned, cristae thick, gradually merging with posterolateral margin marked by thin crista. Front bilobed, margin sinuous, cristate; epibranchial regions striated; epigastric cristae positioned slightly anterior of postorbital cristae; epibranchial tooth large, distinct, pointed but not sharp, separated from distinct orbital angle by a V-shaped cleft. Cervical groove shallow; H-shaped central depression deep, distinct. Eyes well developed. Chelipeds (fig. 1C) unequal, robust, outer surfaces striated but otherwise smooth; fingers with 2 shallow furrows each on outer and inner surfaces; palm rugose, carpus with a prominent spine bordered by 2 small spines at its base; another less prominent spine on inner distal angle, with group of several smaller spines just below it. Ambulatory legs slender, surface gently striated; merus relatively long, with gently serrated margins, blunt spine at ventral aspect; propodus with 2 rows of amber-coloured spines

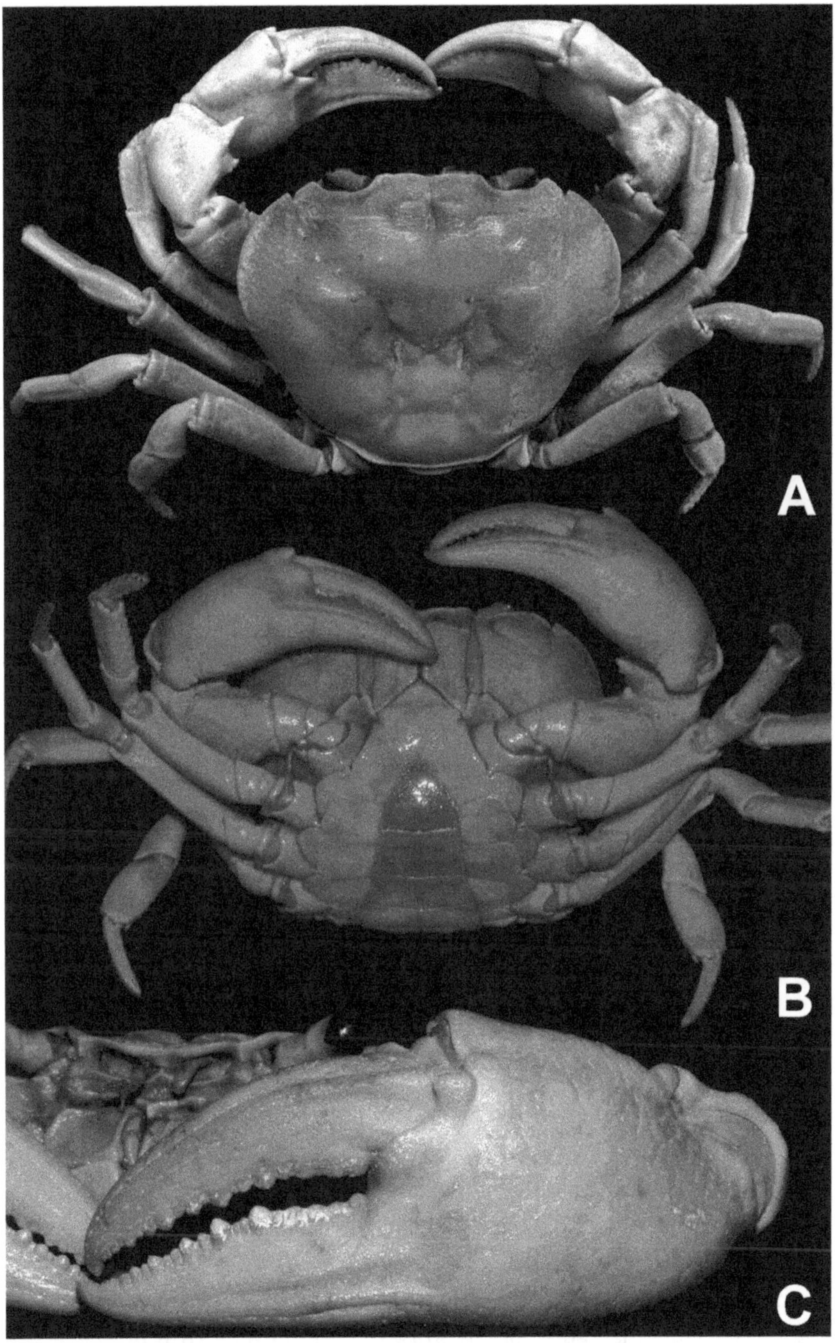

Fig. 1. *Isolapotamon danielae* sp. nov., NMCR 25000, holotype male, cw 52.7 mm, cl 41.9 mm. A, overall dorsal view; B, ventral view of carapace; C, outer view of left chela.

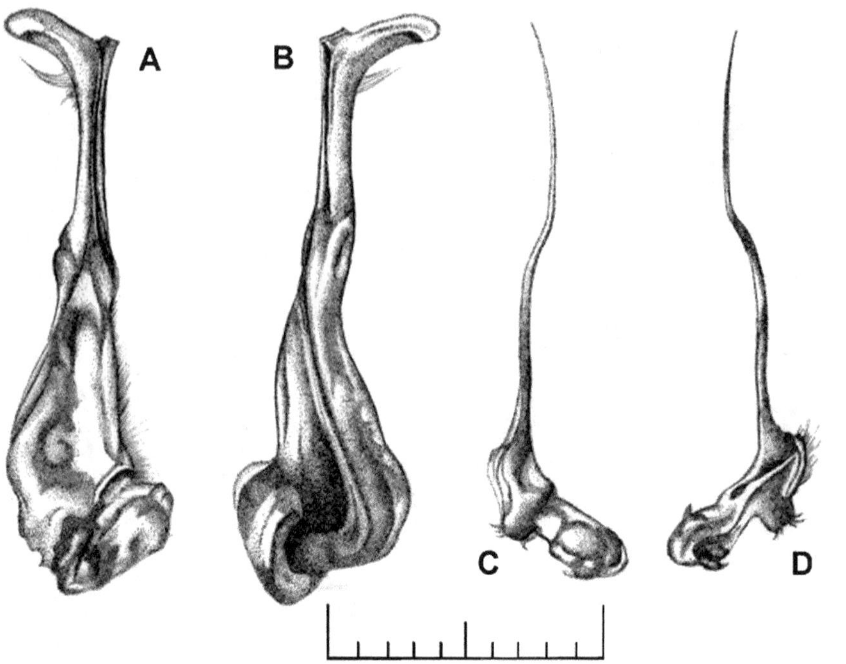

Fig. 2. *Isolapotamon danielae* sp. nov., NMCR 25000, holotype male, cw 52.7 mm, cl 41.9 mm. A, B, left G1; C, D, right G2. Scale = 10 mm.

on ventral margin, spines conspicuous on distal proximal margin surrounding articulation with dactylus; dactylus with 2 rows of amber-coloured spines on dorsal and ventral margins. Male abdomen (fig. 1B) triangular, relatively narrow, third somite widest, narrowing progressively towards telson. G1, G2 long, reaching to half length of thoracic sternite 4; G1 (fig. 2A, B) slightly sinuous, distalmost part inflated, resembling inverted human foot; G2 (fig. 2C, D) as long as G1, distal part slender, tapering.

Paratype female. — The female is similar in all aspects except for the abdomen, which is very broad, occupying almost the entire thoracic sternum.

Local name. — In Tagalog, the national dialect for the Philippines, the crab is known as Alimangong Gubat; whilst in Visaya, it is called Kagang.

Etymology. — The name of the new species honours Madame Danièle Guinot, whose love for crabs and her many contributions on their systematics have been overwhelming.

Remarks. — The new species is the fourth species of *Isolapotamon* collected from Mindanao. It is noteworthy remarking that all four species have been recorded from different parts of the island. *Isolapotamon mindanaoense* from the southern region, *I. sinuatifrons* from the western region, *I. spatha* from the southwestern area, and *I. danielae* sp. nov. from the northern region (see also Ng & Tan, 1998, for discussion of their biogeography).

The G1 of *I. spatha* and *I. mindanaoense* are morphologically similar, especially with regard to the distalmost part of the terminal segment (see Ng & Takeda, 1992, fig. 7; Ng & Tan, 1998, fig. 10A-D). Both species have been recorded from the southern and southwestern regions of Mindanao. *Isolapotamon sinuatifrons* and *I. danielae*, however, are quite different from the other two species. The dilated distalmost part of the terminal segment of these two species is positioned at a sharper right angle relative to the rest of the G1 and is also relatively more rectangular in shape (present fig. 2A, B; see also Ng & Tan, 1998, fig. 10M-P) than in *I. sinuatifrons*. The distalmost part of the G1 of *I. danielae* is nevertheless more similar to that of *I. ingeri* Ng & Tan, 1998, from Sabah, Malaysia, Borneo (see Ng & Tan, 1998, fig. 6E, F). In addition to the differences in carapace characters, *I. danielae* differs markedly from *I. ingeri* in having the distalmost part of the G1 more twisted and angled at 45°, versus about 30° in *I. ingeri* (see Ng & Tan, 1998, fig. 6H) and the distal side being longer than the proximal side in *I. ingeri* (see Ng & Tan, 1998, fig. 6G).

ACKNOWLEDGEMENTS

The author is indebted to Peter K. L. Ng (National University of Singapore) for reviewing the manuscript, for his suggestions, photographs, and untiring logistic and technical support; the Raffles Museum of Biodiversity Research, National University of Singapore, for a visiting fellowship; Jaime C. Bruma, Artist Illustrator II of the National Museum of the Philippines for the drawings; and to the Mindanao State University staff for passing the specimens to the author for study.

REFERENCES

BOTT, R., 1968. Potamiden aus Süd-Asien (Crustacea, Decapoda). Senckenbergiana Biologica, **49**: 119-130.

MILNE EDWARDS, H., 1853. Mémoire sur la famille des Ocypodiens. Annales des Sciences Naturelles, Paris, (Zoologie) (3) **20**: 163-228.

NG, P. K. L., D. GUINOT & P. J. F. DAVIE, 2008. Systema Brachyuorum: Part I. An annotated checklist of extant brachyuran crabs of the world. Raffles Bulletin of Zoology, (Supplement) **17**: 1-286.

NG, P. K. L. & M. TAKEDA, 1992. The freshwater crab fauna (Crustacea, Brachyura) of the Philippines. I. The family Potamidae Ortmann, 1896. Bulletin of the National Science Museum, Tokyo, (A) **18** (4): 149-166.

NG, P. K. L. & S. H. TAN, 1998. A revision of the Southeast Asian freshwater crabs of the genus *Isolapotamon* Bott, 1968 (Crustacea: Decapoda: Brachyura: Potamidae). Proceedings of the Biological Society of Washington, **111** (1): 52-80.

ORTMANN, A., 1896. Das System der Decapoden-Krebse. Zoologische Jahrbücher, (Abtheilung für Systematik, Geographie und Biologie der Thiere) **9**: 409-453.

RATHBUN, M. J., 1904. Les crabes d'eau douce. Nouvelles Archives du Muséum d'Histoire Naturelle, Paris, (4) **6**: 225-312, pls. 9-18.

First received 13 May 2009.
Final version accepted 12 June 2009.

MEDAEUS DANIELITA, A NEW SPECIES OF XANTHID CRAB (DECAPODA, BRACHYURA, XANTHIDAE) FROM THE BOHOL SEA, CENTRAL PHILIPPINES

BY

JOSE C. E. MENDOZA[1,3]) and PETER K. L. NG[1,2,4])

[1]) Department of Biological Sciences, National University of Singapore, 14 Science Drive 4, 117543 Singapore

[2]) Tropical Marine Science Institute, National University of Singapore, 14 Kent Ridge Road S, 119223 Singapore

ABSTRACT

A new species of xanthid crab of the genus *Medaeus* Dana, 1852, is described from the Bohol Sea in the central Philippines. *Medaeus danielita* new species, is similar to *Medaeus aztec* Davie, 1997, but can be easily differentiated from it by the structure of its carapace, ambulatory legs, and male first gonopods. It is only the second species of the genus known from the Philippines.

RÉSUMÉ

Une nouvelle espèce de Xanthidae du genre *Medaeus* Dana, 1852 est décrite de la mer de Bohol aux Philippines. *Medaeus danielita* nouvelle espèce ressemble à *Medaeus aztec* Davie, 1997, mais peut en être facilement différenciée par la structure de sa carapace, ses pattes ambulatoires et le premier gonopode male. Il s'agit de la seconde espèce de ce genre décrite des Philippines.

INTRODUCTION

The genus *Medaeus* was established by Dana (1851) to accommodate a new species of xanthid crab obtained from Lahaina, in the Hawaiian Islands, during the United States Exploring Expedition (1838-1842), although he did

[3]) Corresponding author; e-mail: jcmendoza@nus.edu.sg

[4]) e-mail: dbsngkl@nus.edu.sg

© Koninklijke Brill NV, Leiden, 2010

not name a species then. He later named the included new species as *Medaeus ornatus* in 1852 (Dana, 1852). Subsequently, other authors have described new species of *Medaeus* or transferred previously described taxa into this genus (e.g., A. Milne-Edwards, 1867, 1873; Miers, 1886; Rathbun, 1898, 1930; Ward, 1934; Capart, 1951; Monod, 1956; Sakai, 1965; Crosnier, 1967). Guinot (1967, 1968) eventually restricted *Medaeus* for two Pacific species (*M. ornatus* Dana, 1852, the type species, and *M. elegans* A. Milne-Edwards, 1867) and established several new genera (e.g., *Edwardsium*, *Medaeops*, *Miersiella*, *Monodaeus*, *Paramedaeus*, and *Pseudomedaeus*) to accommodate the other species. As a result of collecting efforts by French scientists in the Pacific, Davie (1993, 1997) added two more new species: *M. grandis* Davie, 1993, from the Tuamotu Archipelago in French Polynesia, and *M. aztec* Davie, 1997, from New Caledonia.

A few beautifully coloured crabs were hand-collected by SCUBA divers off the islands of Panglao and Balicasag in the Bohol Sea, central Philippines, during the PANGLAO 2004 Marine Biodiversity Project. The crabs proved to be a *Medaeus* species and is here described as new. This new species is the second species of *Medaeus* reported from the Philippines, the first one being *M. elegans* A. Milne-Edwards, 1867, reported by Serène & Umali (1972) from the Sulu Archipelago in the southern Philippines.

All measurements of the material examined are expressed as carapace width by carapace length, in millimeters. The nomenclature of carapace regions follows that of Dana (1851) and Serène (1984). The following abbreviations are used: coll., collected by; stn, station; G1, male first gonopod; G2, male second gonopod. The specimens in this report are deposited at the Crustacean Reference Collection of the National Museum of the Philippines, Manila (NMCR), the Muséum national d'Histoire naturelle, Paris (MNHN), and the Raffles Museum of Biodiversity Research, Singapore (ZRC).

TAXONOMIC ACCOUNT

Family XANTHIDAE MacLeay, 1838, sensu Ng et al., 2008

Subfamily EUXANTHINAE Alcock, 1898

Medaeus Dana, 1851

Type species. — *Medaeus ornatus* Dana, 1852, by subsequent monotypy.

Remarks. — *Pilumnus pelagicus* Glassell, 1936, and *Pilumnus spinulifer* Rathbun, 1898, both from the tropical eastern Pacific region, were placed in *Medaeus* by other authors (see Guinot, 1967: 374) on account of the carapace anterolateral margin not meeting the external orbital angle but, instead, descending towards the buccal cavity. However, they were not included in *Medaeus* sensu stricto by Guinot (1967) in her revision of the genus. Ng et al. (2008) included these two species in *Medaeus*, albeit prefixed with a question mark, signifying the authors' reservations about their inclusion, and also, presumably, for the lack of any significant taxonomic action on this problem to date. We agree with Garth's assertion (in litt. to Guinot, 1967: 374) that these two species are congeneric due to similarities in their carapaces, sternums, abdomens, and G1, and conclude that they should be excluded from *Medaeus* as neither of them have a prominent 4M region or a crest or ridge on the 1P region. These features were mentioned as present in *Medaeus* sensu stricto by Guinot (1967), and Davie (1997) affirmed both these features as being diagnostic characters for the genus. Števčić (2005) did propose a new genus, *Stimpsonia*, for *Pilumnus spinulifer* but his name was a nomen nudum because no diagnosis was provided (Ng et al., 2008: 199, 208).

Medaeus danielita sp. nov. (figs. 1, 2A, C, E, 3)

Material examined. — Holotype ♂, 12.3 × 8.0 mm (NMCR 30063), stn R30, reef slope with black coral, 15-37 m, 9°37.1′N 123°46.1′E, Napaling, Panglao Island, Bohol, Philippines, coll. PANGLAO 2004 Marine Biodiversity Project, 8 Jun. 2004. Paratypes: 1 male, 7.7 × 5.4 mm (ZRC 2010.0101), stn B16, coral rubble on sand & gravel, 20 m, 9°37.6′N 123°47.3′E, Bingag, Panglao Island, Bohol, Philippines, coll. PANGLAO 2004 Marine Biodiversity Project, 17 Jun. 2004; 1 female, 7.9 × 5.3 mm (ZRC 2010.0102), stn B41, floor of cave, 17-19 m, 9°30.9′N 123°40.8′E, Balicasag Island, Bohol, Philippines, coll. PANGLAO 2004 Marine Biodiversity Project, 4 Jul. 2004.

Comparative material. — *Medaeus aztec* Davie, 1997: holotype ♂, 9.1 × 6.1 mm (MNHN-B22807), stn DW183, 367 m, 23°18.3′S 168°04.9′E, Aztec Bank, New Caledonia, coll. SMIB 8, 31 Jan. 1993; *Medaeus grandis* Davie, 1993: holotype ♂, 37.2 × 24.6 mm (MNHN-B22218), in trap, 210 m, 21°46.9′S 138°55.4′W, Mururoa, Tuamotu Archipelago, French Polynesia, coll. J. Poupin, 30 Nov. 1989; paratype ♂, 33.4 × 22.3 mm (MNHN-B22219), same data as holotype; paratype ♀, 21.2 × 14.7 mm (MNHN-B22220), in trap, stn 246, 90 m, 18°04.5′S 141°01.6′W, Hao, Tuamotu Archipelago, French Polynesia, coll. J. Poupin, 2 Jun. 1990.

Description. — Carapace (figs. 1, 2A, E) about 1.5 times as broad as long, regions well defined; 2M partially divided longitudinally, inner branch fused with 1M, which is separated by groove from 2F, outer branch indistinctly divided longitudinally; 3M, 4M, 2L, 3L, 5L distinct, entire; 1L indistinct; 4L merging with third anterolateral tooth; 6L partially divided transversely by indentation on medial margin; 1P with smooth, transverse crest; 2P with medially interrupted, granular transverse ridge; 1R, 2R fused, separated from 3R by distinct, oblique groove; grooves bordering regions, particularly 3M, posterior 2M, 6L deep, smooth; dorsal regions of carapace granular, with anterior regions bearing compacted granules; suborbital, subhepatic, pterygostomial re-

Fig. 1. *Medaeus danielita* sp. nov., holotype ♂, 12.3 × 8.0 mm (NMCR 30063), Panglao Island, Bohol, Philippines.

gions similarly granular. Front about 0.3 times carapace width, bilobed, minimally produced beyond internal orbital angle; frontal margin nearly straight, granular; lobes separated by narrow cleft, which continues as median sulcus on frontal region. Supraorbital margin granular, with no clear external orbital tooth, not clearly meeting anterolateral margin. Orbits relatively small, width about 0.1 times carapace width. Anterolateral margin with 4 broadly triangular teeth: first, fourth smaller than second, third, with apices of third, fourth teeth at the point of maximum carapace width; anterior part not clearly meeting orbital margin. Posterolateral margin straight, convergent posteriorly. Median part of posterior carapace margin straight, lined with single row of pearliform granules.

Eyes with short stalks, distal edge with cornea lined with small, tooth-like granules; corneas well developed (fig. 2E). Antennules (fig. 2E) folding transversely. Basal antennal segment long, granular, subrectangular, occupying entire space between antennular fossa, internal orbital angle, filling orbital hiatus; flagellum arising from distal margin, almost reaching outer edge of orbit. Posterior margin of epistome with median projection. Outer surface of third maxillipeds (fig. 2E) granular. Merus subquadrate, anterior margin with median concavity, median length about half that of ischium, with 2 shallow depressions on either side of low, submedian, granular ridge; margins nearly

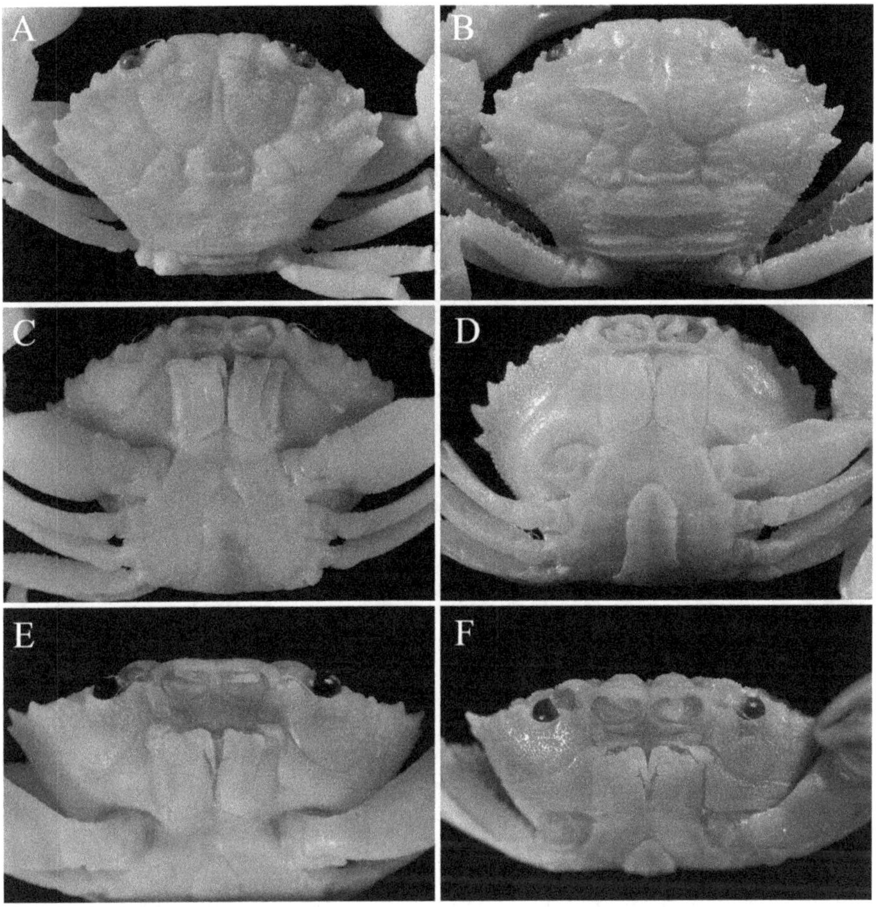

Fig. 2. A, C, E, *Medaeus danielita* sp. nov., holotype ♂, 12.3 × 8.0 mm (NMCR 30063), Panglao Island, Philippines; and, B, D, F, *Medaeus aztec* Davie, 1997, holotype male, 9.1 × 6.1 mm (MNHN-B22807), New Caledonia. A, B, dorsal view of carapace; C, D, ventral view showing thoracic sternum and abdomen; E, F, fronto-ventral view.

straight, lined with small granules. Ischium subrectangular, inner margin with short, stiff setae; with shallow, longitudinal sub-median groove; separated from basis by feeble suture. Exopod granulose, tapering toward distal end, which just reaches anterior edge of merus; flagellum long.

Surface of thoracic sternum (fig. 2C) granular, anterior region elongate. Sternites 1, 2 completely fused into triangular plate without trace of suture, separated from sternite 3 by distinct transverse suture. Sternites 3, 4 partially fused, with lateral notch on either side, which is replaced by distinct groove medially; sternite 4 with short, median, longitudinal furrow just anterior to sterno-abdominal cavity. Sterno-abdominal cavity deep, sternal press-button slightly off-centre on sternite 5, slightly nearer to suture with sternite 6, abdomen almost reaching to imaginary line joining posterior edges of cheliped coxae.

Chelipeds subequal. Merus granular on external surface, distal end exceeding anterolateral margin of carapace in dorsal view. Carpus short; dorsal, ventral surface granular, inner margin with blunt tooth. External surface of palm of both chelae granular, with slight depression forming groove near upper margin, upper, lower margins almost straight, widening distinctly from proximal end to distal end; inner surface relatively smoother. Fingers pigmented, with pigmentation on fixed finger extending slightly onto palm; tips pointed, incurved, crossing each other when fingers closed. Fixed finger slightly deflexed, with submarginal groove on external surface, cutting margin with 5, 6 low teeth. Dactylus longer than upper margin of palm, slightly curved, external surface with 2 grooves, cutting margin with 2 low, simple teeth near base.

Ambulatory legs relatively long, slender; third leg longest, coxa-to-dactylus length about 1.0 times carapace width. Meri subrectangular, subcylindrical in cross-section; anterior edges armed with small conical granules, except in fourth leg, which is more finely granular; distal end of anterior margins with prominent tooth. Carpi narrow proximally, widening distally, dorsal surface granular, with low, submarginal crest. Propodi subrectangular, with larger conical granules on edges, dorsal surface. Dactyli straight, anterior, posterior edges covered with short, stiff setae; distal end with short, chitinous claw.

External surface of male abdomen, telson (fig. 2C) with low granules. Somite 1 longer at lateral edges, forming concave distal margin; larger granules on lateral portions. Somite 2 subtrapezoidal, proximal, lateral margins convex, distal margin concave; with larger granules on lateral portions. Somites 3-5 immovably fused, vestigial sutures seen as shallow grooves; external surface with finer, lower granules; lateral margins markedly concave. Somite 6 rectangular, about 1.4 times as wide as long, central region raised, lateral margins straight. Telson subtriangular with rounded tip, convex lateral margins, median length about 0.9 times basal width, about as long as penultimate somite.

G1 (fig. 3A, B) moderate in length, slender, curving laterally, tapering distally into 2 unequal lobes; external lobe small, narrow; internal lobe longer; with about 7 stiff, simple setae subdistally. G2 (fig. 3C) about half length of G1, curving medially, tip recurved.

Coloration. — The holotype is almost entirely red dorsally, except for a few patches of pinkish white on the third and fourth anterolateral teeth, and on

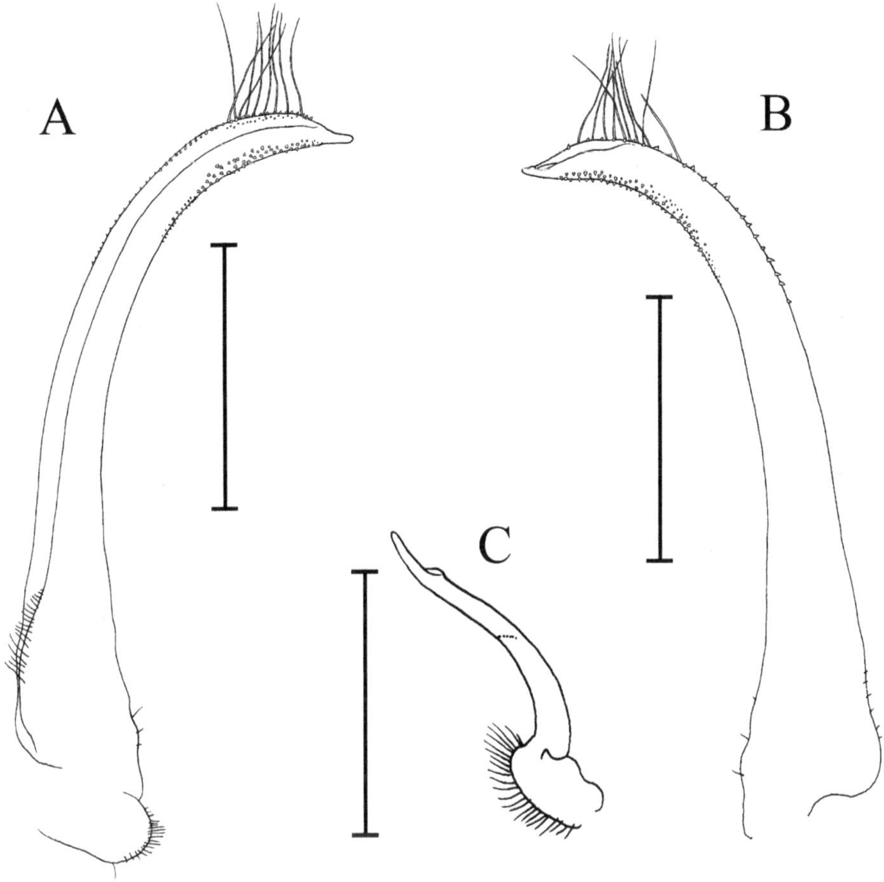

Fig. 3. *Medaeus danielita* sp. nov., holotype ♂, 12.3 × 8.0 mm (NMCR 30063), Panglao Island, Philippines. A, left G1, external view; B, idem, internal view; C, left G2, external view. Scale: A-C = 1.0 mm.

the posterior half of the carapace. The corneas are a brilliant red. The fingers of the chelipeds are brownish black throughout most of their length, and the ambulatory legs have pinkish white bands at the joints between the carpi and propodi and between the propodi and dactyli. In the fourth ambulatory legs, the coxae and ischia are also pinkish white.

Habitat. — *Medaeus danielita* sp. nov. is currently known only from the type locality, Panglao and Balicasag Islands. The crabs were hand-collected by SCUBA divers from a reef slope with good cover of black coral (*Antipathes*), an underwater cave and also among coral rubble, at a depth range of 15-37 m.

Etymology. — We take pleasure in naming this beautiful new species after Danièle Guinot, a good friend and mentor, whose work on brachyuran crabs, particularly the Xanthidae, has been most valuable to our research, and who has always been a most excellent host whenever we visited the MNHN in Paris. The specific epithet *danielita* means "little Danièle", the suffix "lita" being commonly used in Spanish and Filipino as a feminine, diminutive term of endearment. The name is used here as a noun in apposition.

Remarks. — *Medaeus danielita* sp. nov., fits well in the genus due to the prominent 4M region and the presence of a transverse crest or ridge on the cardiac (1P) region. Among the members of this genus, it is morphologically most similar to *M. aztec* (cf. Davie, 1997: 362, figs. 9, 15a, 18d) on account of: (1) the front being minimally produced and almost level with the internal orbital angle, (2) the margins of the frontal lobes being straight rather than concave, (3) the relatively thin and smooth crest on the 1P region of the carapace, (4) the relatively flatter and less rugose dorsal surface of the carapace, (5) the convex lateral margins of the male telson, (6) the slender ambulatory legs, and (7) the absence of plumose subterminal setae on the G1. However, it can be distinguished from *M. aztec* based on the following features: (1) the anterolateral teeth are more broadly triangular (figs. 1, 2A) (acutely triangular in *M. aztec*, fig. 2B), (2) the distal end of the cheliped meri exceeds the anterolateral margin of the carapace (figs. 1, 2A) (does not exceed anterolateral margin of carapace in *M. aztec*, fig. 2B), (3) the anterior margins of the meri of the ambulatory legs are either armed with small conical granules (first-third legs) or are finely granular (fourth leg) (figs. 1, 2A) (armed with uniformly spaced spines in *M. aztec*, fig. 2B); (4) the ambulatory legs are more slender (fig. 1) (relatively stouter in *M. aztec*, cf. Davie, 1997, fig. 15a); and (5) the G1 has fewer subterminal setae (~7), the external terminal lobe is narrowly spatulate and is smaller than and parallel to the internal lobe (fig. 3A, B) (G1 has more subterminal setae (~14), and the external terminal lobe is triangular and is larger than and perpendicular to the internal lobe in *M. aztec*, cf. Davie, 1997, fig. 9g, h).

Medaeus danielita is also similar to *M. grandis* (cf. Davie, 1993: 526, fig. 8, pl. 8) based on the minimally produced front, with straight margins, the broadly triangular anterolateral teeth, the relatively long and slender legs, and the gently curving G1. However, it can be distinguished from *M. grandis* by the following features: (1) the surfaces of the carapace, sternum, and pereopods are more finely granular and the areolation of the regions is not pronounced (figs. 1, 2A) (more coarsely granular carapace, sternum, and pereopods and markedly pronounced areolation of the carapace regions in *M. grandis*, cf. Davie, 1993: 527, fig. 8A, D, pl. 8A-C); (2) the chelipeds are relatively more slender (fig. 1) (stouter and more robust in *M. grandis*, cf. Davie, 1993, pl. 8C); (3) the transverse prominence on 1P is a thin, smooth crest (figs. 1, 2A) (thicker and coarser ridge in *M. grandis*, cf. Davie, 1993, pl. 8A), and (4) the G1 is proportionately more slender and has much fewer subterminal setae and spines (fig. 3A, B) (stouter and with more subterminal setae and spines in *M. grandis*, cf. Davie, 1993: 527, fig. 8E, F). Finally, *M. danielita* was found in relatively shallower waters (about 15-37 m, as opposed to 90-210 m for *M. grandis*).

ACKNOWLEDGEMENTS

We are very grateful to Danièle Guinot and Régis Cleva at the Muséum national d'Histoire naturelle, Paris, for their support of our visits to the museum and for specimen loans. The PANGLAO 2004 Marine Biodiversity Project was made possible through the efforts of the organizers, Philippe Bouchet (MNHN) and Danilo Largo (University of San Carlos, Cebu), and local and international participants; with the permission of the Philippine Department of Agriculture and Bureau of Fisheries and Aquatic Resources and support from the TOTAL Foundation and the ASEAN Regional Center for Biodiversity Conservation. We also thank Peter Castro, Peter Davie, and Paul Clark for their helpful comments. This study was supported by the National University of Singapore through its ARF Grant #R-154-000-334-112.

REFERENCES

ALCOCK, A., 1898. Materials for a carcinological fauna of India. No. 3. The Brachyura Cyclometopa. Part I. The family Xanthidae. Journal of the Asiatic Society of Bengal, Calcutta, **67** (2) (1): 67-233.

CAPART, A., 1951. Crustacés Décapodes Brachyures. Expédition océanographique belge dans les eaux côtières africaines de l'Atlantique sud (1948-1949), **3** (1): 13-205.

CROSNIER, A., 1967. Remarques sur quelques Crustacés Décapodes benthiques ouest-africaines. Description de *Heteropanope acanthocarpus* et *Medaeus rectifrons* spp. nov. Bulletin du Muséum national d'Histoire naturelle, Paris, (2) **39** (2): 320-344.

DANA, J. D., 1851. On the classification of the Cancroidea. American Journal of Science, **12** (34): 121-131.

— —, 1852. Conspectus Crustaceorum quae in Orbis Terrarum circumnavigatione lexit et descripsit. Conspectus of the Crustacea of the Exploring Expedition under Capt. Wilkes U.S.N. Proceedings of the Academy of Natural Sciences of Philadelphia, **6**: 73-86.

DAVIE, P. J. F., 1993. Deepwater xanthid crabs from French Polynesia (Crustacea, Decapoda, Xanthoidea). Bulletin du Muséum national d'Histoire naturelle, Paris, (4) **14** A (2): 501-561, pls. 1-13.

— —, 1997. Crustacea Decapoda: deep water Xanthoidea from the south-western Pacific and the western Indian Ocean. In: A. CROSNIER, Résultats des campagnes MUSORSTOM, 18. Mémoires du Muséum national d'Histoire naturelle, Paris, **176**: 337-387.

GLASSELL, S. A., 1936. The Templeton Crocker Expedition. I. Six new brachyuran crabs from the Gulf of California. Zoologica, **21** (3): 213-218.

GUINOT, D., 1967. Recherches préliminaires sur les groupements naturels chez les Crustacés Décapodes Brachyoures. II. Les anciens genres *Micropanope* Stimpson et *Medaeus* Dana. Bulletin du Muséum national d'Histoire naturelle, Paris, (2) **39** (2): 345-374.

— —, 1968. Recherches préliminaires sur les groupements naturels chez les Crustacés Décapodes Brachyoures. IV. Observations sur quelques genres de Xanthidae. Bulletin du Muséum national d'Histoire naturelle, Paris, (2) **39** (4): 149-166.

MACLEAY, W. S., 1838. Illustrations of the Annulosa of South Africa; being a portion of the objects of natural history chiefly collected during an expedition into the interior of South Africa, under the direction of Dr. Andrew Smith, in the years 1834, 1835, and 1836; fitted out by the "Cape of Good Hope Association for Exploring Central Africa." In: A. SMITH, Illustrations of the zoology of South Africa investigations: 1-75, pls. 1-4. (Smith, Elder and Co., London).

MIERS, E. J., 1886. Part II. Report of the Brachyura collected by H.M.S. Challenger during the years 1873-76. In: Report on the Scientific Results of the Voyage of H.M.S. Challenger during the years 1873-1876 under the command of Captain George S. Nares, N.R., F.R.S. and the late Captain Frank Tourle Thomson, R.N. prepared under the Superintendence of the late Sir C. Wyville Thomson, Knt., F.R.S. & Regis Professor of Natural History in the University of Edinburgh of the civilian scientific staff on board and now of John Murray one of the naturalists of the expedition, (Zoology) **17**: i-l, 1-362, pls. 1-29. (Her Majesty's Stationary Office, London, Edinburgh and Dublin).

MILNE-EDWARDS, A., 1867. Descriptions de quelques espèces nouvelles de Crustacés Brachyures. Annales de la Société entomologique de France, (4) **7**: 263-288.

— —, 1873. Recherches sur la faune carcinologique de la Nouvelle-Calédonie, Deuxième Partie. Nouvelles Archives du Muséum d'Histoire naturelle, Paris, **9**: 155-332, pls. 4-18.

MONOD, T., 1956. Hippidea et Brachyura ouest-africains. Mémoires de l'Institut Française d'Afrique Noire, **45**: 1-674.

NG, P. K. L., D. GUINOT & P. J. F. DAVIE, 2008. Systema Brachyurorum: Part I. An annotated checklist of extant brachyuran crabs of the world. Raffles Bulletin of Zoology, (Supplement) **17**: 1-286.

RATHBUN, M. J., 1898. The Brachyura collected by the U.S. Fish Commission steamer Albatross on the voyage from Norfolk, Virginia, to San Francisco, California, 1887-1888. Proceedings of the United States National Museum, **21**: 567-616.

— —, 1930. The cancroid crabs of America of the families Euryalidae, Portunidae, Atelecyclidae, Cancridae, and Xanthidae. United States National Museum Bulletin, **152**: 1-609, pls. 1-230.

SAKAI, T., 1965. On two new genera and five new species of xanthoid crabs from the collection of His Majesty the Emperor of Japan made in Sagami Bay. Crustaceana, **8** (1): 97-106.

SERÈNE, R., 1984. Crustacés Décapodes Brachyoures de l'océan Indien occidental et de la mer Rouge. Xanthoidea: Xanthidae et Trapeziidae. Avec un addendum par A. Crosnier: Carpilidae et Menippidae. Faune Tropicale, **24**: 1-400, pls. 1-48.

SERÈNE, R. & A. F. UMALI, 1972. The family Raninidae and other new and rare species of brachyuran decapods from the Philippines and adjacent regions. Philippine Journal of Science, **99** (1-2): 21-105, pls. 1-9.

ŠTEVČIĆ, Z., 2005. The reclassification of brachyuran crabs (Crustacea: Decapoda: Brachyura). Natura Croatica, **14** (Supplement 1): 1-159.

WARD, M., 1934. Notes on a collection of crabs from Christmas Island, Indian Ocean. Bulletin of the Raffles Museum, Singapore, **9**: 5-27, pls. 1-3.

First received 17 June 2009.
Final version accepted 17 July 2009.

A NEW SPECIES OF *MICROGONEPLAX* CASTRO, 2007 (DECAPODA, BRACHYURA, GONEPLACIDAE) FROM AMBON, INDONESIA

BY

TOHRU NARUSE[1,3]) and PETER CASTRO[2,4])

[1]) Transdisciplinary Research Organization for Subtropical and Island Studies, University of the Ryukyus, 870 Uehara, Taketomi, Okinawa 907-1541, Japan

[2]) Biological Sciences Department, California State Polytechnic University, Pomona, California 91768-4032, U.S.A.

ABSTRACT

A new species of the genus *Microgoneplax* Castro, 2007 (Goneplacidae sensu stricto), is described based on two male specimens from Ambon, Indonesia. The new species differs from its six congeners by the C-shaped and incurved distal part of the male first gonopod and the lanceolate dactylus of its fourth ambulatory leg. All known species of *Microgoneplax* share an almost identical carapace but have a very different male first gonopod. The male second gonopod is much shorter than the first, setting it apart from most goneplacid genera.

RÉSUMÉ

Une nouvelle espèce de Goneplacidae du genre *Microgoneplax* Castro, 2007, est décrite à partir de deux spécimens males d'Ambon, Indonésie. La nouvelle espèce diffère de ses cinq congénères par la forme en C, incurvée distalement, de son G1 et par la forme lancéolée du dactyle de sa quatrième patte ambulatoire. Toutes les espèces connues de *Microgoneplax* ont une carapace pratiquement identique, mais des G1 très différents. Le G2 de *Microgoneplax* est beaucoup plus court que le G1, ce qui sépare ce genre de la plupart des autres genres de goneplacides.

INTRODUCTION

Serène & Soh (1976) established *Singhaplax* to accommodate the goneplacids *Goneplax ockelmanni* Serène, 1971, and *G. nipponensis* Yokoya, 1933. Castro (2007) revised Goneplacidae, restricted *Goneplax* Leach, 1814, to four

[3]) Corresponding author; e-mail: naruse@lab.u-ryukyu.ac.jp

[4]) e-mail: pcastro@csupomona.edu

© Koninklijke Brill NV, Leiden, 2010

western Atlantic species and described four new genera for the Indo-West Pacific species previously included in the genus. *Singhaplax* was redescribed to include *Goneplax ockelmanni*, *G. nipponensis*, *G. wolffi* Serène, 1964, *Ommatocarcinus orientalis* Tesch, 1918, and four new species, all of which are characterized by their small size, a conspicuous outer orbital tooth but no anterolateral teeth, and a long male second gonopod that is longer, or as long as the male first gonopod. *Microgoneplax* Castro, 2007, was established for *Ommatocarcinus elegans* Chen, 1998, and four new species, all of which are nearly identical to the species of *Singhaplax* except that the male second gonopod is much shorter than the male first gonopod. All 15 species of *Singhaplax* and *Microgoneplax*, including *S. danielae* and *M. guinotae* described by Takeda & Komatsu (2010) and a new *Microgoneplax* species being described here, are morphologically very close to each other and some have been collected from the same stations, suggesting that these species are sympatric. The most obvious difference between these species is the morphology of their respective male first gonopods.

Specimens examined are deposited in the Zoological Reference Collection (ZRC), Raffles Museum of Biodiversity Research, National University of Singapore. Measurements provided are of the carapace length (CL) by the carapace width (CW). The abbreviations G1 and G2 are used for the male first and second gonopods, respectively.

TAXONOMY

GONEPLACIDAE MacLeay, 1838

Microgoneplax Castro, 2007

Microgoneplax danielae n. sp. (figs. 1, 2)

Material examined. — Holotype, male, 3.1 by 6.0 mm, ZRC 2009.0576, Ambon Bay, Indonesia, coarse sand bottom, coll. R. V. "Tirta", 24 Jan. 1975. Paratype, 1 male, 2.6 by 4.7 mm, ZRC 2009.0577, data same as holotype.

Description. — Carapace (figs. 1a, 2) transversely rectangular, widest between outer orbital angles, CW 1.83-1.93 times CL (n = 2), dorsal surface smooth, longitudinally convex, highest between external orbital angles, regions ill-defined. Front directed ventrally, base not constricted. Supraorbital margins divergent posteriorly, thinly cristate, lined with minute granules, lateral third concave; infraorbital margin thinly cristate, lined with minute granules, inner orbital tooth low, rounded, broadly produced, outer two-thirds convex ventroposteriorly, laterally connected to outer orbital angle. Outer orbital

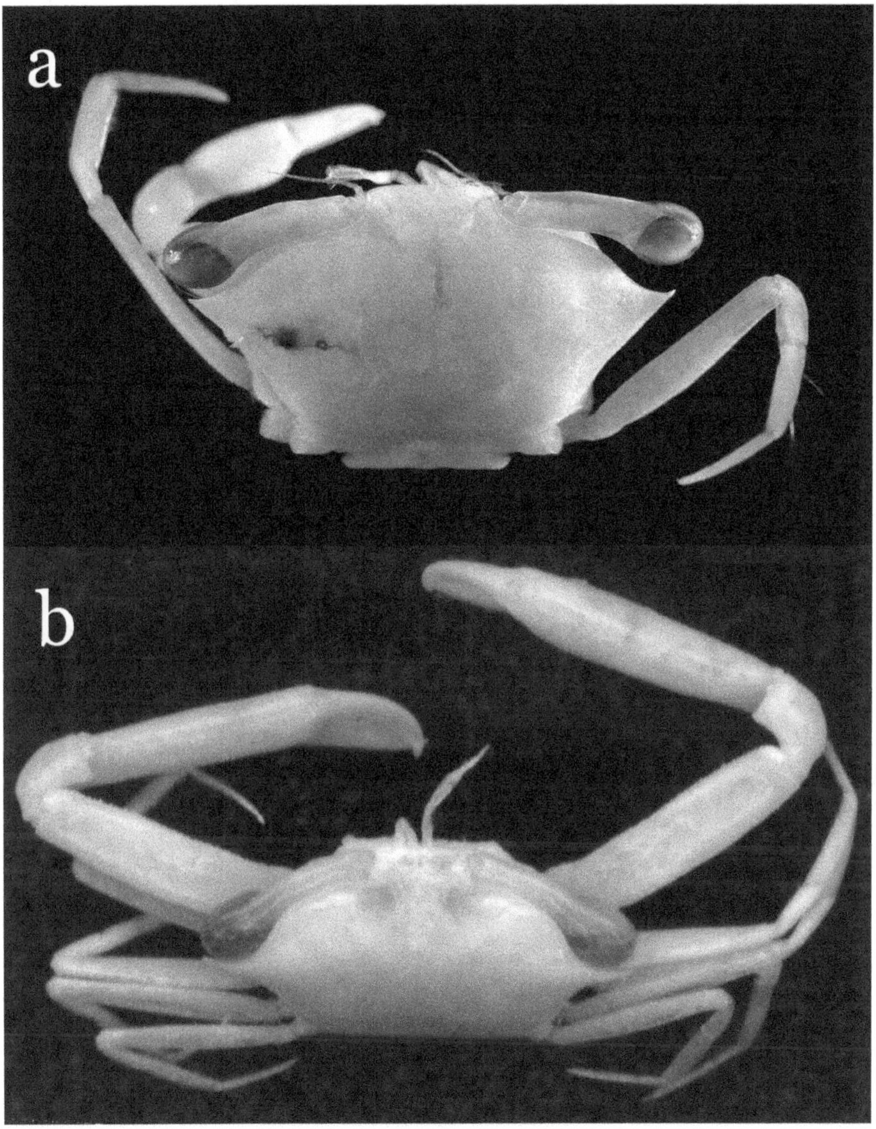

Fig. 1. a, *Microgoneplax danielae* n. sp., holotype male (3.1 × 6.0 mm, ZRC 2009.0576); b, *Ommatocarcinus elegans* Chen, 1998, holotype male, (4.3 × 8.0 mm, SSB III-34).

angle at level of proximal half of carapace, sharp, directed laterally, with tip slightly curved anteriorly, anterolateral tooth absent. Epistome flat, with posterior margin thinly upturned, slightly concave along its breadth.

Antennule developed, long, antennular septum placed behind surfaces of basal antennular segments. Antenna with basal antennal article free in orbital

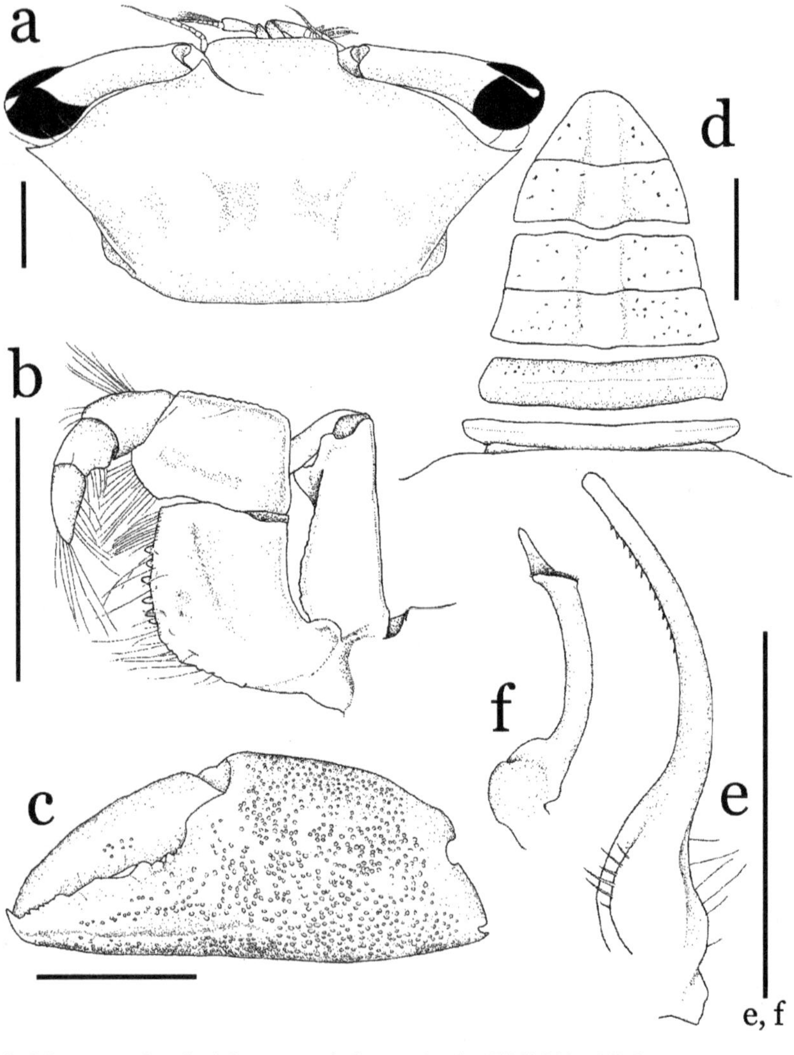

Fig. 2. *Microgoneplax danielae* n. sp., holotype (male, ZRC 2009.0576). a, carapace; b, left third maxilliped; c, left chela; d, male abdomen, telson, and posterior margin of carapace; e, left G1; f, right G2. Scale bars = 1 mm.

hiatus, not reaching front, flagellum short. Eye (fig. 2a) with long peduncle, tip of eye in frontal view exceeding beyond outer orbital angle by distal quarter of cornea, cornea wider than peduncle. Third maxillipeds (fig. 2b) rectangular, leaving wide gape when closed; ischium rectangular, inner margin dentate, lined with setae, mid-length about twice length of merus; merus with distal inner angle concave; exopod wide, distal end reaching external outer angle of merus, with subdistal triangular tooth on inner margin, flagellum long.

Male with wide, triangular abdominal cavity, press-button of abdominal locking mechanism thin, high, placed near anterior margin of thoracic sternite 5. Gonopore coxal, penis emerging from sternite 8 near boundary with thoracic sternite 7.

Male holotype with only relatively slender left cheliped (fig. 2c), male paratype with only relatively stout right cheliped. Meri with cristate anterior margins, lined with granules, produced distally, slightly lobe-like. Carpi small, each with rounded outer surface, inner angle pointed as small tubercle. Chelae rather compressed; outer, inner surfaces covered with minute granules, palm slightly longer than immovable finger in male holotype, about two times as long in male paratype; fingers compressed, with low carina from tip of immovable finger to distoventral part of palm, teeth on fingers small in holotype, those of paratype aligned irregulary, proximal quarter of immovable finger concave.

Ambulatory legs (fig. 1a) slender, long. Meri widest on proximal third, no subterminal tooth on anterior margin. Propodus of fourth leg clearly wider than carpus, flat, inner, outer margins densely lined with plumose setae. Dactylus of fourth leg compressed, wide, lanceolate, as long as or slightly longer than propodus; distal half of inner margin, distal two-thirds of outer margin lined with spines; inner margin also lined with long plumose setae.

All abdominal somites, telson freely articulating (fig. 2d), first somite concealed under carapace, third somite widest, convex laterally but far from coxa of fourth ambulatory leg; sixth somite with sockets for sternal condyle on distal outer angles of sternal side. G1 (fig. 2e) slender, distal two-thirds broadly C-shaped, nearly same width along entire length. G2 (fig. 2f) about half length of G1, ventral margin of distal opening fringed with short setae, dorsal margin distally produced to triangular lobe.

Etymology. — We are honoured to name the new species after Prof. Danièle Guinot, whose pioneering work on the Goneplacidae set the standards for present and future work on this challenging group.

Remarks. — *Microgoneplax danielae*, n. sp. is morphologically most similar to *M. elegans* (Chen, 1998) and *M. prion* Castro, 2007, by each sharing a slender and simple G1. The G1 of *M. danielae*, however, differs from those of *M. elegans* and *M. prion* by its broadly C-shaped and incurved distal part (sinuous in both *M. elegans* and *M. prion*, cf. Chen, 1998, fig. 14 (7); Castro, 2007, fig. 45b). *Microgoneplax danielae* is also distinct from all its congeners by the lanceolate dactylus of its fourth ambulatory leg. In contrast, the fourth ambulatory legs of all other *Microgoneplax* species have a thin dactylus (e.g., Castro,

2007, figs. 42, 46). The dactylus of the fourth ambulatory leg of *M. elegans* has not been described nor illustrated, however. The holotype of *M. elegans* is deposited in the Institute of Oceanology, Qingdao, China. The photograph of the holotype of *M. elegans* (male, SSB III-34, 4.3 by 8.0 mm), kindly sent to us by Dr. Li Xinzheng, clearly shows a non-lanceolate dactylus (fig. 1b).

ACKNOWLEDGEMENTS

We thank Dr. Kasim Moosa who enabled us to examine the present species. Dr. Li Xinzheng (Institute of Oceanology, Qingdao, China) and Nguyen Thanh Son (National University of Singapore) for photographing specimens of *Microgoneplax* species.

REFERENCES

CASTRO, P., 2007. A reappraisal of the family Goneplacidae MacLeay, 1838 (Crustacea, Decapoda, Brachyura) and revision of the subfamily Goneplacinae, with the description of 10 new genera and 18 new species. Zoosystema, **29**: 609-774.

CHEN, H., 1998. The Goneplacidae (Crustacea: Brachyura) from Nansha Islands and adjacent waters. In: The Multidisciplinary Oceanographic Expedition Team of Academia Sinica to Nansha Islands (eds.), Studies on marine fauna and flora and biogeography of the Nansha Islands and neighbouring waters, **3**: 265-316. (Academia Sinica, Beijing). [In Chinese with English abstract.]

LEACH, W. E., 1814. Crustaceology. In: D. BREWSTER (ed.), The Edinburgh Encyclopaedia, **7**: 383-437, pl. 221. (William Blackwood, Edinburgh).

MACLEAY, W. S., 1838. Illustrations of the Annulosa of South Africa; being a portion of the objects of natural history chiefly collected during an expedition into the interior of South Africa, under the direction of Dr. Andrew Smith, in the years 1834, 1835, and 1836; fitted out by the "Cape of Good Hope Association for Exploring Central Africa". In: A. SMITH (ed.), Illustrations of the Zoology of South Africa Investigations: 1-75, pls. 1-4. (Smith, Elder & Co., London).

SERÈNE, R., 1971. Observations préliminaires sur des Brachyoures nouveaux ou mal connus du Sud-Est Asiatique (Crustacea Decapoda). Bulletin du Muséum national d'Histoire naturelle, Paris, (2) **42** (5) [1970]: 903-918.

SERÈNE, R. & C. L. SOH, 1976. Brachyura collected during the Thai-Danish Expedition (1966). Research Bulletin Phuket Marine Biological Center, **12**: 1-37, fig. 28, pls. I-VII.

YOKOYA, Y., 1933. On the distribution of decapod crustaceans inhabiting the continental shelf around Japan, chiefly based upon the materials collected by S. S. Sôyô-Maru, during the years 1923-1930. Journal of the College of Agriculture, Tokyo Imperial University, **12**: 1-226.

First received 22 March 2009.
Final version accepted 31 May 2009.

A NEW SPECIES OF *IHLOPSIS* GUINOT & RICHER DE FORGES, 1995 (DECAPODA, BRACHYURA, HOMOLIDAE) FROM TAIWAN

BY

TOHRU NARUSE[1,3]) and BERTRAND RICHER DE FORGES[2,4])

[1]) Transdisciplinary Research Organization for Subtropical and Island Studies, University of the Ryukyus, 870 Uehara, Taketomi, Okinawa 907-1541, Japan

[2]) Correspondant du Muséum national d'Histoire naturelle, Paris, France

ABSTRACT

A new species of deep-sea homolid crab of the genus *Ihlopsis* Guinot & Richer de Forges, 1995, is described from Taiwan. The new species differs from its two congeners by the shape and position of the spines on the cephalothorax and pereopods. A record of "*Ihlopsis multispinosa* (Ihle, 1912)" from Wakayama, Japan most probably belongs to the new species. This brings the number of known species of *Ihlopsis* to three.

RÉSUMÉ

Une nouvelle espèce de crabe de profondeur Homolidae du genre *Ihlopsis* Guinot & Richer de Forges, 1995, est décrite de Taïwan. Cette nouvelle espèce diffère de ses deux congénères par la forme et la position des épines du céphalothorax et des péréiopodes. La mention d'un « *Ihlopsis multispinosa* (Ihle, 1912) » de Wakayama au Japon correspond très probablement à cette nouvelle espèce. Le genre *Ihlopsis* comprend donc maintenant trois espèces.

INTRODUCTION

Guinot & Richer de Forges (1995) established the genus *Ihlopsis* for their new species, *I. tirardi*, from New Caledonia, and also included *Latreillopsis multispinosa* Ihle, 1912, described from off Kai Island, Indonesia, in their new genus. They distinguished *Ihlopsis* from the closely similar *Moloha* Barnard, 1947, *Latreillopsis* Henderson, 1888, *Dagnaudus* Guinot & Richer

[3]) Corresponding author; e-mail: naruse@lab.u-ryukyu.ac.jp

[4]) e-mail: b.richerdeforges@gmail.com

© Koninklijke Brill NV, Leiden, 2010

de Forges, 1995, and *Homolochunia* Doflein, 1904, mainly on characters like the spinulations of the cephalotholax, spinulation and relative length of the pereopods, as well as the relative length of the ocular peduncle (Guinot & Richer de Forges, 1995: 422).

Recent expeditions in Taiwan, lead by Tin-Yam Chan of the National Taiwan Ocean University, collected two male specimens of an undescribed species of *Ihlopsis*. In addition, "*I. multispinosa* (Ihle, 1912)", recorded from Wakayama, Japan, appears to belong to the new species as well. The present study describes this third *Ihlopsis* species.

Specimens of *Ihlopsis* examined from New Caledonia (Richer de Forges & Ng, 2008) and type specimens of the new species are deposited in Muséum national d'Histoire naturelle (MNHN), Paris; National Taiwan Ocean University (NTOU), Keelung, Taiwan; and the Zoological Reference Collection (ZRC), Raffles Museum of Biodiversity Research, National University of Singapore. Measurements provided are of the carapace length (CL: mid-length from base of rostrum to middle of posterior margin) by the carapace width (CW: excluding spine length). The terminology essentially follows that of Guinot & Richer de Forges (1995). The abbreviations G1 and G2 are used for the male first and second gonopods, respectively.

TAXONOMY

HOMOLIDAE De Haan, 1839

Ihlopsis Guinot & Richer de Forges, 1995

Ihlopsis danieleae n. sp. (figs. 1-3)

Latreillopsis multispinosa — Nagai, 1991: 31, pl. 1 (3).
Ihlopsis multispinosa — Marumura & Kosaka, 2003: 22 (in list, no new material).
Ihlopsis multispinpsa [sic] — Marumura & Kosaka, 2003: pl. 5, fig. 30 (no new material).
(Not *Latreillopsis multispinosa* Ihle, 1912.)

Material examined. — Holotype: male, 21.9 by 14.1 mm, ZRC 2008.0988, station CD385, 322-313 m, 24°39.33′N 122°09.79′E, Taiwan, coll. 25 Jul. 2007. Paratype: 1 male, 18.9 by 18.7 mm, NTOU, station CD386, 301-314 m, 24°37.74′N 122°9.82′E, Taiwan, coll. 25 Jul. 2007.

Comparative material. — *Ihlopsis tirardi* Guinot & Richer de Forges, 1995: 1 male, 23.8 by 14.7 mm, MNHN B 31293, station DW3025, Concalis, New Caledonia, coll. Concalis Cruise, 9 May 2008; 1 ovigerous female, 22.7 by 15.0 mm, ZRC 2008.0987, station DW2984, 450-509 m, 18°0.24′S 163°3.43′E, Concalis, New Caledonia, coll. Concalis Cruise, 5 May 2008.

Description. — Carapace (fig. 2) longer than wide, dorsal, lateral surfaces covered with long, sharp teeth, dorsal surface, bases of teeth covered with short, soft setae. Rostrum long, slender, directed anteroventrally, reaching proximal two-thirds of antennal article 4; pseudorostrum only slightly shorter than CL, branching at distal third, dorsal, ventral branches of similar length, another short dorsal branch from proximal half. Subhepatic, buccal, pterygostomial spines long, buccal spine directed laterally, pterygostomial spine adjacent to buccal cavern.

Eyes (fig. 2) with long peduncle, distally reaching proximal branch of pseudorostrum. Antenna (fig. 2b) with long article, antennal article 4 twice length of basal antennal article, antennal article 5 shorter than basal antennal article. Third maxilliped (fig. 2b) with long spine on each merus, ischium, anterolateral angle of merus with additional short spine, coxa with large mesial lobe holding over telson (fig. 2b).

Chelipeds (figs. 1, 2) equal, subcylindrical in cross section, combined length of basis to merus shorter than that of carpus, chela; coxa with mesial lobe like that of third maxilliped coxa, ischium with 2 teeth on distoventral portion, merus with distal spine, chelae long, palm about 3 times length of fingers. Pereopods 2-5 long, pereopod 3 longest, meri to propodi subcylindrical in cross section, meri of pereopods 2-4 slightly shorter than combined length of carpi, propodi, anterior margin of meri with equidistantly distributed 2 spines, 2 setal tufts, long distal spine (fig. 3a-c), margin entire on pereopod 5 (fig. 3d); dactyli of pereopods 2-4 compressed dorsoventrally, inner margins lined with short spines.

Abdomen oval in outline, second to sixth somites with strong mid-spines, telson longer than sixth somite, strongly sharpened distally. G1 very stout, short, distal part with slender flap (fig. 3e). G2 relatively stout, slightly shorter than G1, tip flattened out, without flagellum (fig. 3f).

Etymology. — The name of the new species honours Dr. Danièle Guinot for her outstanding contributions to carcinology and more specifically to our knowledge of the Homolidae. The name is to be treated as a noun in the genitive singular.

Remarks. — Some of the characters of *Ihlopsis danieleae* n. sp., appear to be intermediate between *I. tirardi* and *I. multispinosa*, although the combination of characters makes it clear that it is a distinct species. The relatively long dorsal spines (fig. 2a) and spinulation of the meri of pereopods 2-4 (fig. 3a-c) of *I. danieleae* are similar to those of *I. multispinosa* (Ihle, 1913, pl. 4 figs. 19, 20; Guinot & Richer de Forges, 1995, fig. 47a, b; Richer de Forges & Ng, 2008, fig. 13), but the shape of the pseudorostrum is closer to that of *I. tirardi* (see Guinot & Richer de Forges, 1995, figs. 46, 47c, d, 48B; Richer de Forges & Ng,

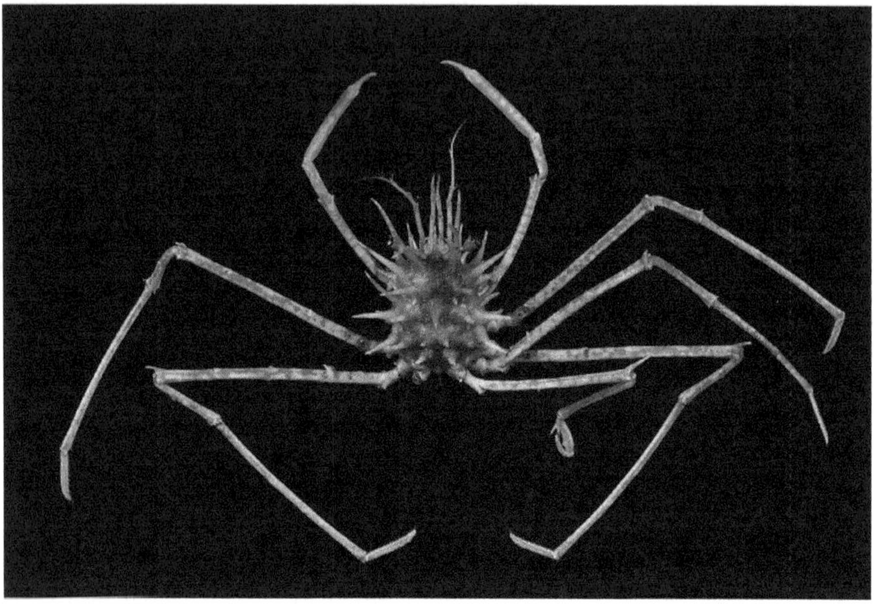

Fig. 1. *Ihlopsis danieleae* n. sp., holotype. Photograph by Tin-Yam Chan.

Fig. 2. *Ihlopsis danieleae* n. sp., cephalothorax of holotype. a, dorsal view; b, ventral view.

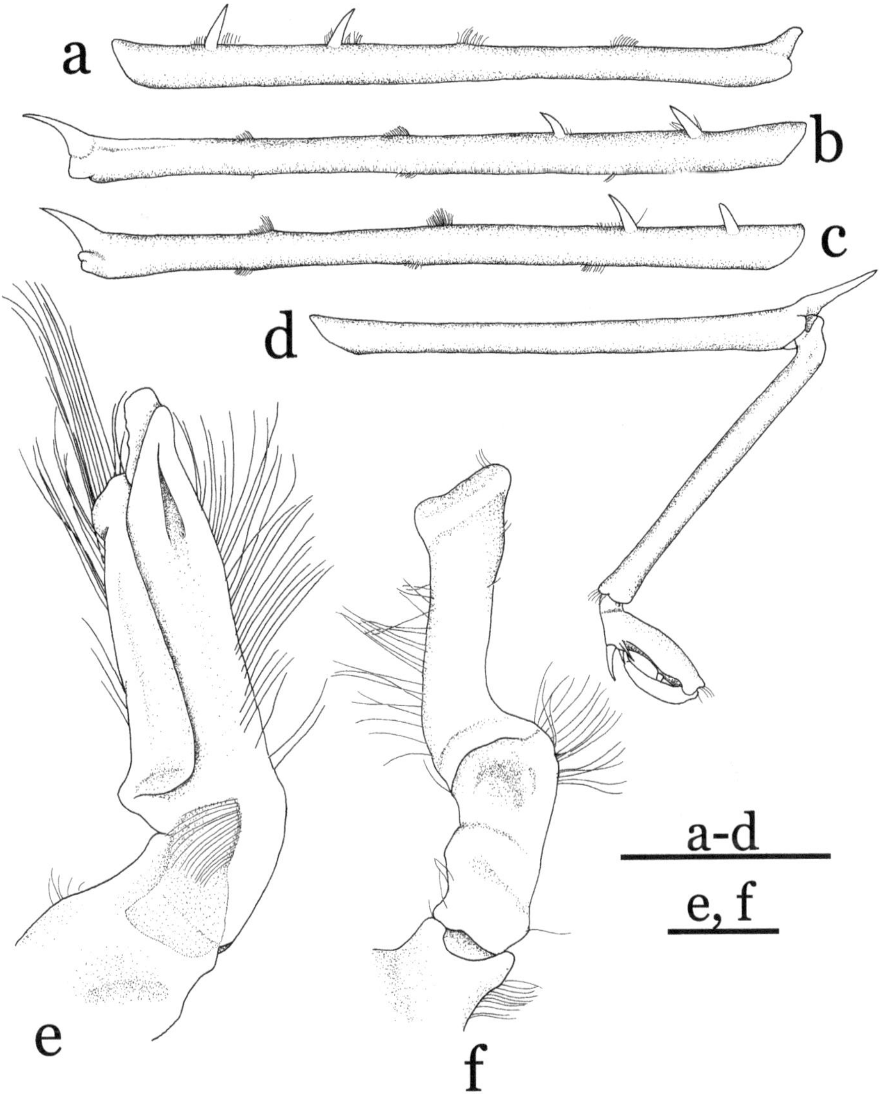

Fig. 3. *Ihlopsis danieleae* n. sp., holotype. a, merus of right pereopod 2; b, merus of left pereopod 3; c, merus of left pereopod 4; d, right pereopod 5; e, left G1, ventral view; f, left G2, ventral view. Scale bars: a-d = 10 mm; e, f = 1 mm.

2008, fig. 13B). In addition to the combination of these characters, *I. danieleae* can be distinguished from both congeners by possessing a buccal spine that is directed laterally (fig. 2b), while those of *I. tirardi* and *I. multispinosa* are directed anterolaterally (Guinot & Richer de Forges, 1995, figs. 46B, 48B; Ihle, 1913, pl. 4 fig. 20). The pterygostomial spine of *I. danieleae* is also

TABLE I
Diagnostic characters and geographical distributions of the species of *Ihlopsis*

	I. danieleae n. sp.	*I. tirardi* Guinot & Richer de Forges	*I. multispinosa* (Ihle)
Pseudorostrum	Branching at distal third, dorsal and ventral branches of same length; another short dorsal branch from proximal half	Branching at distal third, dorsal and ventral branches of same length; another short dorsal branch from proximal half	Short ventral branch from distal third and short dorsal branch from proximal third
Meri of pereopods 2 and 3	2 spines + 2 setal tufts + 1 distal spine distributed equidistantly	3 spines on proximal half + 1 setal tuft + 1 distal spine	2 spines on proximal portion
Merus of pereopod 4	2 spines + 2 setal tufts + 1 distal spine distributed equidistantly	3 spines on proximal half + 1 setal tuft + 1 distal spine	1 spine on proximal portion
Dorsal teeth	Long	Short	Long
Buccal spine	Laterally directed	Anterolaterally directed	Anterolaterally directed
Pterygostomial spine	Placed close to buccal cavern	Placed far from buccal cavern	Placed far from buccal cavern
Mesial lobes of coxae of third maxilliped and first pereopod	Large	Small	Small
Geographical distribution	Taiwan (present study: 301-322 m); Japan (Marumura & Kosaka, 2003: about 300 m)	New Caledonia (Guinot & Richer de Forges, 1995: 390-650 m; Richer de Forges & Ng, 2008: 396-509 m)	Kai Islands, Indonesia (Ihle, 1912: 204 m)

positioned closer to the buccal cavern (fig. 2b), while those of *I. tirardi* and *I. multispinosa* are in a more lateral position (table I; Guinot & Richer de Forges, 1995, figs. 46B, 48B; Ihle, 1913, pl. 4 fig. 20).

Nagai (1991) recorded a female specimen of *I. multispinosa* from off Wakayama, on the Pacific side of mainland Japan. His photograph as well as the description clearly show that the pseudorostrum of the specimen is distally bifurcated. Guinot & Richer de Forges (1995: 424) regarded the Nagai specimen as *I. tirardi* but with hesitation. Nagai's specimen, however, also possesses relatively long dorsal spines as well as long pereopods. The presence of these characters indicates that the Nagai specimen is actually *I. danieleae*, which is also supported by the proximity of the Taiwanese and Japanese locations.

ACKNOWLEDGEMENTS

We are pleased to thank our colleagues, Tin-Yam Chan for collecting this material, and Peter K. L. Ng for his help in many ways to this study. The first author was supported by a postdoctoral position in the Department of Biological Sciences, National University of Singapore; the second author was partially supported by a research fellowship from the Raffles Museum of Biodiversity Research.

REFERENCES

BARNARD, K. H., 1947. Description of new species of South African decapod Crustacea, with notes on synonymy and new records. Annals and Magazine of Natural History, (11) **13** (102), (1946): 361-392.

DOFLEIN, F., 1904. Brachyura. In: Wissenschaftliche Ergebnisse der Deutschen Tiefsee-Expedition auf dem Dampfer "Valdivia", 1898-1899, **6**: i-xiv, 1-314, atlas. (Verlag Gustav Fischer, Jena).

GUINOT, D. & B. RICHER DE FORGES, 1995. Crustacea Decapoda Brachyura: Révision de la famille des Homolidae De Haan, 1839. In: A. CROSNIER (ed.), Résultats des Campagnes MUSORSTOM, **13**. Mémoires du Muséum national d'Histoire naturelle, Paris, **163**: 283-517.

HAAN, W. DE, 1833-1849. Crustacea. In: P. F. VON SIEBOLD (ed.), Fauna Japonica, sive descriptio animalium, quae in itinere per Japoniam, jussu et auspiciis superiorum, qui summum in India Batavia imperium tenent, suscepto, annis 1823-1830 collegit, notis, observationibus a adumbrationibus illustravit, **1-8**: i-xxi, vii-xvii, ix-xvi, 1-243, pls. 1-55, A-Q, circ. pl. 2. (Lugduni Batavorum [=Leiden]).

HENDERSON, J. R., 1888. Report on the Anomura collected by H.M.S. Challenger during the years 1873-1876. In: Report on the scientific results of the voyage of H.M.S. Challenger during the years 1873-1876 under the command of captain George S. Nares, N.R., F.R.S. and the late captain Frank Tourle Thomson, R.N. prepared under the superintendence of the late Sir C. Wyville Thomson, Knt., F.R.S. &c. Regius professor of natural history in the University of Edinburgh of the civilian scientific staff on board and now of John Murray one of the naturalists of the expedition, (Zoology) **27** (1): i-xi, 1-221. (Her Majesty's Stationary, London, Edinburgh and Dublin).

IHLE, J. E. W., 1912. Über einige neue, von der Siboga-Expedition gesammelte Homolidae. Tijdschrift der Nederlandsche Dierkundige Vereeniging, Leiden, (2) **12**: 206-214.

— —, 1913. Die Decapoda Brachyura der Siboga-Expedition. I. Dromiacea. Siboga Expeditie Monographie, **39b** (71): 1-96, figs. 1-38, pls. 1-4.

MARUMURA, M. & A. KOSAKA, 2003. Catalogue of the brachyuran and anomuran crabs donated by the late Mr. Seiji Nagai to the Wakayama Prefectural Museum of Natural History: 1-73. (Wakayama Prefectural Museum, Wakayama, Japan). [In Japanese.]

NAGAI, S., 1991. Some remarkable crabs of Wakayama Prefecture. I. Nankiseibutu, **33** (1): 31-34. [In Japanese.]

RICHER DE FORGES, B. & P. K. L. NG, 2008. New western Pacific records of Homolidae De Haan, 1839, from the western Pacific, with descriptions of new species of *Homolochunia* Doflein, 1904, and *Latreillopsis* Henderson, 1888 (Crustacea: Decapoda: Brachyura). Zootaxa, **1967**: 1-35.

First received 16 June 2009.
Final version accepted 24 July 2009.

TWO NEW SPECIES OF GRAPSOID CRABS OF THE GENUS *EUCHIROGRAPSUS* (DECAPODA, BRACHYURA, GRAPSOIDEA, PLAGUSIIDAE) FROM THE PHILIPPINES AND THE HAWAIIAN ISLANDS

BY

N. K. NG[1,3]) and JOEL W. MARTIN[2,4])

[1]) Department of Biological Sciences, National University of Singapore, Science Drive 4, Singapore 117543, Republic of Singapore

[2]) Natural History Museum of Los Angeles County, Research and Collections/Crustacea, 900 Exposition Boulevard, Los Angeles, CA 90007, United States of America

ABSTRACT

Two recent expeditions to the Philippines and the Hawaiian Islands have resulted in the collection of two new species of grapsoid crabs that are included in the rare deep-sea genus *Euchirograpsus* H. Milne Edwards, 1853. The two new species are different from their congeners in the structures of their carapaces, chelipeds, ambulatory legs, male abdomens, and male first gonopods. Both species are described, bringing the number of species in the genus to ten.

RÉSUMÉ

Deux expéditions récentes aux îles Philippines et aux îles Hawaii ont permis la récolte de deux nouvelles espèces de crabes Grapsidae qui font parties du rare genre de profondeur *Euchirograpsus* H. Milne Edwards, 1853. Ces deux nouvelles espèces diffèrent de leurs congénères par les structures de leurs carapaces, par les pinces, les pattes ambulatoires, les abdomens males et le premier gonopode male. Ces deux espèces sont décrites, portant le nombre d'espèces de ce genre à dix.

INTRODUCTION

The genus *Euchirograpsus* H. Milne Edwards, 1853, currently includes eight species of unusual deep-sea grapsoid crabs from the Pacific, Atlantic,

[3]) Corresponding author; e-mail: dbsngnk@nus.edu.sg

[4]) e-mail: jmartin@nhm.org

© Koninklijke Brill NV, Leiden, 2010

TABLE I
Known species of the genus *Euchirograpsus* H. Milne Edwards, 1853, and their geographical distributions

Species	Geographical distribution	Depth (m)	Reference
Euchirograpsus americanus H. Milne Edwards	Eastern Atlantic	33-137	A. Milne-Edwards, 1880
Euchirograpsus antillensis Türkay	Western Atlantic	192-430	Türkay, 1975
Euchirograpsus danielae n. sp.	Hawaiian Islands (French Frigate Shoals)	178-267	Present study
Euchirograpsus guinotae n. sp.	Philippines (Balicasaq Island)	unknown	Present study
Euchirograpsus liguricus H. Milne Edwards	Eastern Atlantic	45-180	H. Milne Edwards, 1853
Euchirograpsus madagascariensis Türkay	Western Indian Ocean	105-195	Türkay, 1978
Euchirograpsus pacificus Türkay	Galápagos Islands	58	Türkay, 1975
Euchirograpsus (?) *polyodous* (Stebbing)	Natal, South Africa	91	Stebbing, 1921
Euchirograpsus timorensis Türkay	Southwest Pacific	35-360	Türkay, 1975, Crosnier, 2001
Euchirograpsus tuerkayi Crosnier	French Polynesia	200-290	Crosnier, 2001

and Indian oceans: *E. liguricus* H. Milne Edwards, 1853 (type species); *E. americanus* H. Milne Edwards, 1880; *E. polyodous* (Stebbing, 1921) (tentative inclusion); *E. antillensis* Türkay, 1975; *E. madagascariensis* Türkay, 1978; *E. pacificus* Türkay, 1975; *E. timorensis* Türkay, 1975, and *E. tuerkayi* Crosnier, 2001 (see table I and Ng et al., 2008). Two recent expeditions to the Philippines (2003) and the northwestern Hawaiian Islands (2006) have resulted in the collection of two new species of the genus, which differ from their congeners in the forms of their carapaces, chelipeds, ambulatory legs, male abdomens, and first male gonopods.

MATERIAL AND METHODS

The two Hawaiian Islands specimens, both of which are males, were collected as part of a 2006 biodiversity survey of French Frigate Shoals, northwestern Hawaiian Islands, that was conducted using NOAA's research vessel "Oscar Elton Sette." The *Euchirograpsus* specimens were collected from commercial lobster traps baited with dead mackerel and set at depths of 240-267 m and 178-198 m, and left overnight.

The two Philippines specimens, also both males, were purchased from local fishermen in July of 2003 on Balicasaq Island, Panglao Islands, Bohol.

Measurements provided are of carapace width (listed first) and carapace length. The abbreviations G1 and G2 are used for the male first and second pleopods, respectively. The overall length to width of the chela refers to the ratio of the length, measured from the tip of the dactylus to the proximal margin of the palm, against the chela width, measured from the highest to lowest part of the palm (propodus). The overall length to width ratio of the male abdomen refers to the length, measured from the tip of the distal end of the telson to the proximal end of the second abdominal somite, against the width, measured as the widest part of the second abdominal somite. The abbreviations P2, P3, P4, and P5 refer to the first, second, third, and fourth ambulatory legs, respectively. Specimens examined are deposited in Natural History Museum of Los Angeles County (LACM), Los Angeles, California; Museum national d'Histoire naturelle (MNHN), Paris; the National Museum of the Philippines Manila (NMCR); Forschunginstitut Senckenberg (SNG), Frankfurt am Main; United States National Museum of Natural History (USNM), Smithsonian Institution, Washington, D.C.; and the Zoological Reference Collection (ZRC), Raffles Museum of Biodiversity Research, National University of Singapore.

TAXONOMY

Family PLAGUSIIDAE Dana, 1851

Subfamily PLAGUSIINAE Dana, 1851

Euchirograpsus H. Milne Edwards, 1853

Diagnosis. — Carapace subquadrate; surface with small round granules, densely setose with short stiff setae; regions relatively well-defined. Frontal margin about 0.4 times carapace width, with 2 distinct lobes separated by deep, U-shaped notch; each lobe further divided into 2 smaller lobes by antennular fossae cleft. Anterolateral margin with 4 anterolateral teeth including external orbital tooth. Basal antennal segment covered with small granules, immobile, laterally locked against inner orbital tooth. Third maxillipeds with inner margins meeting medially, not completely closing buccal cavern, palp of exopod articulating near antero-external angle of merus. Chelipeds long, robust, granular, margins lined with sharp, distinct granules, anterior margin with 2 distinct sub-distal teeth. Chelae proportionally long, dorsal surface with 3 rows of distinct, rounded granules; posterior region of chela with row of granules extending from proximal margin of chela to tip of immobile finger; posterior margin densely granular. Ambulatory legs with all segments long, slender, flattened; P4 longest. Male abdomen broadly triangular, with somites 3-6 fused, but sutures may still be evident, entirely covering sternum between last pair of legs, locking mechanism present. Male thoracic sternum with penis emerging at outer edge of thoracic sternite 8. G1 broad, short. G2 usually one-third the length of G1. [After Türkay, 1975: 103; and Crosnier, 2001: 785.]

Remarks. — Of the eight known species of *Euchirograpsus*, only the identity of *E. polyodous* (Stebbing, 1921) is problematic. The species was described as *Pachygrapsus polyodous* based on one male specimen from Natal, South Africa. The sole specimen is unfortunately lost (Ng et al., 2008). Based on the descriptions, figures (Stebbing, 1921: 458, pl. 16), and the depth from which the specimen was collected (see table I), the species might belong to *Euchirograpsus*. The identity of this species, however, will remain in question until new material can be collected from the type locality. *Euchirograpsus polyodous* can be easily distinguished from all congeners by: (a) the shape of the merus of the third maxilliped is trapezoidal with numerous low spines along the lateral margins (Stebbing, 1921, pl. 16) (subquadrate with granular lateral margins in other *Euchirograpsus* species); (b) the frontal margin of the carapace bears numerous distinct, sharp teeth (granular frontal margin in other species); (c) the frontal margin is almost straight with a median notch (two distinct lobes separated by a deep U-shaped notch, each lobe further divided into two smaller lobes by the antennular fossae cleft in other species); (d) numerous sharp teeth on the dorsal margin of the merus of the P5 (serrated dorsal margin of the merus of the P5 in other species); (e) corneous spines on the ventral margin of the propodus of the P5 (corneous spines absent in other species); (f) the presence of five pairs of distinct, sharp corneous spines on the ventral margin of the dactylus of the P5 (six pairs of spines in other species).

Euchirograpsus danielae n. sp. (figs. 1A, 2, 3A, C, 4, 5)

Material examined. — Holotype: 1 male, 25.3 × 24.0 mm (figs. 1a, 2, 3a, c) (LACM), French Frigate Shoals (23.888°N 166.336°W – 23.887°N 166.333°W), northwest Hawaiian

Islands, reef slope, 198-178 m, baited trap, coll. R. Moffitt, A. Collins, C. Pittman, J. Martin, K. Coontz, S. Middleton, 17 Oct. 2006. Paratype: 1 male, 29.6 × 27.8 mm (ZRC 2009.0844), French Frigate Shoals (23.903°N 166.288°W – 23.902°N 166.286°W), northwest Hawaiian Islands, reef slope, 267-240 m, baited trap, coll. R. Moffitt, J. Martin, C. Pittman, G. Paulay, T. Lotufo, E. Soto, K. Coontz, 18 Oct. 2006.

Description. — Carapace subquadrate (figs. 1A, 2, 3A), as wide as long or slightly broader than long (width to length ratio 1.1), dorsal surface covered with minute round granules, short stiff setae, regions relatively well defined; cervical grooves indistinct, shallow, broad; gastric grooves distinct; epigas-

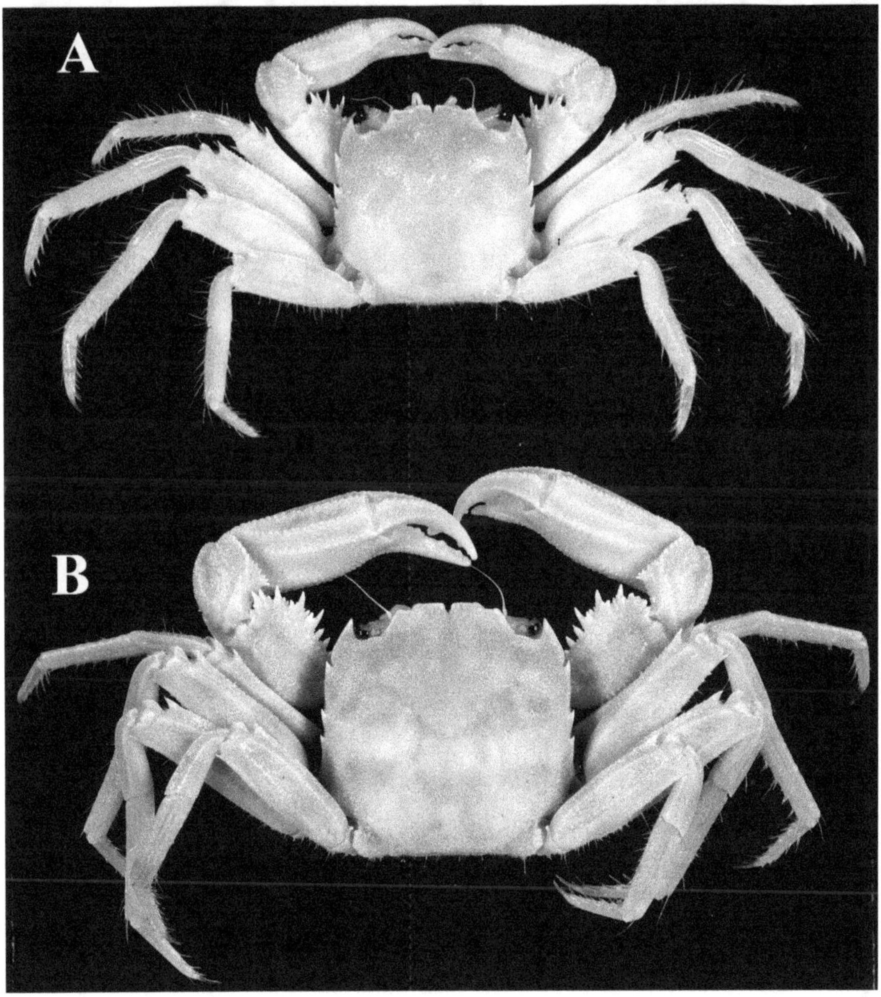

Fig. 1. Habitus. A, *Euchirograpsus danielae* n. sp., holotype male, 25.3 × 24.0 mm (LACM); B, *Euchirograpsus guinotae* n. sp., holotype male (37.8 × 37.8 mm) (ZRC 2009.0845).

Fig. 2. *Euchirograpsus danielae* n. sp., holotype male, 25.3 × 24.0 mm (LACM). [Also copyrighted by J. W. Martin.]

tric, postorbital cristae absent; low ridge present at region adjacent to junction between posterolateral margin, posterior carapace margin. Anterolateral margin with 4 anterolateral teeth including external orbital tooth, decreasing in size posteriorly; lined with small, rounded granules; not clearly demarcated from posterolateral margin; external orbital tooth broad, separated from second tooth by broad U-shaped cleft; third anterolateral tooth separated from second tooth by broad U-shaped cleft; fourth tooth narrowest, separated from third tooth by narrow V-shaped notch; posterolateral margins gently convex before converging towards almost straight posterior margins. Frontal margin about 0.4 times carapace width, with 2 distinct lobes separated by deep, U-shaped notch; each lobe further divided into 2 smaller lobes by antennular fossae cleft. Orbit deep sunken cavity, almost complete. Lower margin of external orbital tooth extending medially as ridge to form new lower orbital margin, separated from large, broad, rounded infra-orbital tooth by narrow fissure; sub-orbital crest extending laterally beyond orbit as slightly sinuous granular row, but not appearing to be stridulatory. Supraorbital margin granular, strongly sinuous, subparallel with frontal margin; frontal margin (figs. 1A, 3A, 4A) straight, strongly deflexed, with broad shallow sinus between 2 lobes, broad frontal median triangle present, formed by 3 granuliform ridges; infra-orbital margins granular, complete, congruent with anterolateral margin; distinct row of rounded granules present below infra-orbital margin on suborbital region, extending to subbranchial region (fig. 3A). Antennal segments entering orbital

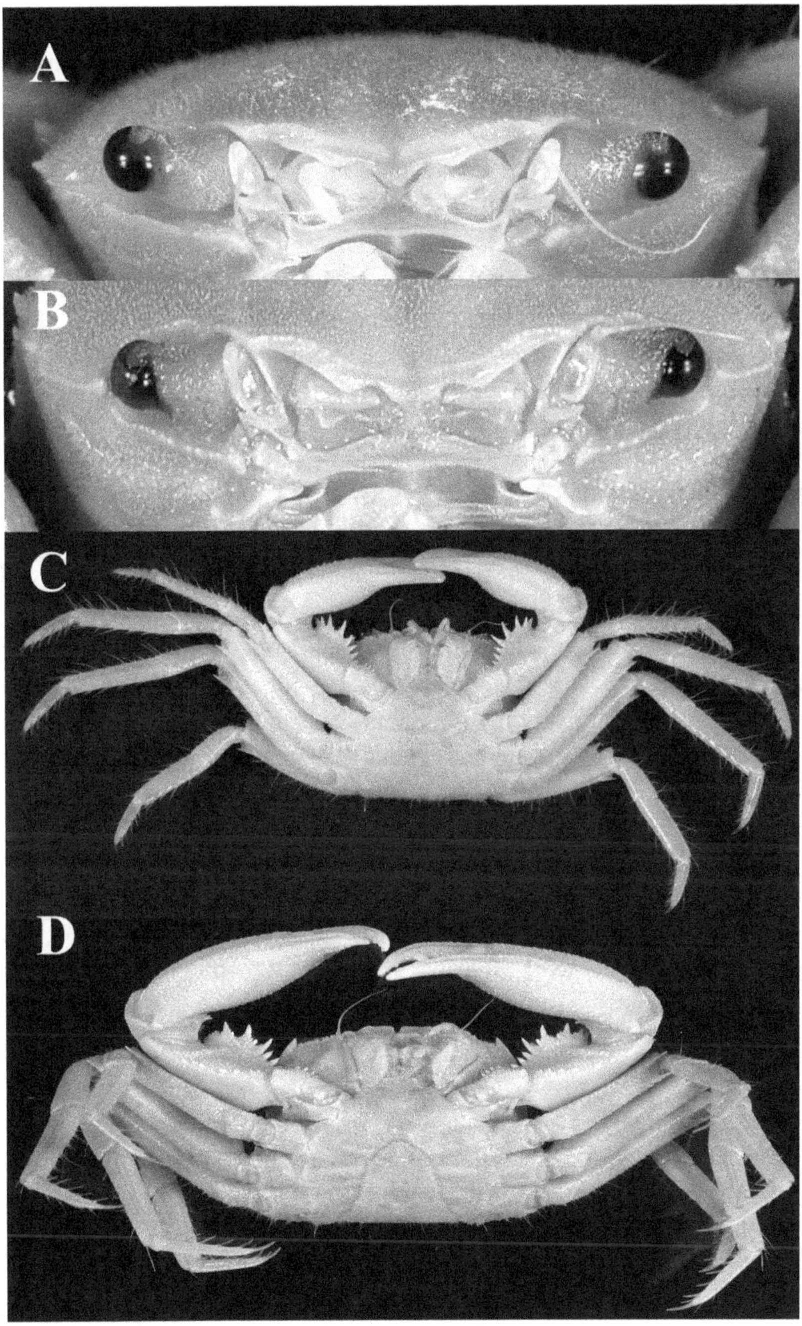

Fig. 3. A, C, *Euchirograpsus danielae* n. sp., holotype male, 25.3 × 24.0 mm (LACM); B, D, *Euchirograpsus guinotae* n. sp., holotype male, 37.8 × 37.8 mm (ZRC 2009.0845). A, B, frontal view; C, D, ventral view.

hiatus. Basal antennal segment covered with small granules, immobile, laterally locked against inner orbital tooth; antennal flagellum short. Antennules broad, folded up transversely, slightly oblique. Eyes well developed, cornea pigmented.

Third maxillipeds (fig. 4B) with merus quadrate in shape, inner margins meeting medially, not completely closing buccal cavern, merus broad (length to width ratio 1.2), ischium elongated, rectangular in shape (length to width ratio 1.8), without an oblique setose crest; exopod narrow, slender, with flagellum; palp articulating near antero-external angle of merus.

Epistome (fig. 4C) narrow, anterior margin strongly sinuous, posterior margin with 2 gentle lateral lobes, median part broadly triangular, margin with rounded granules.

Male chelipeds long, robust, granular. Merus long, slender (length to width ratio ca. 2.1), both anterior, posterior margins lined with sharp, distinct granules, anterior margin with 2 distinct sub-distal teeth (fig. 3D). Carpus (fig. 4E) slightly elongated, dorsal surface granular, ventral margin rounded, without spine (length to width ratio 1.2). Chelae proportionally long (fig. 4F-H) (ratio of chela length to cheliped length 2.1), dorsal surface with 3 rows of distinct, rounded granules; posterior region of chela with row of granules extending from proximal margin of chela to tip of immobile finger; posterior margin densely granular; movable finger slender, 0.6 times as long as chelae, dorsal surface with 3 rows of distinct, rounded granules; inner surface of immobile finger covered with granules. Chelae of female unknown. Ambulatory legs with all articles long, slender, flattened; P4 longest. All meri with broad, thick subdistal spine on each dorsal margin; both dorsal, ventral margins serrated, ventral margin interspersed with long, stiff setae; all carpi elongated, both margins serrated, dorsal margin interspersed with long, stiff setae, dorsal surfaces with 3 ridges; all propodi with dorsal, ventral margins serrated, interspersed with long, stiff setae, slightly depressed along median length; dactyli short, broad, each ventral margin with six pairs of short, corneous spines. P2 with merus slender (length to width ratio 2.9), propodus more slender (length to width ratio 3.9); dactylus long and curveous (length to width ratio 5.5); P3 with merus even more slender than first (length to width ratio 3.6), propodus longer (length to width ratio 5.3), proportionally long dactylus (length to width ratio 6.2); P4 (fig. 5A) with merus less slender than in P3 (length to width ratio 3.1), propodus broad, short (length to width ratio 2.9), dactylus long, slender (length to width ratio 6.9); P5 with merus (fig. 5B) long, slender (length to width ratio 2.8); propodus longer than in P4 (length to width ratio 4.3); dactylus shorter than in P4 (length to width ratio 6.2).

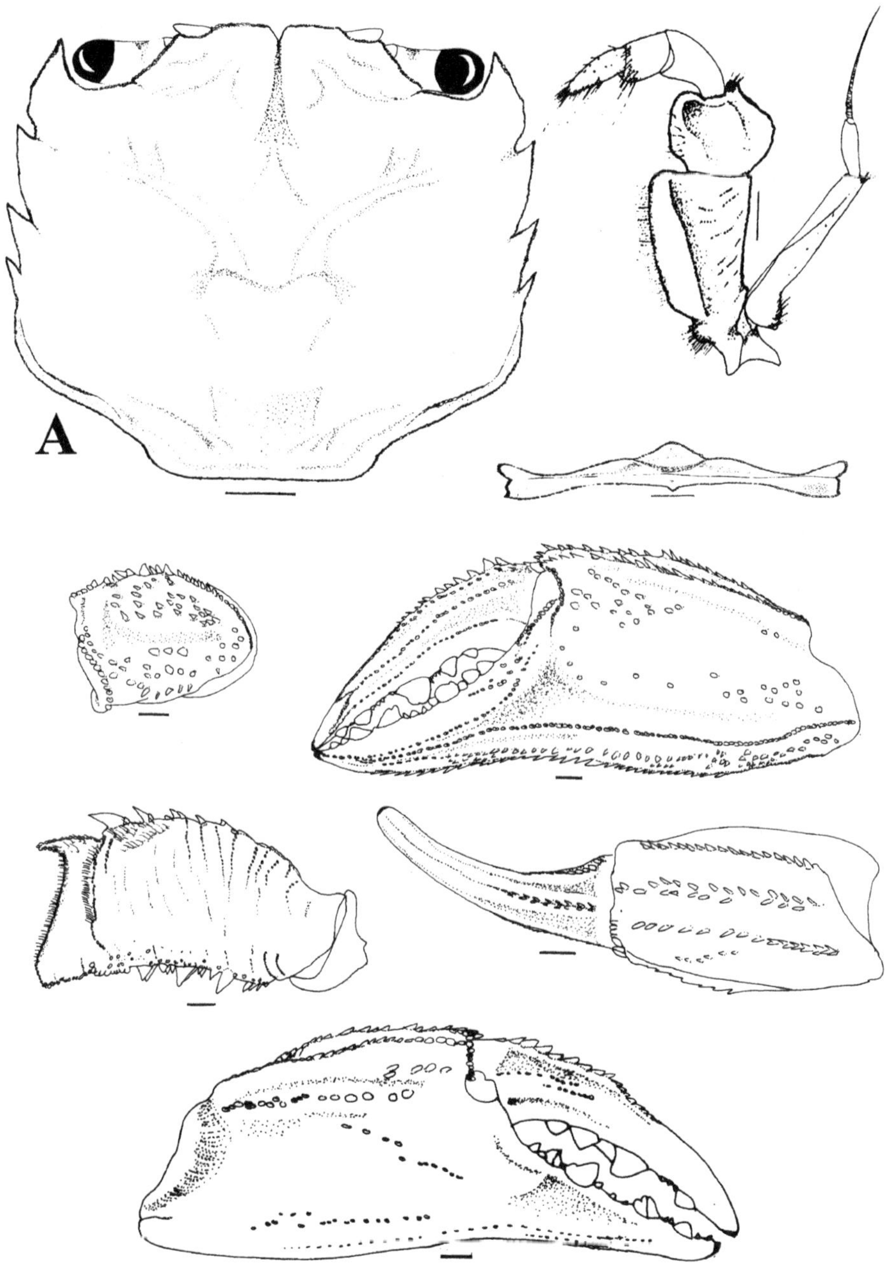

Fig. 4. *Euchirograpsus danielae* n. sp., paratype male, 29.6 × 27.8 mm (ZRC 2009.0844). A, carapace; B, left third maxilliped; C, epistome; D, dorsal view of merus of left cheliped; E, dorsal view of carpus of left cheliped; F, dorsal view of chela of left cheliped; G, anterior view of chela of left cheliped; H, inner view of chela of left cheliped. Scale bars: A, 2 mm; B-H, 1 mm.

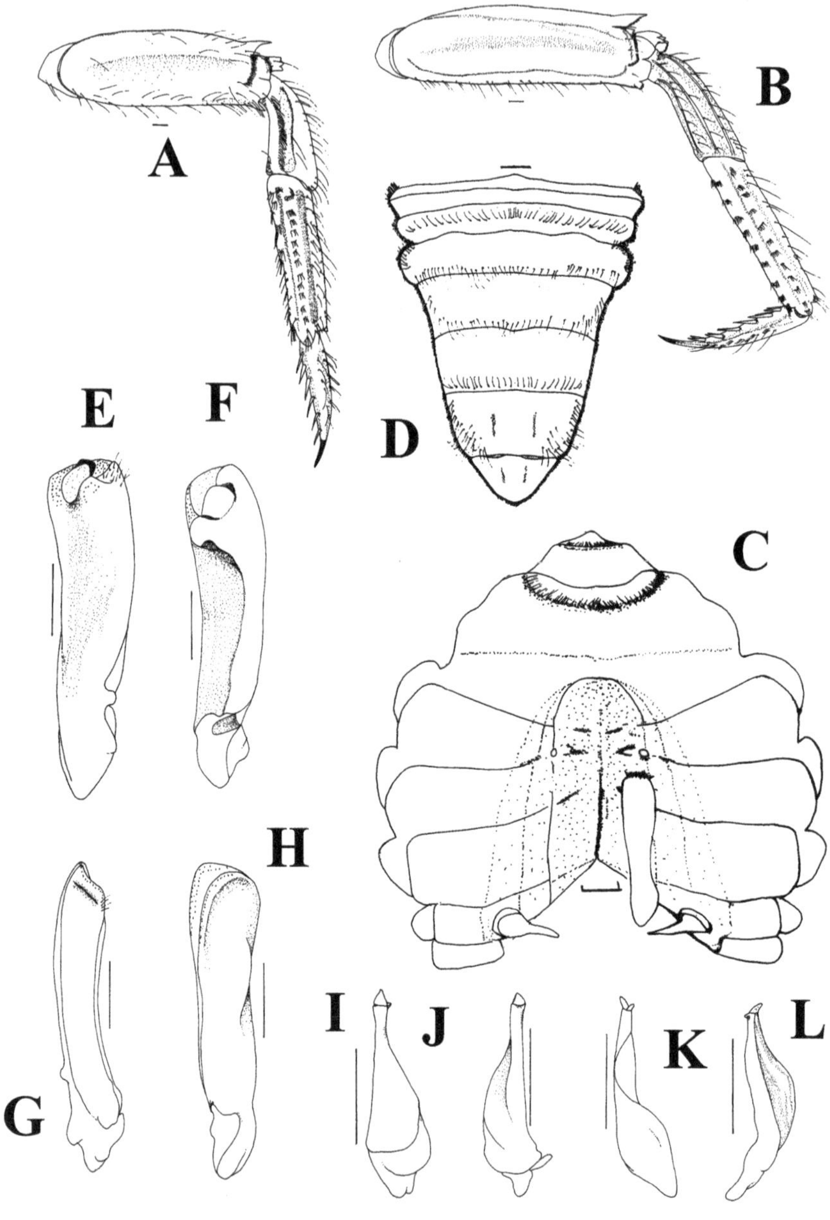

Fig. 5. *Euchirograpsus danielae* n. sp., paratype male, 29.6 × 27.8 mm (ZRC 2009.0844). A, right P5; B, right P4; C, sternite; D, male abdomen; E, dorsal view of G1; F, right view of G1; G, ventral view of G1; H, left view of G1; I, left view of G2; J, ventral view of G2; K, dorsal view of G2; L, right view of G2. Scale bars: A-L, 1 mm.

Male abdomen broadly triangular (figs. 3C, 5D), with somites 3-6 fused, but sutures are still evident, entirely covering sternum between last pair of legs, locking mechanism present, surface setose, interspersed with long stiff setae; lateral margins covered with short, dense setae; first abdominal somite weakly arched, with transverse ridge; second abdominal somite narrow, short; third abdominal somite broadest (width to length ratio 4.3) compared to fourth, fifth, and sixth abdominal, lateral margins swollen; telson broadly triangular (length to width ratio 1.7), lateral margins weakly concave, distal margin rounded.

Male thoracic sternum (figs. 3C, 5C) with lateral margins of first 3 thoracic sternites finely granulated; suture between sternites 2 and 3 slightly concave; lateral margins of sternites 3 and 4 sinuous, with deep, broad notch, demarcating edge of suture; with low granulated ridge parallel to lateral margins of sternites 3 and 4; median groove between sternites 5 and 6 broad, shallow, with distinct rounded locking button between the margins of sternites 5 and 6; deep narrow median longitudinal groove between sternites 7 and 8. Thoracic sternum of female unknown. Penis emerging at outer edge of thoracic sternite 8.

G1 broad, short (length to width ratio 3.0), gently curving outwards, reaching slightly beyond posterior margin of sternite 6; anterior end of G1 densely setose, gently bi-lobed, outer lobe with short, broad corneous tip, inner lobe with distinct, rounded tip; genital opening lateral to base of lobes (fig. 5E-H). G2 short (0.33 length of G1), small, narrow with broad base; tip distinctly bi-lobed with 2 equal-sized lobes (fig. 5I-L).

Female unknown.

Etymology. — The new species is named in honour of Professor Danièle Guinot, Museum national d'Histoire naturelle, Paris, in recognition of her enormous contributions to the science of carcinology, and for being a wonderful mentor to junior carcinologists all over the world. The name consequently is a noun in the genitive singular.

Colour. — Fresh colour of the carapace is mottled bright orange on light orange background; chelipeds and ambulatory legs with bright orange and light orange-coloured bands, seen as distinct bands in fig. 2. Preserved specimens are dark cream throughout.

Remarks. — The new species is most similar to *E. pacificus* (see Türkay, 1975: 119, figs. 8-9) in the form of the carapace, but it can be easily separated from *E. pacificus* by: (a) a stouter exopod of the third maxilliped, length to width ratio 6.2 (more slender third maxilliped exopod, length to width ratio 7.3, in *E. pacificus*); (b) stouter cheliped merus, length to width ratio 2.0 (more slender merus, length to width ratio 2.2, in *E. pacificus*); (c) propodus of the P2 is more slender, length to width ratio 2.9 (stouter propodus of the P2, length to width ratio 3.4, in *E. pacificus*); (d) dactylus of the first leg is stouter, length to width ratio 5.5 (more slender dactylus, length to width ratio 6.0, in *E. pacificus*); (e) dactylus of the P3 is more slender, length to width ratio 6.2 (stouter dactylus, length to width ratio 5.1, in *E. pacificus*); (f) relatively short and stout propodus of the P4, length to width ratio 3.0 (longer and more slender propodus, length to width ratio 5.6, in *E. pacificus*); (g) a longer and more slender dactylus of the P4, length to width ratio 6.9 (stouter dactylus, length to width ratio 5.3, in *E. pacificus*); (h) longer and more slender dactylus of the P5, length to width ratio 6.2 (stouter dactylus, length to width ratio 5.0, in *E. pacificus*); and (i) the tip of G1 is almost rectangular in shape (triangular tip in *E. pacificus* cf. Türkay, 1975: 119, fig. 18).

Euchirograpsus danielae is also similar to *E. timorensis* (see Türkay, 1975: 117, figs. 10, 11) in the form of the chelipeds, but it can be easily distinguished from the latter by: (a) the antennular fossae cleft on each lobe of the frontal margin is about half the distance from the middle of the frontal margin (two-thirds from the middle of the frontal margin in *E. timorensis*); (b) the teeth on the anterolateral margin are acute (obtuse in *E. timorensis*); (c) spines on the dorso-anterior margin of the cheliped merus are more acute and longer (broader and shorter spines in *E. timorensis*); (d) dactylus of the P2 is more slender, length to width ratio 5.5 (stouter dactylus, length to width ratio 4.3, in *E. timorensis*); (e) dactylus of the P3 is more slender, length to width ratio 6.2 (stouter dactylus, length to width ratio 5.8, in *E. timorensis*); (f) short and stout propodus of the P4, length to width ratio 3.0 (long and slender propodus, length to width ratio 5.3, in *E. timorensis*); (g) more slender dactylus of the P4, length to width ratio 6.9 (stouter dactylus, length to width ratio 6.2, in *E. timorensis*); (h) third abdominal somite is broader and stouter, width to length ratio 4.3 (more slender and longer, width to length ratio 4.8, in *E. timorensis*); (i) the telson is distinctly narrower, length to width ratio 1.7 (broader telson, length to width ratio 2.1, in *E. timorensis*); (j) tip of the G1 is short and broad, almost rectangular in shape (bi-lobed, narrowly triangular terminal tip in *E. timorensis*); and (k) the presence of two equally-sized lobes on the tip of the G2 (one small and one big lobe on the tip of the G2 in *E. timorensis* cf. Türkay, 1975, fig. 21).

Euchirograpsus danielae is also similar to *E. tuerkayi* (cf. Crosnier, 2001: 791, figs. 3B, 4B-7) in the form of the carapace but can be easily distinguished from the latter by the following characters: (a) more acute and longer spines on the dorso-anterior margin of the cheliped merus of the cheliped (broader and shorter spines in *E. tuerkayi*); (b) slender exopod of the third maxilliped, length to width ratio 6.2 (stouter exopod, length to width ratio 5.2, in *E. tuerkayi*); (c) dactylus of the P3 is proportionately more slender, length to width ratio 6.2 (stouter dactylus, length to width ratio 5.7, in *E. tuerkayi*); (d) proportionately short and stout propodus of the P4, length to width ratio 2.9 (long and slender propodus, length to width ratio 4.2, in *E. tuerkayi*); (e) stout merus of the P5, length to width ratio 2.8 (more slender merus, length to width ratio 3.4, in *E. tuerkayi*); (f) slender and long dactylus of the P5, length to width ratio 6.2 (stouter and shorter dactylus, length to width ratio 5.0, in *E. tuerkayi*); (g) short, broad, almost rectangular tip of the G1 (short, narrowly triangular in *E. tuerkayi*); and (h) two equal lobes on the tip of G2 (two dissimilar lobes in *E. tuerkayi* cf. Crosnier, 2001, fig. 6).

Euchirograpsus guinotae n. sp. (figs. 1B, 3B, D, 6, 7)

Material examined. — Holotype: 1 male (37.8 × 37.8 mm) (NMCR), Balicasaq Island, Panglao Islands, Bohol, Philippines, purchased from local fishermen, leg. P. K. L. Ng. 25-30 Jul. 2003. Paratype: 1 male (33.9 × 32.0 mm) (ZRC 2009.0845), Balicasaq Island, Panglao Islands, Bohol, Philippines, purchased from local fishermen, leg. P. K. L. Ng. 25-30 Jul. 2003.

Description. — Carapace distinctly quadrate (figs. 1B, 3B) as wide as long, dorsal surface covered with minute round granules and short stiff setae, regions relatively well defined; cervical grooves indistinct, shallow, broad; gastric grooves distinct; epigastric cristae absent, postorbital cristae absent; low ridge present at region adjacent to junction between posterolateral margin and posterior margin. Anterolateral margins with 4 anterolateral teeth including external orbital tooth, decreasing in size posteriorly; lined with small, rounded granules; not clearly demarcated from posterolateral margins; external orbital tooth broad, separated from second tooth by broad U-shaped cleft; third anterolateral tooth separated from second tooth by a broad U-shaped cleft; fourth tooth narrowest, separated from the third tooth by narrow V-shaped notch; posterolateral margins gently convex before converging towards almost straight posterior margin. Frontal margin about 0.5 times carapace width, with 2 distinct lobes separated by deep U-shaped notch; each lobe is further divided into 2 smaller lobes by antennular fossae cleft. Orbit deeply sunken cavity, almost complete. Lower margin of external orbital tooth extending medially as ridge to form new lower orbital margin, separated from large, broad, rounded infra-orbital tooth by narrow fissure; suborbital crest extending laterally beyond orbit as slightly sinuous granular row, but not appearing to be stridulatory. Supraorbital margin granular, strongly sinuous, sub-parallel with frontal margin; frontal margin (figs. 1B, 3B, 6A) sinuous, strongly deflexed, with broad

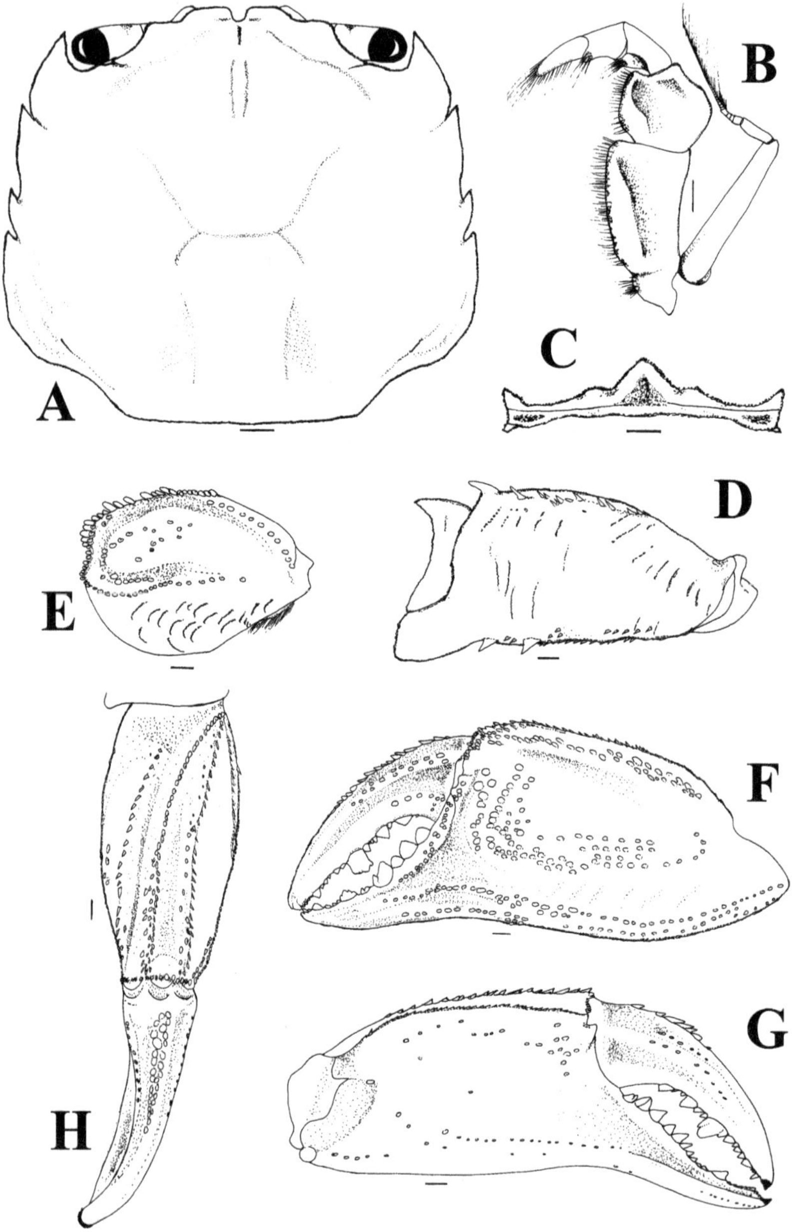

Fig. 6. *Euchirograpsus guinotae* n. sp., holotype male, 37.8 × 37.8 mm (ZRC 2009.0845). A, carapace; B, left third maxilliped; C, epistome; D, dorsal view of merus of left cheliped; E, dorsal view of carpus of left cheliped; F, dorsal view of chela of left cheliped; G, inner view of chela of left cheliped; H, anterior view of chela of left cheliped. Scale bars: A, 2 mm; B-H, 1 mm.

shallow sinus between 2 lobes, broad frontal median triangle present, formed by 3 granuliform ridges; infra-orbital margins granular, complete, congruent with anterolateral margin; distinct row of rounded granules below infra-orbital margin on suborbital region, extending to sub-branchial region (fig. 3B). Antennal segments entering orbital hiatus. Basal antennal segment covered with small granules, immobile, laterally locked against inner orbital tooth; antennal flagellum short. Antennules broad, folded up transversely, slightly oblique. Eyes well developed, cornea pigmented.

Third maxillipeds (fig. 6B) with merus quadrate in shape, inner margins meeting medially, not completely closing buccal cavern, merus broad (length to width ratio 1.1), ischium elongated, rectangular in shape (length to width ratio 2.0), without an oblique setose crest; exopod narrow, slender, with flagellum; palp articulating near antero-external angle of merus.

Epistome (fig. 6C) narrow, anterior margin strongly sinuous, posterior margin with 2 gentle lateral lobes, median part broadly triangular, margin with rounded granules.

Male chelipeds long, robust, all surfaces covered with granules. Merus long, slender (length to width ratio 2.0), both anterior, posterior margins lined with sharp, distinct granules, anterior margin with 2 distinct subdistal teeth (fig. 6D). Carpus (fig. 6E) slightly elongated, dorsal surface covered with granules, ventral margin rounded, without spine (length to width ratio 1.3). Chelae relatively long (ratio of chela length to cheliped length 2.3), dorsal surface with 3 rows of distinct, rounded granules; posterior region of chela with row of granules extending from proximal margin of chela to tip of immobile finger (fig. 6F, G); posterior margin densely granular; movable finger slender, 0.5 times as long as chelae, dorsal surface with 3 rows of distinct, rounded granules; inner surface of immobile finger covered with granules (fig. 6H).

Ambulatory legs with all different articles long, slender, flattened; P4 longest. All meri with broad, thick sub-distal spine on each dorsal margin; both dorsal, ventral margins serrated, ventral margins interspersed with long, stiff setae; all carpi elongated, both margins serrated, dorsal margin interspersed with long, stiff setae, each dorsal surface with 3 ridges; all propodi with dorsal, ventral margins serrated, interspersed with long, stiff setae, slightly depressed along median length; dactyli short, broad, ventral margins with 6 pairs of short, corneous spines. P2 with merus slender (length to width ratio 3.5), propodus long, slender (length to width ratio 4.2), dactylus long, curved (length to width ratio 5.6); P3 with merus slightly longer, more slender than first (length to width ratio 3.6), propodus more slender than P2 (length to width ratio 5.1), dactylus more slender than first (length to width ratio 6.7); third leg (fig. 7A) with merus stouter then second (length to width ratio 3.4), propodus slightly longer, more slender than second (length to width ratio 5.3); dactylus long, slender, but not as long, slender as P3 (length to width ratio 6.3); P5 (figs. 1B, 7B) with merus stouter, shorter than third (length to width ratio 3.1); propodus proportionally long (length to width ratio 4.2); dactylus shorter and broader than P4 (length to width ratio 5.8).

Male abdomen broadly triangular (figs. 3D, 7D); somites 3-6 fused, but sutures may still be evident, entirely covering sternum between last pair of legs, locking mechanism present, surface setose, interspersed with long stiff setae; lateral margins covered with short, dense setae; first abdominal somite weakly arched, with transversal ridge; second somite narrow, short; third somite long, narrow (width to length ratio 5.4), compared to fourth, fifth, sixth abdominal somites, lateral margins swollen; telson broadly triangular (length to width ratio 2.1), lateral margins weakly concave, distal margin rounded.

Male thoracic sternum (figs. 3D, 7C) with lateral margins of first 3 thoracic sternites finely granular; suture between sternites 2, 3 slightly concave; lateral margins of sternites 3, 4 sinuous, with deep, broad notch, demarcating edge of suture; with low granular ridge parallel to lateral margins of sternites 3, 4; median groove between sternites 5, 6 broad, shallow, with distinct rounded locking button between margins of sternites 5, 6; deep, narrow median longitudinal groove between sternites 7, 8. Penis emerging at outer edge of thoracic sternite 8.

G1 broad, short (length to width ratio 3.2), gently curving outwards, reaching slightly beyond posterior margin of sternite 6; anterior end of G1 densely setose, slightly bi-lobed, outer lobe with short, narrow corneous tip, inner with distinct rounded tip; opening lateral to base of lobes (fig. 7E-H). G2 short (0.33 length of G1), small, narrow with broad base; tip distinctly bi-lobed with two dissimilar lobes (fig. 7I-L).

Female unknown.

Etymology. — This species is named in honour of Professor Danièle Guinot, Museum national d'Histoire naturelle, Paris, using her family name as a noun in the genitive singular. Professor Guinot is an inspiration to many young carcinologists all over the world.

Colour. — Live colour not known. All preserved specimens dark cream throughout.

Remarks. — The new species, *Euchirograpsus guinotae*, is closely similar to *E. timorensis* (see Türkay, 1975: 117, figs. 10, 11) but can be separated by

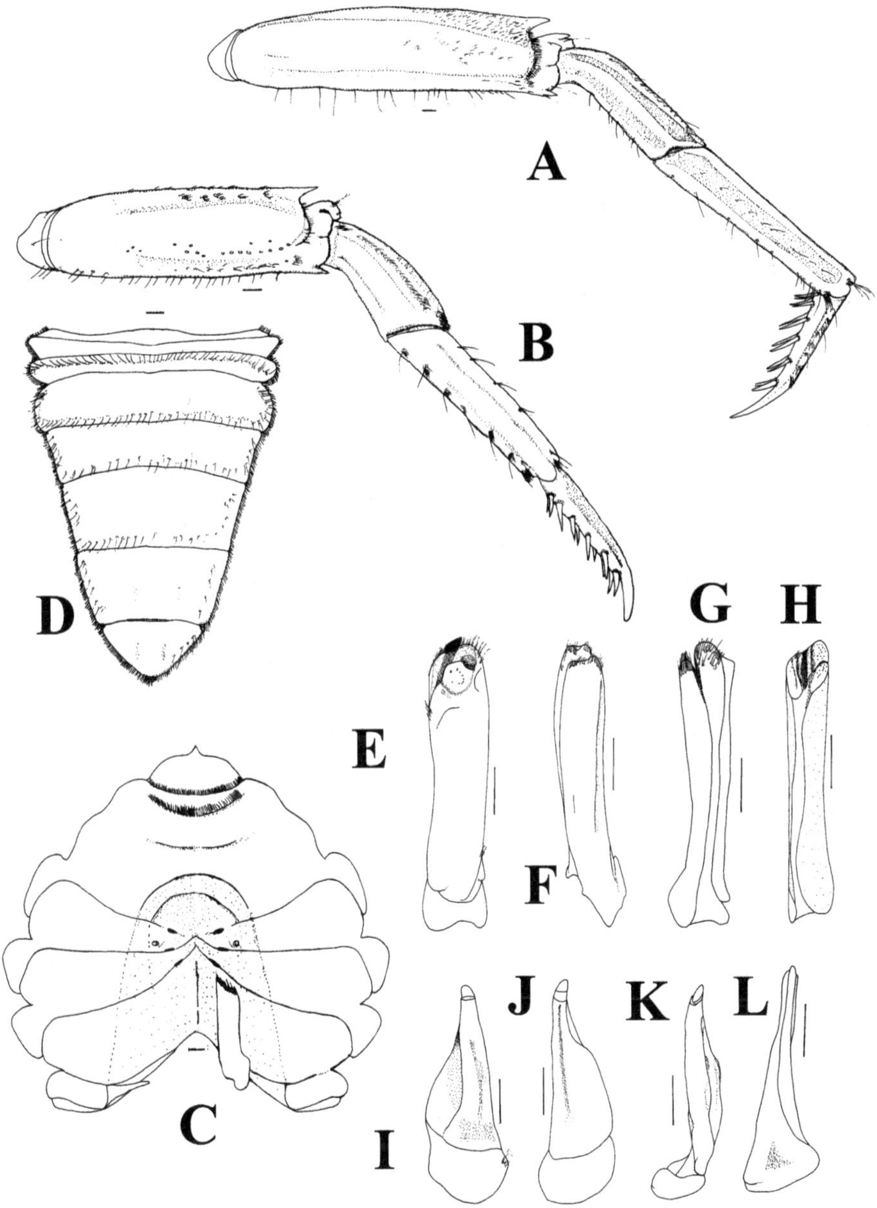

Fig. 7. *Euchirograpsus guinotae* n. sp., holotype male, 37.8 × 37.8 mm (ZRC 2009.0845). A, right P4; B, right P5; C, male abdomen; D, sternite; E, dorsal view of G1; F, right view of G1; G, ventral view of G1; H, left view of G1; I, left view of G2; J, ventral view of G2; K, dorsal view of G2; L, right view of G2. Scale bars: A-L, 1 mm.

the following characters: (a) orbit is narrower, orbit to carapace ratio 0.2 (wider orbit, orbit to carapace ratio 0.3, in *E. timorensis*); (b) gap between the first and second of the anterolateral teeth is narrow (broad in *E. timorensis*); (c) narrow third maxilliped exopod, length to width ratio 6.4 (broader exopod, length to width ratio 5.2, in *E. timorensis*); (d) merus of the P2 is more slender, length to width ratio 3.5 (broader merus, length to width ratio 2.6, in *E. timorensis*); (e) dactylus of the P2 is longer and more slender, length to width ratio 5.6 (broader and shorter dactylus, length to width ratio 4.3, in *E. timorensis*); (f) merus of the P3 is slightly more slender, length to width ratio 3.6 (broader merus, length to width ratio 2.8, in *E. timorensis*); (g) propodus of the P3 is longer and more slender, length to width ratio 5.1 (broader propodus, length to width ratio 4.6, in *E. timorensis*); (h) dactylus of the P3 is long and slender, length to width ratio 6.7 (much stouter and broader dactylus, length to width ratio 5.8, in *E. timorensis*); (i) merus of the P5 is proportionately more slender and narrow, length to width ratio 3.1 (broad merus, length to width ratio 2.6, in *E. timorensis*); (j) the propodus of the P5 is more slender, length to width ratio 4.2 (slightly stouter propodus, length to width ratio 3.9, in *E. timorensis*); (k) the dactylus of the P5 is proportionately more slender, length to width ratio 5.8 (slightly stouter dactylus, length to width ratio 5.3, in *E. timorensis*); (l) G1 is relatively short and stout, length to width ratio 3.2 (much longer and slender G1, length to width ratio 5.6, in *E. timorensis*); and (m) short, narrow rectangular corneous tip on the tip of the G1 (high, narrow triangular corneous tip on the G1 in *E. timorensis*, cf. Türkay, 1975, fig. 21).

Euchirograpsus guinotae is also closely similar to *E. tuerkayi* (cf. Crosnier, 2001: 791, figs. 3B, 4B-7) but can be easily distinguished from *E. tuerkayi* by: (a) exopod of the third maxilliped is long and slender, length to width ratio 6.4 (stouter exopod, length to width ratio 5.2, in *E. tuerkayi*); (b) propodus of the P3 is longer and more slender, length to width ratio 5.1 (broader propodus, length to width ratio 4.0, in *E. tuerkayi*); (c) dactylus of the P3 is long and slender, length to width ratio 6.7 (much stouter and broader dactylus, length to width ratio 5.7, in *E. tuerkayi*); (d) propodus of the P4 is more sender, length to width ratio 5.3 (stouter propodus, length to width ratio 4.2, in *E. tuerkayi*); (e) dactylus of the P4 is slightly stouter and broader, length to width ratio 6.3 (slightly more slender dactylus, length to width ratio 6.7, in *E. tuerkayi*); (f) dactylus of the P5 is more slender, length to width ratio 5.8 (broader and stouter dactylus, length to width ratio 5.0, in *E. tuerkayi*); (g) slightly stouter and shorter G1, length to width ratio 3.2 (longer and more slender G1, length to width ratio 4.0, in *E. tuerkayi*); and (h) short, narrow rectangular corneous tip of the G1 (short triangular tip of the G1 in *E. tuerkayi* cf. Crosnier, 2001, fig. 5A-D).

Comparative material examined. — *Euchirograpsus liguricus* H. Milne Edwards, 1853: 7 males (8.8-20.4 × 7.9-17.8 mm), 4 females (8.3-10.9 × 7.4-9.2 mm) (SNG-4741), off Morocco coast (33°19′N 9°W), stn 8-13a, 120-180 m, F. S. "Meteor", 23 Jan. 1967; 1 female (23.5 × 20.8 mm) (SNG), off Morocco (31°35′N 10°10.5′W), stn 9c-82a, 45-180 m, F. S. "Meteor", 18-19 Jun. 1967; 1 female (ovig.) (MNHN-B12588), **Maroe** coast, coll. "De Belloc", 1935; 1 female (15.2 × 14.3 mm) (MNHN-B12587), Pte. Parga, Gabon, 70 m, 25 Jul. 1960; 1 female (13.4 × 12.0 mm) (MNHN-B12489), stn 64-219, "President Theodore Tissen", 29 May 1955; 1 female (9.5 × 8.8 mm) (MNHN-B12591), 1 female (13.6 × 12.1 mm) (MNHN-B12592), **Le Chavaillenp**, 1982.

Euchirograpsus americanus H. Milne Edwards, 1880: Paratype male (10.8 × 10.5 mm) (MNHN-B12606); 1 male (13.4 × 12.7 mm), 1 female (ovig.) (16.4 × 15.3 mm) (SNG), Florida, U.S.A., leg. W. L. Schmitt, 4 Aug. 1931; 1 female (ovig.) (13.5 × 13.9 mm) (USNM 155628), Bahamas, Straits of Florida, 29 Mar. 1975; 1 male (11.5 × 11.2 mm), 1 female (ovig.) (9.6 × 9.4 mm) (USNM 232638), off South Carolina, U.S.A., 9 Mar. 1981; 1 male (8.7 × 8.1 mm), 1 female (6.3 × 5.8 mm) (USNM 202806), off North Carolina, 99 m, 14 May 1981; 1 female (10.0 × 6.4 mm) (USNM 276579), off Florida, Gulf of Mexico, 10 Dec. 1983; 1 female (18 × 12.4 mm) (USNM 276578), same data; 1 male (12.4 × 12.2 mm) (USNM 98673), U.S. Fish & Wildlife, "Oregon", stn 1343, Gulf of Mexico, 22°89′N 79°17′W, Jul. 1955; 2 females (13.9-14.1 × 13.4-13.6 mm) (USNM 202808), 33°48′18″N 76°34′18″W, 100 m, off North Carolina, U.S.A., coll. Duke University, 11 Nov. 1981; 1 female (2.0 × 2.0 mm) (USNM 202802), 33°32′18″N 76°25′89″W, 33 m, off North Carolina, U.S.A., coll. Duke University, 5 May 1981; 1 female (8.4 × 7.9 mm) (USNM 242728), Florida, U.S.A., Gulf of Mexico; 1 male (7.2 × 6.9 mm) (USNM 242730), 26°16′44″N 83°42′49″W, off Florida, Gulf of Mexico, coll. 30 Apr. 1989; 1 female (6.3 × 6.2 mm) (USNM 242729), same data; 3 females (7.0-8.7 × 7.0-8.6 mm) (USNM 242726), 26°16′43″N 83°46′49″W, off Florida, Gulf of Mexico, 4 Nov. 1980; 1 female (ovig.) (USNM 273305), same data; 1 female (6.5 × 6.4 mm) (USNM 242727), same data; 1 male (6.0 × 6.2 mm) (USNM 242633), 26°16′40″N 84°04′05″W, 137 m, off Florida, Gulf of Mexico, 25 Jul. 1981; 2 females (ovig.) (USNM 24263), same data; 1 male (10.9 × 10.5 mm) (USNM 242631), same data; 1 male (13.3 × 12.4 mm) (USNM 267396), 29°05′06″N 80°08′24″W, 76 m, off North Carolina, 17 Nov. 1968; 1 female (USNM 242627), 25°16′50″N 83°52′21″W, 127 m, off Florida, Gulf of Mexico, 2 Aug. 1981; 2 males (9.0-9.8 × 8.4-9.5 mm) (USNM 273477), 25°16′53″N 83°37′47″W, 70 m, off Florida, Gulf of Mexico, 1 Feb. 1981; 5 males (6.7-11.0 × 6.6-10.2 mm), 1 female (9.4 × 8.9 mm) (USNM 232639), 31°3′48″N 79°44′36″W, depth 56 m, off coast of South Carolina, North Atlantic Ocean, U.S.A., 4 May 1981; 1 female (8.3 × 7.4 mm) (USNM 276573), 25°16′55″N 83°37′47″W, depth 74 m, off Florida, Gulf of Mexico, North Atlantic Ocean, U.S.A., 5 Mar. 1984; 1 male (9.2 × 9.0 mm) (USNM 276575), same data; 1 female (10.0 × 9.5 mm) (USNM 276577), same data; 1 female (8.0 × 7.6 mm) (USNM 276576), same data; 1 male (6.5 × 4.9 mm) (USNM 276574), same data.

Euchirograpsus madagascariensis Türkay, 1978: Holotype male (10.3 × 10.3 mm) (MNHN-MPB-12607), dredging 71, Fort Dauphin, Madagascar (25°13.1′S 47°17.8′E), 105-115 m, coll. A. Crosnier, 3 Mar. 1973. Paratypes: 2 females (8.7-10.8 × 8.8-10.3 mm) (MNHN-B12607) same data as holotype; 1 male (8.4 × 8.0 mm), 1 female (10.0 × 9.4 mm) (MNHN-B12607), same data as holotype; 1 male (11.5 × 11.0 mm), 1 female (8.1 × 8.0 mm) (MNHN-B12609), same data as holotype; 1 male (9.4 × 9.4 mm), 1 female (10.0 × 9.7 mm) (SNG 7794), same data as holotype. Other material: 1 female (ovig.) (MNHN-B10542), La Réunion; 1 female (18.9 × 17.0 mm) (MNHN-B10540), La Réunion, stn DC 176, 195 m, 8 Sep. 1982.

Euchirograpsus timorensis Türkay, 1975: 1 male (18.1 × 17.1 mm), 3 females (12.8-23.5 × 12.7-22.5 mm) (USNM 308983), New Caledonia, 18 Oct. 1992; 2 males (15.4-17.2 × 15.0-16.7 mm), 3 females (18.8-23.3 × 18.5-22.2 mm) (USNM 308984), New Caledonia, 29 Jan. 1993; 1 male (26.0 × 23.8 mm) (MNHN-B12597), off Dahomey, 35 m, coll. A. Crosnier, Oct. 1963; 1 male (8.8×8.7 mm), 1 female (15.5×14.2 mm) (MNHN-B6825), stn CCB102 (22°48′S 167°03′E), 360 m, ORSTROM, 13 Apr. 1978; 1 male (13.1 × 12.5 mm) (MNHN-B10541), La Réunion, stn CP57, 225-227 m.

ACKNOWLEDGEMENTS

The first author thanks Lady McNeice Yuen Ping for her extremely generous financial support, which allowed research visits to European museums. The French Merlion-Singapore Fellowship and research grant R-154-00-044-113 from the National University of Singapore are also acknowledged for financial support.

This work was partially supported by grant DEB 0531616 from the National Science Foundation's "Assembling the Tree of Life" program to the second author. Special thanks to P. K. L. Ng (National University of Singapore), his staff and students for hosting a visit of the second author in October of 2008. The 2006 expedition to French Frigate Shoals was supported by the Census of Marine Life's Census of Coral Reefs (CReefs) project, National Oceanic and Atmospheric Administration, National Science Foundation and fifteen institutions, government agencies and organizations. Fig. 2 is also copyrighted by J. W. Martin. Special thanks to our fellow scientists and colleagues who made the expedition possible and who collected many of the crabs, and to the officers and crew of the NOAA vessel "Oscar Elton Sette."

We are also indebted to D. Guinot and R. Cleva (MNHN), M. Türkay (SMF), R. Lemaitre and C. Walter (USNM), and Tan Swee-Hee and Tan Xiong-Kiat (Raffles Museum of Biodiversity Research, National University of Singapore) for the loans of comparative material.

LITERATURE CITED

CROSNIER, A., 2001. Grapsidae (Crustacea, Decapoda, Brachyura) d'eau profonde du Pacifique sud-ouest. Zoosystema, **23** (4): 783-796.

DANA, J. D., 1851. Conspectus Crustaceorum quæin Orbis Terrarum circumnavigatione, Carolo Wilkes e Classe ReipublicæFæderatæDuce, Lexit et descripsit J. D. Dana. Pars VIII. Proceedings of the Academy of Natural Sciences of Philadelphia, **5** (10): 247-254.

KINGSLEY, J. S., 1880. Carcinological notes. No. IV: Synopsis of the Grapsidae. Proceedings of the Academy of Natural Sciences of Philadelphia, **1880**: 187-224.

MIERS, E. J., 1885. The Brachyura. In: C. WYVILLE THOMSON & J. MURRAY (eds.), Narrative. Report on the scientific results of the voyage of H.M.S. Challenger during the years 1873-76 under the command of Captain George, S. Nares, R.N., F.R.S. and the late Captain Frank Tourle Thomson, R.N., **1** (2): 585-592.

MILNE-EDWARDS, A., 1880. Reports on the results of dredging under the supervision of Alexander Agassiz, in the Gulf of Mexico, and in the Caribbean Sea, 1877, '78, '79, by the U.S. coast survey steamer Blake, Lieut.-Commander, C. D. Sigsbee, U.S.N., and Commander, J. R. Bartlett, U.S.N., commanding. VIII: Études preliminaries sur les Crustacés. Bulletin of the Museum of Comparative Zoology, Harvard, **8**: 1-68, pls. 1, 2.

MILNE EDWARDS, H., 1853. Mémoire sur la famille des Ocypodiens. Annales de Sciences naturelles, (3) **20** (4): 163-226, pls. 6-11.

NG, P. K. L., D. GUINOT & P. J. F. DAVIE, 2008. Systema Brachyurorum: Part 1. An annotated checklist of extant brachyuran crabs of the world. Raffles Bulletin of Zoology, (Supplement) **17**: 1-286.

STEBBING, T. R. R., 1921. South African Crustacea. (Part XI of South African Crustacea, for the Marine Investigations in South Africa). Annals of the South African Museum, **18** (4): 453-468, pls. 13-20.

TÜRKAY, M., 1975. Zur Kenntnis der Gattung *Euchirograpsus* mit Bemerkungen zu *Brachygrapsus* und *Litocheira* (Crustacea: Decapoda). Senckenbergiana Biologica, **56** (1/3): 103-132.

— —, 1978. Zwei neue Grapsidaen-Arten aus dem Indopazifik (*Euchirograpsus madagascariensis* n. sp. und *Miersiograpsus australiensis* n. sp.) mit Einfürung von *Miersiograpsus* n. gen. (Crustacea: Decapoda: Grapsidae). Senckenbergiana Biologica, **59** (1/2): 133-141.

First received 2 July 2009.
Final version accepted 19 August 2009.

A NEW GENUS FOR *CLEISTOSTOMA EDWARDSII* MACLEAY, 1838 (OCYPODOIDEA, CAMPTANDRIIDAE) FROM SOUTH AFRICA

BY

PETER K. L. NG[1,3]), PAUL F. CLARK[2,4]) and CHERYL G. S. TAN[1])

[1]) Tropical Marine Science Institute and Raffles Museum of Biodiversity Research, Department of Biological Sciences, National University of Singapore, Kent Ridge, Singapore 119260, Republic of Singapore

[2]) Department of Zoology, The Natural History Museum, Cromwell Road, London SW7 5BD, England

ABSTRACT

The camptandriid crab *Cleistostoma edwardsii* MacLeay, 1838, was assigned to *Paratylodiplax* Serène, 1974, but it differs from the type species of the latter, *P. blephariskios* (Stebbing, 1924), with respect to the structures of the frontomedial margin, infraorbital margin, epistome, pterygostomian region, and male abdomen. The present paper reviews the unresolved generic status of *C. edwardsii* and establishes a new genus to accommodate the species.

RÉSUMÉ

Cleistostoma edwardsii MacLeay, 1838, a été attribué au genre *Paratylodiplax* Serène, 1974, mais il diffère de *P. blephariskios* (Stebbing, 1924) par les structures du bord frontal, de la marge infraorbitale, de l'épistome, de la région ptérygostomienne et de l'abdomen mâle. Le présent article fait le bilan des caractères génériques de *C. edwardsii* et établit un nouveau genre pour accueillir cette espèce.

INTRODUCTION

The Camptandriidae Stimpson, 1858, currently contain 39 species assigned to 19 genera (Ng et al., 2008) including *Cleistostoma* De Haan, 1833. The type species, *Ocypode* (*Cleistostoma*) *dilatata* De Haan, 1833, was subsequently

[3]) e-mail: dbsngkl@nus.edu.sg

[4]) e-mail: pfc@nhm.ac.uk

© Koninklijke Brill NV, Leiden, 2010

designated by De Man (1888) and although at one time eight species were assigned to *Cleistostoma*, a number of authors have attempted to revise this taxon. Serène (1974) moved *C. edwardsii* MacLeay, 1838, and *C. algoense* Barnard, 1954, to his new genus *Paratylodiplax* (type species *Cleistostoma blephariskios* Stebbing, 1924) (see also Guinot & Crosnier, 1963) while Manning & Holthuis (1981) transferred *C. wardi* Rathbun, 1926, *C. dotilliforme* Alcock, 1900, and *C. mcneilli* Ward, 1933, to *Paracleistostoma* De Man, 1895 (type species *P. depressum* De Man, 1895). Manning (1991) subsequently transferred *C. dotilliforme* to a new genus, *Nasima*; and Al-Khayat & Jones (1996) placed *C. kuwaitense* Jones & Clayton, 1983, in a new monotypic genus, *Leptochryseus*.

Ng et al. (2008) listed the known taxa of Camptandriidae and commented that *Cleistostoma mcneilli* Ward, 1933, is not a *Cleistostoma* species and although some workers have referred it to *Paracleistosoma*, it is also misplaced. Peter J. F. Davie (Queensland Museum) is currently revising the Australian camptandriids and will assign this taxon to a new genus along with two other new species indigenous to Australia. Ng et al. (2008: 234) also commented that *Leptochryseus* Al-Khayat & Jones, 1996, was a junior subjective synonym of *Cleistostoma*, and the two should be synonymized. Furthermore, Ng et al. (2008: 234) considered that *C. edwardsii* MacLeay, 1838, was not really a *Paratylodiplax* species and merited its own genus. Ng et al. (2009) recently clarified the taxonomy of *Nasima dotilliformis* and showed that *Leptochryseus* was in fact a valid genus. This present paper deals with the unresolved generic status of *C. edwardsii*.

Abbreviations used: The Natural History Museum, London (NHM); South African Museum, Cape Town (SAM); National Museum of Natural History, Smithsonian Institution (USNM), Washington, D.C.; National Museum of Natural History (Naturalis), Leiden (RMNH); and Zoölogisch Museum, Amsterdam (ZMA); G1 and G2 for the male first and second pleopods, respectively. Measurements are of the carapace length and width, respectively.

TAXONOMY

Family CAMPTANDRIIDAE Stimpson, 1858

Danielita n. gen.

Type species. — *Cleistostoma edwardsii* MacLeay, 1838, by present designation.

Diagnosis. — Male. Carapace subrectangular, broader than long, surface without prominent ridges or tubercles, gastro-cardiac groove deep, lateral margins with short setae; anterolateral margins sometimes forming low lobes; posterolateral margin straight, not distinctly demarcated from anterolateral margin; posterior margin straight, with thickened border. Front bilobed from dorsal view, deflexed, margin thickened, lateral angles well developed, frontomedial tooth acute, triangular. Antennal flagellae and eyestalks not setose. Orbits sparsely setose; epibranchial lobe well developed; supraorbital margin sinuous, granular; infraorbital margin with inner two-thirds cristate, forming high ridge, outer third produced, visible from dorsal view; inner infraorbital tooth well developed, distinctly separated from rest of margin; suborbital ridge distinct, produced, visible from dorsal view. Anterior half of epistome short, slightly sunken into buccal cavity; posterior half with thickened posterior margin, postero-medial tooth long, narrow, posterior margin on either side weakly concave or straight. Third maxillipeds with inner margins thickened, setose; merus with outer margin rounded, not lobiform; ischium longer than merus, subrectangular, inner distal angle slightly produced. Chelipeds large, robust; carpus may have large teeth on inner surface; palm about twice length of fingers; fingers with submarginal row of setae on inner surface, tips spatulate, movable finger with molariform tooth close to proximal end, immovable finger may have denticulate lobe on cutting edge in larger males, outer margin curved outwards. All ambulatory legs with thick tomentum on merus, carpus, and propodus only, second pair longest, last pair shortest. Abdominal somite 1 wider than somite 2; somites 1-3 free; somites 3-5 fused, sutures between these somites not discernible except for faint, median, transverse groove; constriction present at level of somite 5, telson longer than broad. G1 with small setose lobe just before recurved portion, junction between recurved and unrecurved portions with triangular lobe, long, slender distal process present, apex with strong, short spines.

Female. Supraorbital margin almost straight, not sinuous as in male. Chelipeds small, fingers slender, tips spatulate, sparsely setose. Ambulatory legs relatively less setose than in male. Abdomen with all somites free, telson broadly triangular. Other non-sexual characters similar to those of male.

Remarks. — The similarity in the basic structure of the G1, the structure of the epistome (long, acute postero-medial projection) and the male abdominal structure, seem to indicate a close relationship between *Danielita* n. gen., and *Paratylodiplax*. However, the two genera are easily to be distinguished. In *Danielita*, an acute triangular projection is present on the frontomedial margin, but in *Paratylodiplax*, this margin is entire, without any projection. The eyestalk of *Danielita* lacks the long setae commonly found on the eyestalk of *Paratylodiplax* species. In *Danielita*, the infraorbital ridge is well developed, being visible from dorsal view; but in *Paratylodiplax*, the ridge is weak and not produced much beyond the infraorbital margin and, thus, not visible from dorsal view. The anterior part of the pterygostomian region is rather bare in *Danielita*, but in *Paratylodiplax*, this region bears a distinct, dense row of long setae. The telson of the male abdomen is generally broader than long in *Paratylodiplax* species whilst in *Danielita*, it is more elongate, being longer than broad. In *Danielita*, the epistome is not deeply sunken into the buccal cavity and the posterior margin on either side of the postero-medial tooth is concave, whereas in *Paratylodiplax*, the epistome is sunken deeply into the buccal cavity and the posterior margin is strongly sinuous or lobiform on either side.

Etymology. — The name *Danielita* is an arbitrary name that is derived from Danièle Guinot's first name and the Latin word "sapienta" for proficiency in philosophy and science. This name alludes to her outstanding work on brachyuran reproductive structures, which has contributed significantly to their systematics. The gender is feminine.

Danielita edwardsii (MacLeay, 1838) new comb. (figs. 1, 2)

Cleistotoma edwardsii MacLeay, 1838: 64; Krauss, 1843: 40.
Cleistostoma edwardsii — Stebbing, 1910: 328; Buitendijk, 1939: 76, figs. 1-5; Barnard, 1950: 105, fig. 21a-f; 1954: 121, fig. 2e-h; 1955: 3 (list); Ng et al., 2008: 234.
Cleistostoma edwarsi [sic] — Serène, 1968: 101.
Paratylodiplax edwarsi [sic] — Serène, 1974: 65.
Paratylodiplax edwardsii — Manning & Holthuis, 1981: 211, fig. 54f-h.

Material examined. — South Africa: 9 ♂♂ (5.2-7.6 mm × 6.8-10.6 mm), 3 ♀♀ (5.3-6.9 mm × 7.0-9.7 mm), Zwartkops, Cape of Good Hope, NHM 1956.11.12.7-12; male (7.5 × 9.7 mm), Knysna Estuary, Cape Province, USNM 170883; 1 ♂ (7.1 × 9.2 mm), 11 ♂♂, 3 ♀♀, RMNH D 3274, 18 Sep. 1938; 1 ♂ (8.0 × 11.0 mm), 1 ♀ (6.1 × 7.8 mm), Langebaan Lagoon, SAM A10775; 4 ♂♂, 4 ♀♀, 1 juvenile, Knysna, Laguna, coll. Engel, 16 Sep. 1948, ZMA.

Neotype. — ♂ (7.6 × 10.6 mm), Zwartkops, Cape of Good Hope, NHM 1956.11.12.7-12.

Comparative material. — *Paratylodiplax blephariskios* (Stebbing, 1924): 1 ♂ (5.4 × 8.8 mm), 1 ♀ (4.7 × 7.0 mm), Cape Town, South Africa, leg. J. Day, Sep. 1957, USNM 137247.

Type locality. — Zwartkops, Eastern Cape Province, South Africa, Indian Ocean.

Description. — Male. Carapace 1.3-1.4× as broad as long, sparsely setose, smooth, clear transverse groove in middle of carapace, carapace moderately inflated, lateral margins subparallel, weakly convex, anterolateral margin with large, outer orbital angle, which curves inwards slightly; small tooth present posterior to outer orbital angle, dorsolateral margin with line of granules, posterior margin straight. Front weakly bilobed from dorsal view, weakly thickened, dorsal region just posterior to frontal edge slightly depressed, epigastric

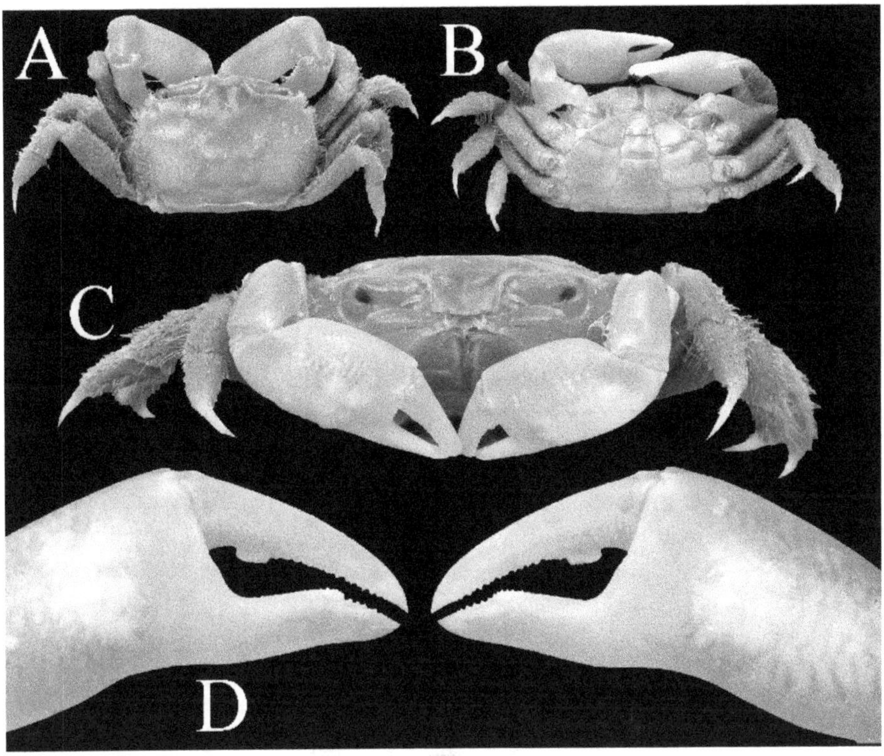

Fig. 1. *Danielita edwardsii* (MacLeay, 1838) new comb., neotype ♂, NHM 1956.11.12.7-12. A, dorsal view; B, ventral view; C, frontal view; D, chelae. [Photo: Harry Taylor NHM Photo Unit.]

ridges relatively flat, rounded; frontomedial projection acutely triangular. Orbits sparsely setose; supraorbital margin beaded, weakly sinuous from dorsal view; infraorbital margin well developed into ridge along inner two-thirds, margin of ridge with large, coarse granules, outer portion of ridge extended forwards to form projecting lobe, lobe visible from dorsal view; inner infraorbital tooth truncate, with straight upper edge, distinctly separated from rest of infraorbital margin; suborbital ridge distinct, produced, visible from dorsal view, defined by row of large pointed granules. Anterior half of epistome sunken in, posterolateral margin of epistome separated from anterior pterygostomian margin by clear, closed fissure, postero-medial tooth narrow, elongated, lateral margins on either side weakly concave; pterygostomial sulci faint, visible as shallow groove. Third maxilliped inner edge thickened, ischium 1.4× as long as merus, measured along inner edge, projection of inner distal angle of ischium only slightly further than outer distal angle, outer edge of merus expanded, rounded.

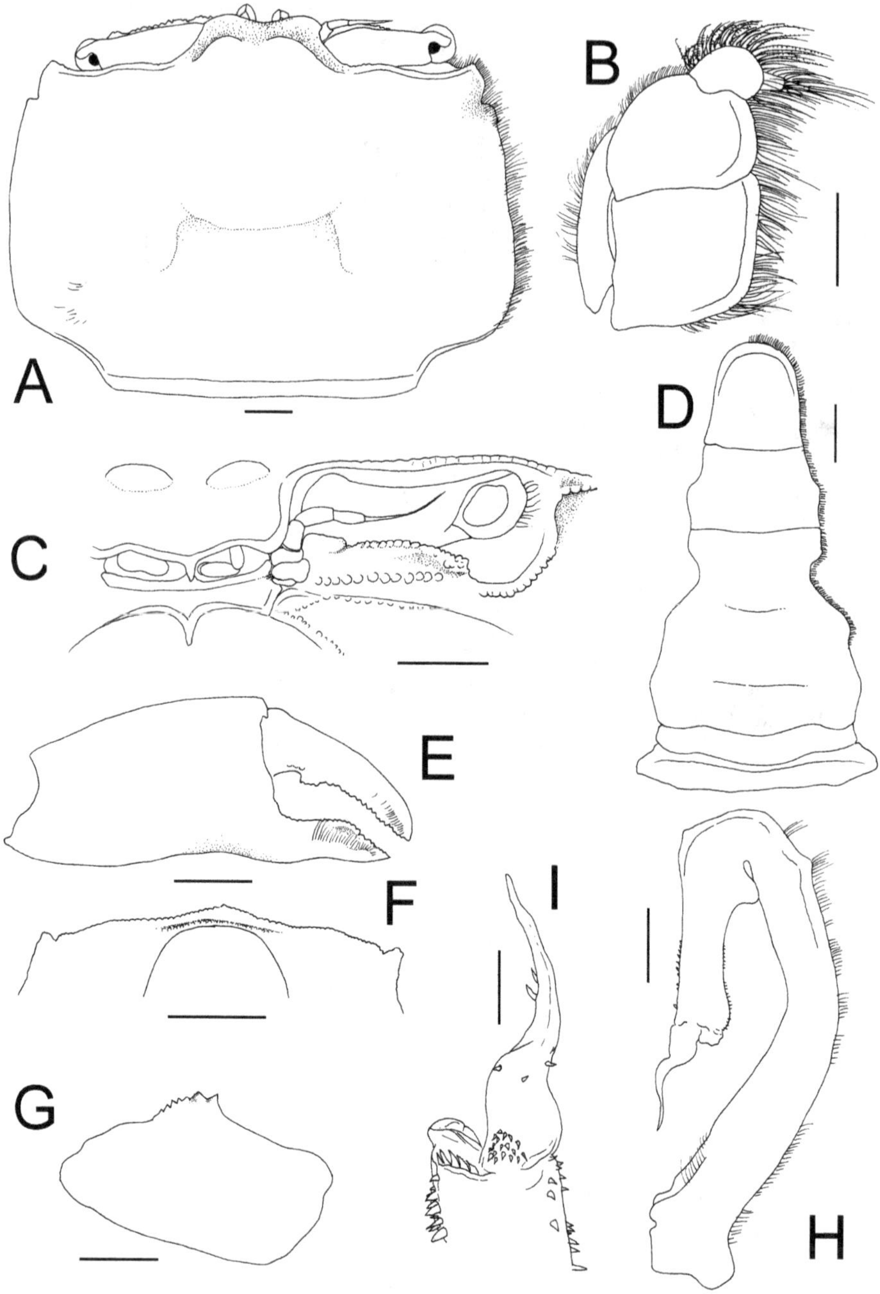

Fig. 2. *Danielita edwardsii* (MacLeay, 1838) new comb., neotype ♂, NHM 1956.11.12.7-12. A, dorsal view of carapace; B, third maxilliped; C, anterior view of orbit; D, male abdomen; E, first three somites of sternum; F, right chela; G, carpus of chela; H, G1; I, distal part of G1. Scale bars: A-G = 1.0 mm; H = 0.5 mm; I = 0.15 mm.

Inner edge of cheliped carpus with projecting lobe with row of small teeth, palm 2.4 times as long as immovable finger, cutting edge of immovable finger with irregularly shaped denticles, outer edge slightly sinuous, movable finger with truncate molariform tooth close to proximal end, cutting edge of tooth with small denticles, rest of cutting edge of movable finger with irregularly shaped denticles.

Ambulatory legs with long, thin setae especially prominent on dorsal margin of merus, carpus, and propodus and short, thick tomentum on merus, carpus, and propodus naked.

Thoracic sternum smooth, anterior margin broadly triangular, margin lined with large rounded granules, sternites 1-3 appear compressed, separated from sternite 4 by clear groove that does not reach lateral margins of sternal plastron, anterolateral angles of sternite 4 small, rounded, directed forwards, anterior margin of sternal concavity lined with large, rounded granules. Sutures between abdominal somites deep, reaching to lateral margin of abdomen; somites 1-3 free, mobile; somites 3-5 fused; somite 6 with median lobe on lateral margin, shorter than telson; telson longer than broad.

G1 apex inner edge lined with distinct row of strong spines, small lobe present just adjacent to base of distal appendage, distal appendage with bulbous base covered sparsely with short, robust spines, cluster of spines present close to base, thinner end of distal appendage with 2 strong, subequal spines.

Female. — Infraorbital margin even, without median projecting lobe, granules less sharp. Chelipeds slender, movable finger without molariform tooth, tips spatulate. Ambulatory legs with less long setae as compared to males. Abdomen with 6 free somites and telson. Other, non-sexual aspects similar to males.

Remarks. — A good series of specimens from the various depositories was examined. The G1 fits Barnard's (1954, fig. 2e-h) description well. However, in the specimens examined, the second, more posterior notch on the anterolateral margin of the carapace, which is shown clearly in Barnard's figure (Barnard, 1950, fig. 21a) is absent. Also, Barnard (1950) omitted the median transverse groove on the carapace, which is evident in all the specimens examined.

A unique feature found in *D. edwardsii* is the presence of a projecting lobe on the median part of the infraorbital margin in males of the species. This is visible even from dorsal view. In all male specimens examined during this study, the suture between the second and third somites of the abdomen is clear, complete, and mobile.

Barnard (1950: 106) remarked that "Mr Melbourne Ward tells me (4/3/37) that the type of this species is not amongst the McLeay collection of South African Crustacea in the Australian Museum". Ng & Shane (2001), who photographed and clarified the identities of MacLeay's extant types also did not record the species among the existing material. The types are no longer considered to be extant. As *Cleistostoma edwardsii* MacLeay, 1838, is now made the type of a new genus, a male specimen (7.6 × 10.6 mm) (NHM 1956.11.12.7-12) is hereby designated as the neotype of the species to stablize the taxonomy of these taxa.

Distribution. — Known only from South Africa.

ACKNOWLEDGEMENTS

With much respect we dedicate this study to our great friend, colleague, and carcinologist, Danièle Guinot. We would also like to thank Harry Taylor, NHM Photo Unit, for taking the photographs presented in this paper.

REFERENCES

AL-KHAYAT, J. A. & D. A. JONES, 1996. Two new genera, *Manningis* and *Leptochryseus* (Decapoda: Camptandriinae), and descriptions of the first zoea of six brachyurans from the Arabian Gulf. Journal of Crustacean Biology, **16** (4): 797-813.

ALCOCK, A., 1900. Materials for a carcinological fauna of India. No. 6. The Brachyura Catometopa or Grapsoidea. Journal of the Asiatic Society of Bengal, Calcutta, **69** (3): 279-486.

BARNARD, K. H., 1950. Descriptive catalogue of South African decapod Crustacea (crabs and shrimps). Annals of the South African Museum, **38**: 1-837, figs. 11-154.

— —, 1954. New records and new species of Crustacea from South Africa. Annales du Musée Royale de Congo Belge, Tervuren, (Série in Quarto, Zoologie) **1**: 120-131, figs. 1-8.

— —, 1955. Additions to fauna-list of South African Crustacea and Pycnogonida. Annals of the South African Museum, **43** (1): 1-107.

BUITENDIJK, A. M., 1939. Biological results of the Snellius Expedition. 5. The Dromiacea, Oxystomata and Oxyrhyncha of the Snellius Expedition. Temminckia, **4**: 223-276, pls. 7-11.

CLARK, P. F., K. HARRISON & S. L. GOODMAN, 1990. Brock's Indian Archipelago Decapoda and Stomatopoda by De Man: 1887 or 1888? Archives of Natural History, London, **17** (1): 79-80.

GUINOT, D. & A. CROSNIER, 1963. Remarques sur les genres *Cleistostoma, Paracleistostoma* et *Tylodiplax* et description de *Tylodiplax derijardi* sp. nov. (Crust. Decap. Brachyura). Bulletin du Muséum national d'Histoire naturelle, Paris, (2) **35** (6): 606-619.

HAAN, W. DE, 1835. Crustacea. Fauna Japonica, sive Descriptio animalium, quae in itinere per Japoniam, jussu et auspiciis superiorum, qui summum in India Batava imperium tenent, suscepto, annis 1823-1830 collegit, notis, observationibus et adumbrationibus illustravit P. F. de Siebold. Conjunctis studiis C. J. Temminck et H. Schlegel pro Vertebratis atque W. de Haan pro Invertebratis elaborata Regis aupicus edita. Decas II: 25-64, pls. 9-15, 17, C, D. (P. F. v. Siebold, Lugduni-Batavorum [= Leiden]).

JONES, D. A. & D. CLAYTON, 1983. The systematics and ecology of crabs belonging to the genera *Cleistostoma* De Haan and *Paracleistostoma* De Man on Kuwait mudflats. Crustaceana, **45** (2): 183-199.

KRAUSS, G., 1843. Die Südafrikanischen Crustaceen. Eine Zusammenstellung aller bekannten Malacostraca, Bemerkungen über deren Lebensweise und geographische Verbreitung, nebst Beschreibung und Abbildung mehrerer neuen Arten: 1-137, pls. 1-4. (E. Schweizerbart'sche Verlagshandlung, Stuttgart).

MACLEAY, W. S., 1838. On the brachyurous decapod Crustacea. Brought from the Cape by Dr. Smith. In: A. SMITH (ed.), Illustrations of the Zoology of South Africa; consisting chiefly of figures and descriptions of the objects of natural history collected during an expedition into the interior of South Africa, in the years 1834, 1835, and 1836; fitted out by 'The Cape of Good Hope Association for Exploring Central Africa': together with a summary of African zoology, and an inquiry into the geographical ranges of species in that quarter of the globe, published under the Authority of the Lords Commissioners of Her Majesty's Treasury, Invertebratae. **IV** [1849]: 53-71, pls. 2, 3. (Smith, Elder & Co., London).

MAN, J. G. DE, 1888. Bericht über die im indischen Archipel von Dr. J. Brock gesammelten Decapoden und Stomatopoden. Archiv für Naturgeschichte, Berlin, **53** (I. Band 3. Heft): 289-600, Taf. XI-XXIIa [issued 1888]. [For dates of publication, see Clark et al., 1990.]

— —, 1895. Bericht über die von Herrn Schiffscapitän Storm zur Atjeh, an den westlichen Küsten von Malakka, Borneo und Celebes sowie in der Java-See gesammelten Decapoden und Stomatopoden. Zoologische Jahrbücher, (Abtheilung für Systematik, Geographie und Biologie der Thiere) **8**: 485-609, pls. 12-14.

MANNING, R. B., 1991. The status of *Cleistostoma dotilliforme* Alcock, 1900 (Crustacea: Decapoda: Ocypodidae), with the description of a new genus. Raffles Bulletin of Zoology, **39** (2): 299-306.

MANNING, R. B. & L. B. HOLTHUIS, 1981. West African brachyuran crabs (Crustacea: Decapoda). Smithsonian Contribution of Zoology, **306**: 1-379.

NG, P. K. L. & S. T. AHYONG, 2001. Brachyuran type specimens (Crustacea: Decapoda) in the Macleay Collection, University of Sydney, Australia. Raffles Bulletin of Zoology, **49** (1): 83-100.

NG, P. K. L., D. GUINOT & P. J. F. DAVIE, 2008. Systema Brachyurorum: Part I. An annotated checklist of extant brachyuran crabs of the world. Raffles Bulletin of Zoology, (Supplement) **17**: 1-286.

NG, P. K. L., D. L. RAHAYU & M. D. NASER, 2009. The Camptandriidae of Iraq, with description of a new genus and notes on *Leptochryseus* Al-Khayat & Jones, 1996 (Crustacea: Decapoda: Brachyura). Zootaxa, **2312**: 1-26.

RATHBUN, M. J., 1926. Brachyuran crabs from Australia and New Guinea. Records of the Australian Museum, **15** (2): 177-182, pls. 14-16.

SERÈNE, R., 1968. The Brachyura of the Indo Pacific Region. In: Prodromus for a check list of the non-planctonic marine fauna of South East Asia. Special Publication of the Singapore National Academy of Science, **1**: 33-120.

— —, 1974. Note on the genera and species of the Camptandriinae Stimpson, 1858 (Decapoda, Brachyura: Ocypodidae). Treubia, **28** (3): 59-117.

STEBBING, T. R. R., 1910. General catalogue of South African Crustacea (Part V of South African Crustacea, for the Marine Investigations in South Africa). Annals of the South African Museum, **6** (4): 281-593, pls. 15-22.

— —, 1924. South African Crustacea (Part XII of South African Crustacea, for Marine Investigations in South Africa). Annals of the South African Museum, **19**: 235-248, pls. 1-7 (pls. 116-122 of Crustacea).

STIMPSON, W., 1858. Prodromus descriptionis animalium evertebratorum, quae in expeditione ad Oceanum Pacificum Septentrionalem, a republica federata missa, Cadwaladaro Ringgold et Johanne Rodgers ducibus, observavit et descripsit. Pars V. Crustacea Ocypodoidea. Proceedings of the Academy of Natural Sciences of Philadelphia, **10**: 93-110 [39-56].

WARD, M., 1933. New genera and species of marine decapod Brachyura. From the coasts of New South Wales and Queensland. Australian Zoologist, **7**: 377-394, pls. 21-23.

First received 27 June 2009.
Final version accepted 23 July 2009.

ELAMENOPSIS GUINOTAE, A NEW SPIDER CRAB FROM BASS STRAIT, AUSTRALIA (BRACHYURA, HYMENOSOMATIDAE)

BY

GARY C. B. POORE[1])
Museum Victoria, GPO Box 666, Melbourne, Victoria 3001, Australia

ABSTRACT

Elamenopsis guinotae sp. nov. is described from a temperate shallow sandy environment in Bass Strait, Australia. The new species is most similar to *E. rotunda*, one of four other species in the genus, all from the tropical western Pacific. The two species share a subcircular carapace with the carapace rim width about 1.2 times as wide as long, and long, slender walking legs. It differs in the absence of an anterolateral notch on the carapace and the broader female abdomen. Features differentiating all five species of *Elamenopsis* are tabulated.

RÉSUMÉ

Elamenopsis guinotae sp. nov. est décrite du Détroit de Bass, dans un environnement d'eaux tempérées, peu profondes et de fonds sableux. La nouvelle espèce est très similaire à *E. rotunda*, l'une des quatre autres espèces du genre, toutes du Pacifique ouest. Les deux espèces ont en commun une carapace semi-circulaire, 1,2 fois plus large que longue et des pattes marcheuses longues et fines. Cette nouvelle espèce se différencie par l'absence d'une encoche antérolatérale sur la carapace et un abdomen femelle plus large. Les caractères séparants des cinq espèces d'*Elamenopsis* sont comparés.

INTRODUCTION

The soft sediment fauna of eastern Bass Strait has a notably high diversity (Coleman et al., 1997) and features many apparently rare (narrowly distributed or rarely caught) species among the more than 800 documented species. The decapods of this area are now well documented (Poore, 2004) and, of the known fauna, only a single species of hymenosomatid crab remains

[1]) e-mail: gpoore@museum.vic.gov.au

© Koninklijke Brill NV, Leiden, 2010

undescribed (Poore, 2004). The species was featured in O'Hara's (2002) discussion of endemism, rarity, and vulnerability along a temperate coastline of Australia because it is represented by few individuals taken from four grab samples along a 25 km long 14-15 metre depth contour. The species, which was then incorrectly referred to the genus *Halicarcinus* White, 1846 is described here as a new species of *Elamenopsis* A. Milne-Edwards, 1873.

Hymenosomatid spider crabs are represented in Australia by 30 species (Lucas, 1980; Davie, 2002). In the Bass Strait region between Victoria and Tasmania four species are known thus far: *Halicarcinus ovatus* Stimpson, 1859, *H. rostratus* (Haswell, 1881), *Hymenosoma hodgkini* Lucas, 1980, and *Trigonoplax longirostris* McCulloch, 1908 (cf. Poore, 2004).

The diagnosis of the new species is based on the characters in Naruse & Ng's (2007) key to species of *Elamenopsis*. Type material is lodged in Museum Victoria, Melbourne (NMV) and the Queensland Museum, Brisbane (QM). Three dimensions are given: carapace length along the median line from the rostrum to the posterior margin (cl); carapace width measured at its widest point (cw); and carapace rim width (often the only measurement available from comparative illustrations) measured transversely at the widest point of the dorsal rim (crw).

Elamenopsis A. Milne-Edwards, 1873

Remarks. — *Elamenopsis* was diagnosed and discussed by Naruse & Ng (2007). The genus comprises only four species, all from the western Pacific: *E. lineata* A. Milne-Edwards, 1873 from New Caledonia, the Philippines, Sulawesi, Singapore, and north-eastern Australia (Ng & Richer de Forges, 1996), *E. ariakensis* (Sakai, 1969) from Japan and China (Kosuge et al., 2002), *E. comosa* Ng & Chuang, 1996 from Indonesia, and *E. rotunda* Naruse & Ng, 2007 from Singapore. Naruse & Ng (2007) compared them all and provided a key. The new species described here extends the range of the genus into temperate southern Australia.

Elamenopsis guinotae sp. nov. (figs. 1, 2)

Material examined. — Holotype: Australia, Victoria, Bass Strait, 1 km off Woodside Beach (38°33′S 146°57′E), 15 m, 08 Mar. 1989 (stn MSL-LV 2W4), NMV J59277 (female, cl 2.16 mm, cw 2.95 mm).

Paratypes (many lacking some or all ambulatory legs): collected with holotype, NMV J31879 (3 males, cw 1.6 mm); QM W28877 (male, cw 1.6 mm; female, 1.7 mm). One (1) km off Woodside Beach (38°33′S 146°57′E), 14.5 m, 08 Mar. 1989 (stn MSL-LV 1W5), NMV J31881 (juvenile male, cw 1.6 mm); (stn MSL-LV 2W5), NMV J31878 (6 juvenile males, cw

1.2 mm; 2 females, cw 1.6 mm), NMV J59646 (male, cw 2.0 mm). One (1) km off Delray Beach (38°14′S 147°22′E), 14.8 m, 24 Mar. 1989 (stn MSL-LV 3T3), NMV J31880 (female, cw 3.0 mm).

Diagnosis. — Carapace ovoid, flat or barely convex, widest at about three-quarters length, cw : cl ratio of adult female 1.36, crw : cl ratio of adult female 1.21; anterolateral margin evenly curved; subhepatic region broadly swollen; rostrum triangular in anterior view, with lateral ridges, bilobed in dorsal view; maxilliped 3 ischium mesial margin 3 times as long as merus mesial margin; ambulatory legs slender (merus of third leg 4.3 times as long as wide), long (length of third leg 1.5 times cw); male abdomen 6-segmented, uropods prominent; gonopod 1 stout, spindle-shaped, 4 times as long as greatest thickness, tapering distally.

Description. — Female (cw 2.95 mm). Carapace (fig. 1a) ovoid, widest at about three-quarters length, cw : cl ratio of adult female 1.35, crw : cl ratio of adult female 1.2; lateral surfaces around bases of second and third legs expanded laterally; dorsally flat or barely convex, dorsal rim without angles, regions well defined; cervical groove branching anteriorly to weakly defined hepatic region, connected posteriorly with gastro-cardiac and thoracic grooves, thoracic groove connecting to cardiac-intestinal groove, branchial region divided by weak oblique-transverse groove into pre- and post-branchial regions, intestinal region with weak median groove. Rostrum triangular, apically bifid in dorsal view, strongly deflexed, with prominent lateral ridges not confluent with dorsal rim but running over orbit. Subhepatic region swollen in upper half, with weakly curved ridge defining lateral margin of orbit; with slight boss mesially under orbit. Dorsal rim, lateral margins of carapace with scattered short, stout setae. Epistome (fig. 1b) with sharp medial anterior projection, not separating antennules, with rounded posterior projection. Eyestalk short, dilating; cornea rounded, visible in dorsal view. Maxilliped 3 (fig. 2e) meri covering 0.6 of width of mouthfield, leaving wide gape when closed; ischium triangular, mesial margin 3 times as long as merus mesial margin; merus ovoid, 1.2 times as long as wide; ischium, merus with row of long mesial setae; exopod base 3.8 times as long as wide, distal end not reaching distal outer angle of merus, flagellum about one-third length of base.

Chelipeds (fig. 1d, e) unequal, with setose outer surfaces; propodus oval in cross-section, deepest near midpoint, total length 2.1 times depth in larger cheliped, 2.4 times in smaller cheliped; propodus-dactylus articulation oblique; cutting edges of fixed finger, dactylus unevenly dentate; dactylus 0.55 times total propodus length, 3.0 times as long as wide in larger cheliped, 3.6 times as long as wide in smaller cheliped.

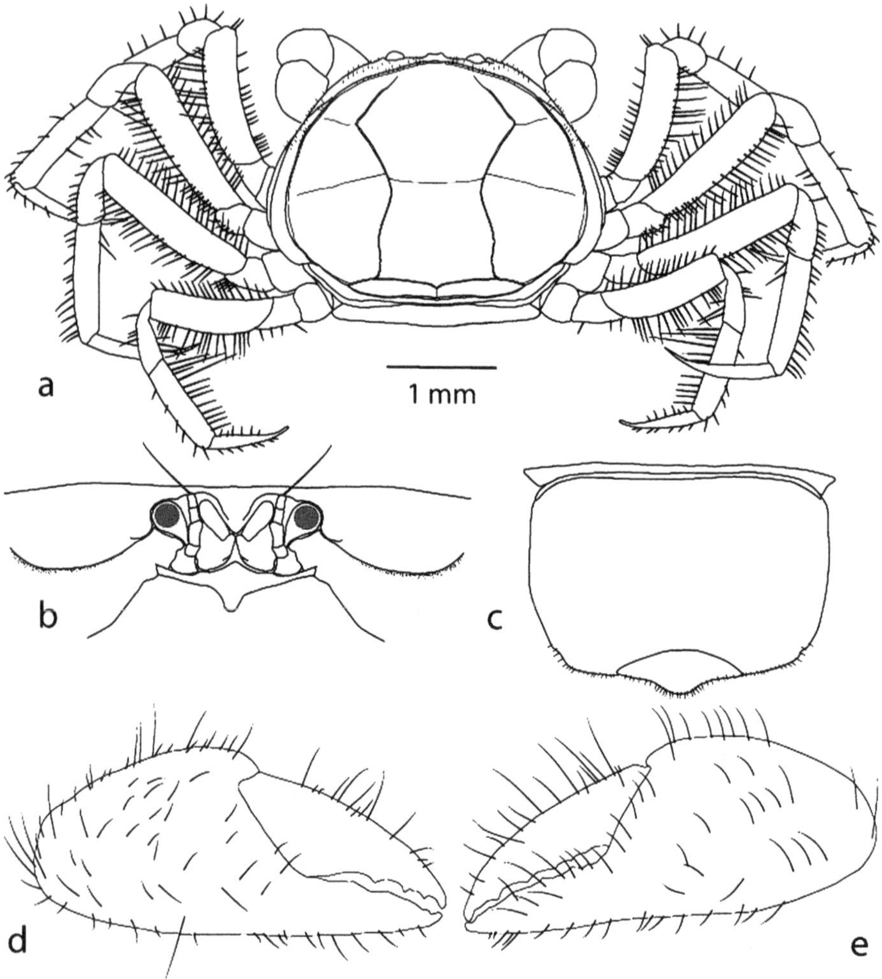

Fig. 1. *Elamenopsis guinotae* sp. nov., female holotype. a, habitus; b, front; c, abdomen; d, e, face views of right and left chelipeds. Scale bar refers to fig. 1a only.

Ambulatory legs (fig. 2f) long, second leg the longest (third leg 1.5 times as long as cw, 1.7 times as long as crw), slender (merus of third leg 4.3 times as long as wide), flattened, flexor, extensor margins of meri densely fringed with long setae, carpus-dactylus with fewer setae scattered over all faces, dactylus about as long as propodus, evenly tapering, curved.

Abdomen (fig. 1c) with 2 short free somites, fused somites 3-5 are 1.6 times as wide as long, with lateral margins strongly converging distally, somites not evident laterally; telson free, width 0.4 abdomen width. Pleopods 1-4 biramous.

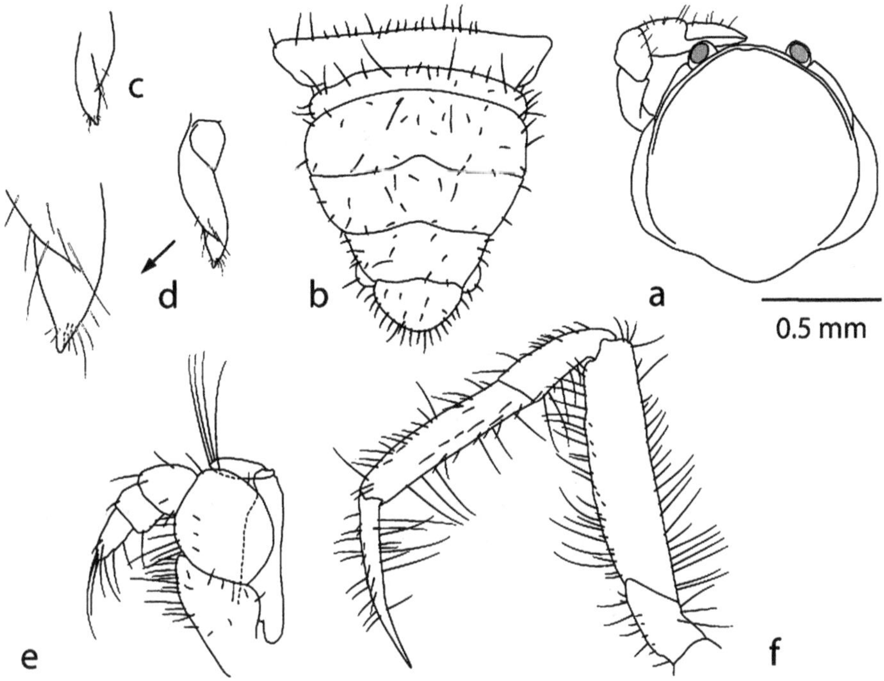

Fig. 2. *Elamenopsis guinotae* sp. nov.: a, juvenile male NMV J31878. b-d, adult male NMV J59646: b, abdomen; c, gonopod 1 (posterior view); d, gonopod 1 (anterior view and detail). e, female, NMV J31880: maxilliped 3; f, female holotype: walking leg 3. Scale bar refers to fig. 2a only; abdomen and gonopod 1 to same scale.

Male (cw 2.00 mm). — Abdomen (fig. 2b) 6-segmented, 1.06 times as long as greatest width, third somite widest, fifth somite with prominent uropods. Gonopod 1 (fig. 2c, d) stout, spindle-shaped, 4 times as long as greatest thickness, twisting, tapering distally, with 6 distal setae along anterior ridge, 7 setae plus setules on triangular apex.

Juvenile male (cw 1.2 mm). — Carapace (fig. 2a) narrower than in adults, subcircular, cw : cl = 1.1, crw : cl = 0.95.

Distribution. — Australia, Victoria, eastern Bass Strait, 14-15 m depth, coarse sandy-shelly sediments.

Etymology. — The species name recognizes the lifetime of insightful contributions to the systematics of brachyuran crabs by Danièle Guinot, and her friendship.

Remarks. — *Elamenopsis guinotae* sp. nov. is most similar to *E. rotunda*, sharing a subcircular carapace with the carapace rim width about 1.2 times as wide as long, and long, slender walking legs (table I). It differs in the absence of an anterolateral notch on the carapace and the broader female abdomen.

TABLE I
Comparison of the five species of *Elamenopsis* (? indicates unknown)

	E. lineata A. Milne-Edwards, 1873	*E. comosa* Ng & Chuang, 1996	*E. ariakensis* (Sakai, 1969)	*E. rotunda* Naruse & Ng, 2007	*E. guinotae* sp. nov.
Distribution	New Caledonia, Philippines, Sulawesi, Singapore, north-eastern Australia	Indonesia	Japan, China	Singapore	SE Australia
crw : cl	1.3	1.4	1.2	1.2	1.2
Gastric region	smooth	smooth	with 3 tubercles	smooth	smooth
Anterolateral notch on carapace	absent	absent	present	present	absent
Maxilliped 3 merus inner margin	equal to ischium	longer than ischium	longer than ischium	shorter than ischium	shorter than ischium
Walking leg 3 length : crw	1.2	?	1.3	1.6	1.7
Walking leg 3 merus length : width	3	?	3	4.0	4.3
Male abdominal somites	5 (somites 4 and 5 fused)	6	4 (somites 3-5 fused)	?	6
Uropods	absent	absent	absent	?	present
Male gonopod 1	slender, tapering, twisted	slender, apex sharply curved	slender, tapering, evenly curved, twisted	?	stout, apex not curved
Female fused abdomen somites width : length	1.2	1.0	?	1.2	1.6
Female abdominal segmentation	laterally evident	laterally evident	?	not evident	not evident

The male of *E. rotunda* is unknown. In *E. ariakensis*, a third species with a subcircular carapace, there is a median notch on the dorsal rim and swellings on each side; the hepatic region has a clear notch and the gastric region has three tubercles; maxilliped 3 has a longer merus; the male abdominal somites 3-5 are fused and the gonopod 1 is slender. The other two species, *E. lineata* and *E. comosa*, have a broader carapace, stouter walking legs (at least in *E. lineata*), and the somites of the female abdomen are laterally indicated (table I).

No author has previously reported on female pleopods in species of *Elamenopsis*. The new species has four pairs of biramous pleopods. Uropods, present in this species as lateral protrusions on the penultimate abdominal somite of the male, have not been illustrated for the three species where males are known (table I).

Elamenopsis lineata has been reported from estuaries in Indonesia and Queensland, Australia (Ng & Chuang, 1996), *E. rotunda* from sublittoral mudflats off Singapore, and *E. ariakensis* from a mudflat off Nagasaki, Japan, commensal with a common holothurian, *Protankyra bidentata* (Woodward & Barrett, 1858). In contrast to these tropical and muddy shallow environments, the new species, *Elamenopsis guinotae*, is known from a temperate shelf environment with coarse sandy-shelly sediments. No associates are known.

ACKNOWLEDGEMENTS

I thank the organizers of this volume, Bertrand Richer de Forges, Peter Castro, Peter J. F. Davie, and Peter K. L. Ng, for the invitation to contribute. I thank Peter Ng for comments on a first draft of the manuscript, the late Noel Coleman of the Victorian Marine and Freshwater Research Institute, Queenscliff, for the opportunity to examine the collections from eastern Bass Strait, and Helen Lew Ton for much of the sorting and preliminary identifications.

REFERENCES

COLEMAN, N., A. S. H. GASON & G. C. B. POORE, 1997. High species richness in the shallow marine waters of south-east Australia. Marine Ecology Progress Series, **154**: 17-26.

DAVIE, P. J. F., 2002. Crustacea: Malacostraca: Eucarida (Part 2): Decapoda — Anomura, Brachyura: i-xiv, 1-641. (CSIRO Publishing, Melbourne).

HASWELL, W. A., 1881. On some new Australian Brachyura. Proceedings of the Linnean Society of New South Wales, **6**: 540-551.

KOSUGE, I., Y. KOSHIISHI & N. SUYAMA, 2002. Rediscovery of *Elamenopsis ariakensis* (Brachyura, Hymenosomatidae) from Ariake-kai Sound, Kyushu, western Japan. Nanki-seibutsu, **44**: 103-105. [In Japanese with English summary.]

LUCAS, J. S., 1980. Spider crabs of the family Hymenosomatidae (Crustacea; Brachyura) with particular reference to Australian species: systematics and biology. Records of the Australian Museum, **33**: 148-247.

MCCULLOCH, A. R., 1908. Studies in Australian Crustacea. No. 1. Records of the Australian Museum, **7**: 51-59, pl. 12.

MILNE-EDWARDS, A., 1873. Recherches sur la faune carcinologique de la Nouvelle-Calédonie. Nouvelles Archives du Muséum d'Histoire Naturelle, Paris, **9**: 155-322, pls. 4-18.

NARUSE, T. & P. K. L. NG, 2007. On a new species of *Elamenopsis* from Singapore, with notes on *Crustaenia palwanensis* (Sèrene, 1971) (Crustacea: Decapoda: Brachyura: Hymenosomatidae). Raffles Bulletin of Zoology, **55**: 121-125.

NG, P. K. L. & C. T. N. CHUANG, 1996. The Hymenosomatidae (Crustacea: Decapoda: Brachyura) of southeast Asia, with notes on other species. Raffles Bulletin of Zoology, (Supplement) **3**: 1-82.

NG, P. K. L. & B. RICHER DE FORGES, 1996. The Hymenosomatidae (Crustacea: Decapoda: Brachyura) of New Caledonia, with descriptions of two new genera and two new species. Memoirs of the Queensland Museum, **39**: 263-276.

O'HARA, T. D., 2002. Endemism, rarity and vulnerability of marine species along a temperate coastline. Invertebrate Systematics, **16**: 671-684.

POORE, G. C. B., 2004. Marine decapod Crustacea of southern Australia. A guide to identification (with chapter on Stomatopoda by Shane Ahyong): 1-574. (CSIRO Publishing, Melbourne).

SAKAI, T., 1969. Two new genera and twenty-two new species of crabs from Japan. Proceedings of the Biological Society of Washington, **82**: 243-280.

STIMPSON, W., 1859. Prodromus descriptionis animalium evertebratorum, quae in expeditione ad Oceanum Pacificum septentrionalem, a Republica Federata missa, Cadwaladaro Ringgold et Johanne Rodgers ducibus, observavit et descripsit. Pars VII. Crustacea Anomura. Proceedings of the Academy of Natural Science of Philadelphia, **1858**: 225-251.

WHITE, A., 1846. Notes on four new genera of Crustacea. Annals and Magazine of Natural History, (1) **18**: 176-178, pl. 2.

First received 23 February 2009.
Final version accepted 8 May 2009.

NOTONYX GUINOTAE, A NEW SPECIES OF GONEPLACID CRAB (BRACHYURA, GONEPLACIDAE) FROM LOMBOK ISLAND, INDONESIA

BY

DWI LISTYO RAHAYU[1,3]) and PETER K. L. NG[2,4])

[1]) Marine Bio-Industry Implementation Unit — Research Center for Oceanography, Indonesian Institute of Sciences (LIPI), P.O. Box 1124, Mataram 83000, NTB, Indonesia

[2]) Tropical Marine Science Institute and Department of Biological Sciences, National University of Singapore, Lower Kent Ridge Road, Singapore 119260, Republic of Singapore

ABSTRACT

A new species of goneplacid crab of the genus *Notonyx* A. Milne-Edwards, 1873, is described from Lombok, Indonesia. The new species, *N. guinotae*, resembles *N. kumi* Naruse & Maenosono, 2009, but differs in having a relatively more slender male abdomen, proportionately longer ambulatory legs, as well as male first and second gonopods that are different in morphology. The discovery of this new species brings the number of species now known to six.

RÉSUMÉ

Une nouvelle espèce de crabe Goneplacidae du genre *Notonyx* A. Milne-Edwards, 1873, est décrite de Lombok en Indonésie. Cette nouvelle espèce, *N. guinotae* ressemble à *N. kumi* Naruse & Maenosono, 2009, mais en diffère par un abdomen male relativement plus étroit, des pattes ambulatoires relativement plus longues et également par les structures des premier et second gonopodes mâles qui sont de forme différentes. La découverte de cette nouvelle espèce porte à six le nombre d'espèces connues de ce genre.

INTRODUCTION

Members of the genus *Notonyx* A. Milne-Edwards, 1873 (type species *Notonyx nitidus* A. Milne-Edwards, 1873), are defined principally by their

[3]) Corresponding author; e-mail: dwilistyo@yahoo.com

[4]) e-mail: dbsngkl@nus.edu.sg

© Koninklijke Brill NV, Leiden, 2010

quadrangular shaped and smooth carapace, without clear indication of regions, the absence of teeth or spines on the anterolateral margins, a relatively stout male first gonopod, as well as a male second gonopod which is long, whip-like, and longer than the male first gonopod. As has been discussed at length by Clark & Ng (2006), Castro (2007), and Ng & Clark (2008), *Notonyx* is clearly a member of the Goneplacidae MacLeay, 1838, despite its unusual carapace form.

This genus has been well studied in the last few years, with the identity of *Notonyx nitidus* A. Milne-Edwards, 1873, clarified and several new species described (Clark & Ng, 2006; Ng & Clark, 2008; Naruse & Maenosono, 2009). Currently, five species are included in the genus: *Notonyx nitidus* A. Milne-Edwards, 1873, *N. vitreus* Alcock, 1900, *N. gigacarcinicus* Clark & Ng, 2006, *N. latus* Ng & Clark, 2008, and *N. kumi* Naruse & Maenosono, 2009 (see also Ng et al., 2008). During ecological studies of the seagrass beds of Lombok Island, Indonesia, a good number of *Notonyx* specimens were collected. While they superficially resemble *N. kumi* in having a swollen subdistal part of the male first gonopod, they nevertheless differ in many other morphological characters. This species is considered as new and it is described herein as *N. guinotae*.

The material examined is deposited in the Museum Zoologi Bogor, Indonesian Institute of Sciences, Cibinong, Indonesia; the former Rijksmuseum van Natuurlijke Historie (RMNH), Leiden, The Netherlands [now National Museum of Natural History — Naturalis]; and the Zoological Reference Collection (ZRC) of the Raffles Museum of Biodiversity Research, National University of Singapore, Rep. of Singapore. Size is indicated by the carapace width and length (carapace width × carapace length) in mm. The terminology used in the description generally follows that of Clark & Ng (2006). The abbreviations G1 and G2 refer to the male first gonopod and male second gonopod, respectively.

TAXONOMY

Family GONEPLACIDAE MacLeay, 1838

Notonyx A. Milne-Edwards, 1873

Remarks. — The systematics of this genus was explained at length by Clark & Ng (2006), Ng & Clark (2008), and Castro (2007: 754). The characters of the present new species conform with all the generic and familial features recorded by these authors.

Notonyx guinotae n. sp. (figs. 1-5)

Material. — Holotype, 1 male (15.5 × 11.5 mm) (MZB Cru 2583), Sira, northern Lombok Island, 08°21.641′S 116°06.272′E, coll. D. L. Rahayu, 13 May 2007. Paratypes: 1 female (6.5 × 4.7 mm) (MZB Cru 2584), 1 female (7.5 × 5.6 mm) (ZRC 2009.0746), same locality as holotype; 2 males (5.9 × 4.9 mm, 11.4 × 8.2 mm) (ZRC 2009.0745), Kuta, central Lombok Island, coll. D. L. Rahayu, 18 August 2006; 1 male (7.8 × 6.0 mm) (MZB Cru 2585), Medana, northern Lombok Island, 08°21.791′S 116°07.608′E, coll. D. L. Rahayu, 12 June 2007; 1 male (7.0 × 5.8 mm) (MZB Cru 2586), Sekotong, western Lombok Island, coll. D. L. Rahayu, 16 May 2007.

Comparative material. — *Notonyx kumi* Naruse & Maenosono, 2009, paratype male (9.3 × 6.8 mm) (ZRC 2008.0022), Oh-jima Island, near Kume-jima Island, Ryukyu Islands, Japan, coll. T. Maenosono, 1 April 2008; 1 male (6.0 × 4.6 mm) (RMNH), station 47, Bay of Bima, north coast of Sumbawa, Lesser Sunda Islands, Indonesia, 13-31 m deep, coll. SIBOGA Expedition.

Etymology. — The new species is named in honour of Danièle Guinot, a respected colleague and cherished friend, for her many contributions to carcinology, especially towards a better understanding of the Goneplacidae. The name is a noun in the genitive singular.

Diagnosis. — Holotype: carapace subquadrilateral, about 1.4 times as wide as long (fig. 1A); frontal margin gently convex, about 0.3 times width of carapace (fig. 1B). Anterolateral margin short, entire, curved, without teeth; posterolateral margin gently concave, slightly diverging posteriorly (fig. 3A); dorsal surface of carapace convex anteriorly, posteriorly; smooth, without indication of regions except for shallow impression in gastro-cardiac area. Antennules long, terminal part overreaching cornea; basal antennal segment small, short, positioned on edge of orbital hiatus. Eye peduncle moderately long, 0.6 times frontal width; cornea small, pigmented. Gape present between third maxillipeds when closed; merus and ischium subequal, quadrate; antero-external angle of merus forming right angle (fig. 3B). Chelipeds subequal, right one larger (fig. 2B, C); fingers relatively stout, dactylus with longitudinal shallow groove; fixed finger with slight longitudinal keel on ventral margin extending almost to tip of finger; palm smooth; inner distal edge of carpus with low spine, inner margin finely denticulate, with clump of setae (fig. 3E); merus with long, stiff setae on anterior, posterior margins. Ambulatory legs relatively long, smooth; dactylus longer than propodus, with carina on lateral surface; all segments with long, relatively dense setae on dorsal and ventral margins (fig. 3C). Thoracic sternum relatively broad; surface lightly pitted; sternites 1, 2 fused, separated from sternite 3 by distinct transverse suture; sternites 3, 4 fused with lateral incision marking somites (fig. 1C). Male abdomen triangular (figs. 1C, 2A, 3D), relatively narrow with 7 movable somites including telson;

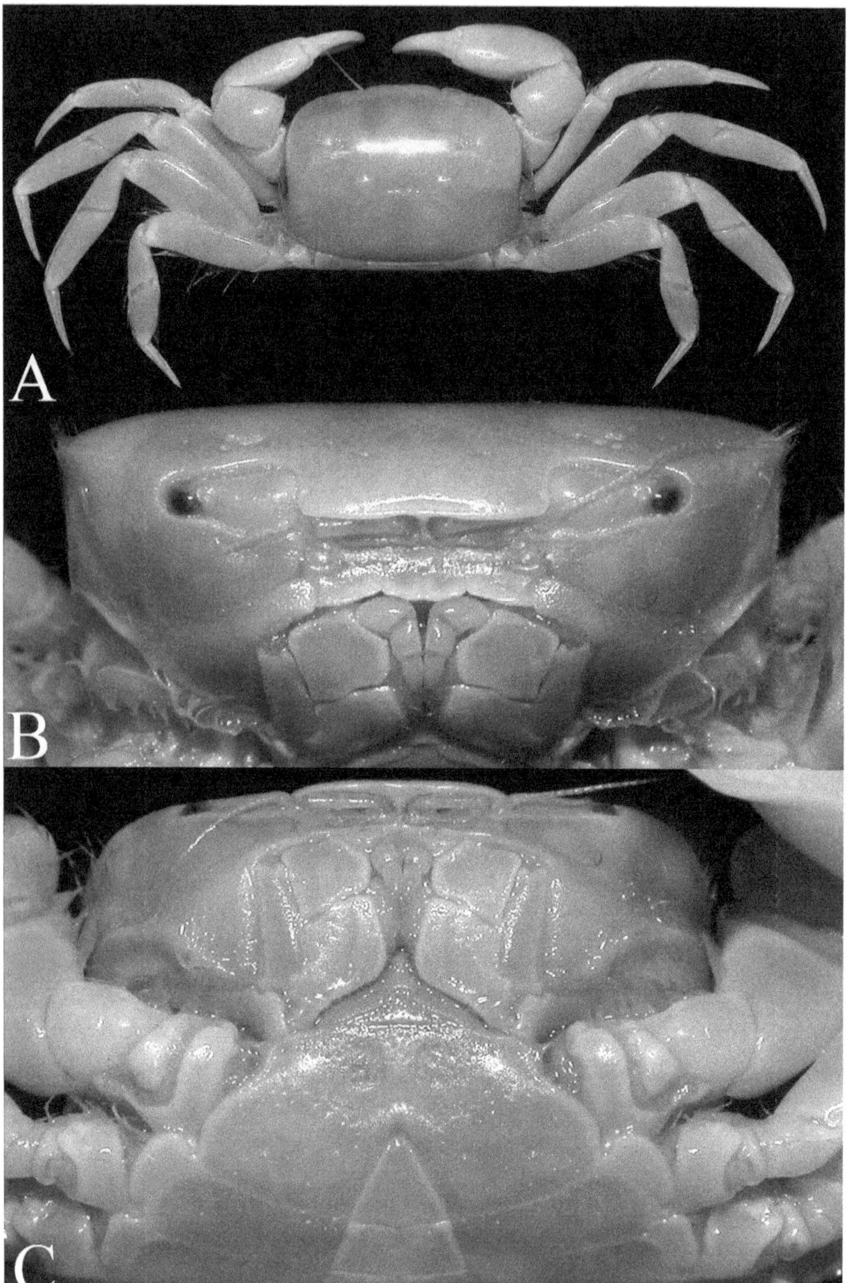

Fig. 1. *Notonyx guinotae* n. sp., holotype, male (15.5 × 11.5 mm) (MZB Cru 2583). A, overall dorsal view; B, frontal view of carapace; C, ventral view showing anterior part of thoracic sternum and abdomen.

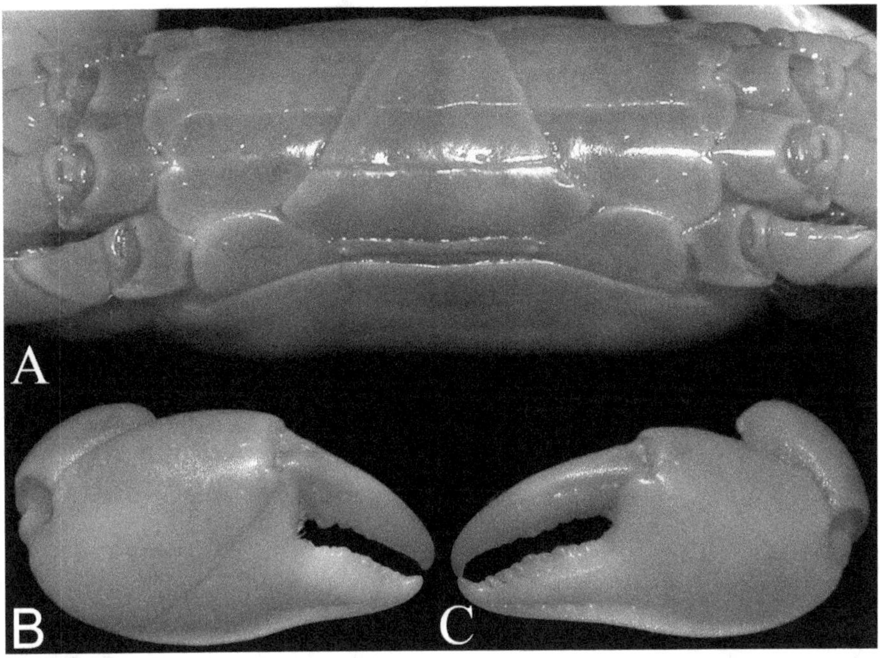

Fig. 2. *Notonyx guinotae* n. sp., holotype, male (15.5 × 11.5 mm) (MZB Cru 2583). A, posterior part of thoracic sternum and abdomen; B, outer surface of right cheliped; C, outer surface of left cheliped.

somite 1 underneath carapace, somite 2 narrow, short, somite 3 widest, somites 4 to 6 tapering to triangular telson; telson longer than somite 6 (figs. 1C, 3D). G1 relatively long (fig. 4), curved, distal part produced, elongate, tip truncate; ventral subdistal part swollen. G2 long (fig. 5), tip slightly twisted, spoon-shaped.

Females. — Chelipeds equal. Abdomen not rounded, ovate, not covering entire surface of thoracic sternum.

Colour. — Small specimens less than 11.4 × 8.2 mm are a uniform ivory white. Large specimens, like the holotype (15.5 × 11.5 mm, MZB Cru 2583), have a dark brown carapace with faint pinkish-orange markings, the chelipeds are pink with white fixed finger and the ambulatory legs are pinkish-white.

Remarks. — The morphologies of the G1 and G2 of *N. guinotae* n. sp. vary according to the size of the specimens. The smaller paratype male (7.0 × 5.8 mm, MZB Cru 2586) has an almost straight G1 with no significant ventral swelling subdistally (fig. 4A) and the tip of the G2 is simple and not twisted (fig. 5A). In a larger paratype specimen (7.8 × 6.0 mm, MZB Cru 2585), the ventral subdistal part of the G1 is moderately swollen and the tip slightly

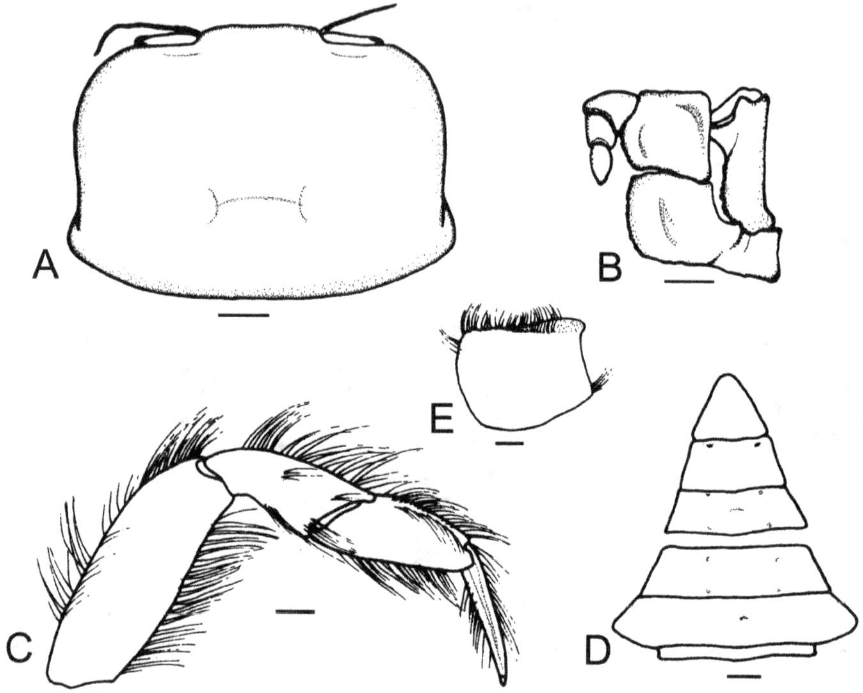

Fig. 3. *Notonyx guinotae* n. sp., holotype, male (15.5 × 11.5 mm) (MZB Cru 2583). A, carapace; B, third maxilliped; C, right fifth pereopod; D, abdomen; E, carpus of right cheliped. Scale: A & E = 2 mm; B-D = 1 mm.

protruded (fig. 4C), with the tip appearing somehow bifurcate when viewed sublaterally (fig. 4D). The tip of the G2 is also slightly twisted (fig. 5C, D). In the 11.4 × 8.2 mm paratype specimen (ZRC 2009.0745), the G1 is relatively longer, more curved, and the ventral subdistal part is prominently swollen with the tip clearly protruded (fig. 4E, F), and the tip of the G2 is twisted and directed upwards (fig. 5E, F). In the largest male (holotype, 15.5 × 11.5 mm, MZB Cru 2583), the G1 is relatively stouter, distinctly curved, with the ventral subdistal part significantly swollen, armed with more spines, and the tip protrudes very prominently (fig. 4G, H). The tip of G2 is now slightly curved and distinctly spoon-shaped (fig. 5G-I). Despite the changes in form, the small spines on the ventral surface of the subdistal part are always in the same position, albeit there are fewer of them in small specimens. The setae on the ambulatory legs also vary. The setae are short and scarce in smaller individuals, becoming gradually longer and denser in larger specimens.

The discovery of a third species of *Notonyx* in the Lesser Sunda Islands in Indonesia is noteworthy. Previously known from Indonesia are *N. latus*

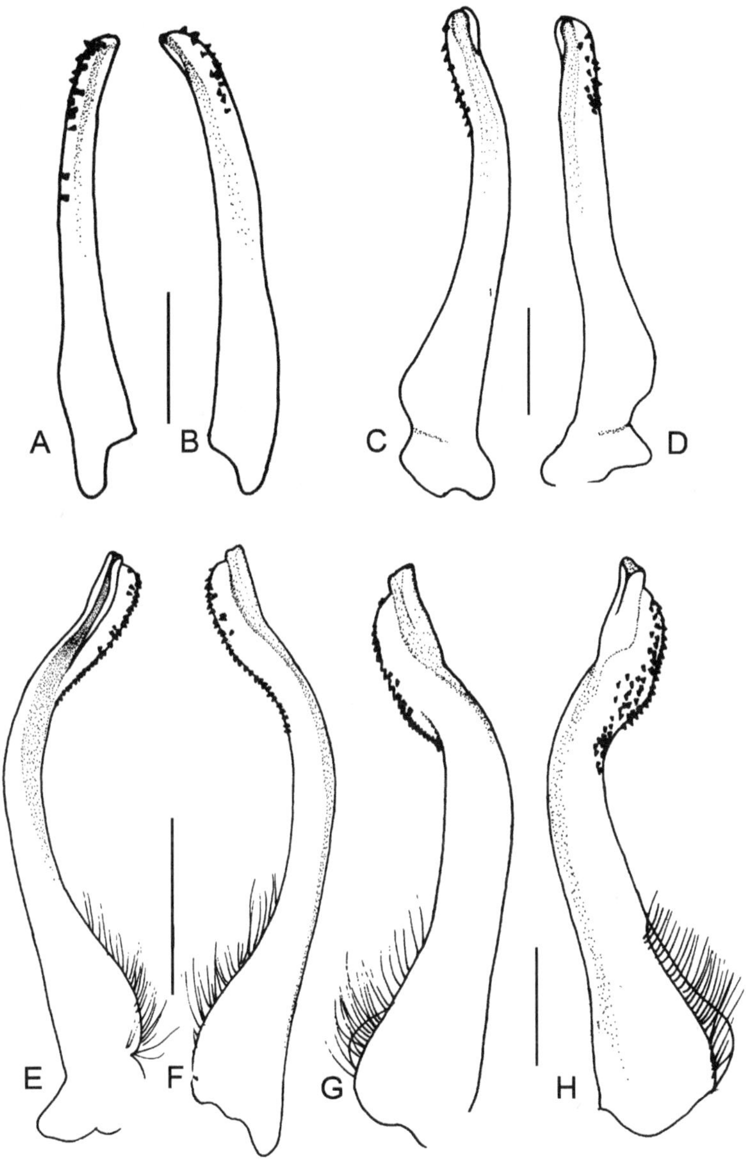

Fig. 4. *Notonyx guinotae* n. sp. Male first gonopod. A, B, paratype male (7.0 × 5.8 mm) (MZB Cru 2586); C, D, paratype male (7.8 × 6.0 mm) (MZB Cru 2585); E, F, paratype male (11.4 × 8.2 mm) (ZRC 2009 0745); G, H, holotype male (15.5 × 11.5 mm) (MZB Cru 2583). Scales: A-D = 0.5 mm; E-H = 1.0 mm.

(Kei Islands) and *N. kumi* (Sumbawa) with the latter also occurring in Japan (type locality). *Notonyx guinotae* is superficially most similar to *N. kumi* but can immediately be distinguished by its adult colour. *Notonyx kumi* has a

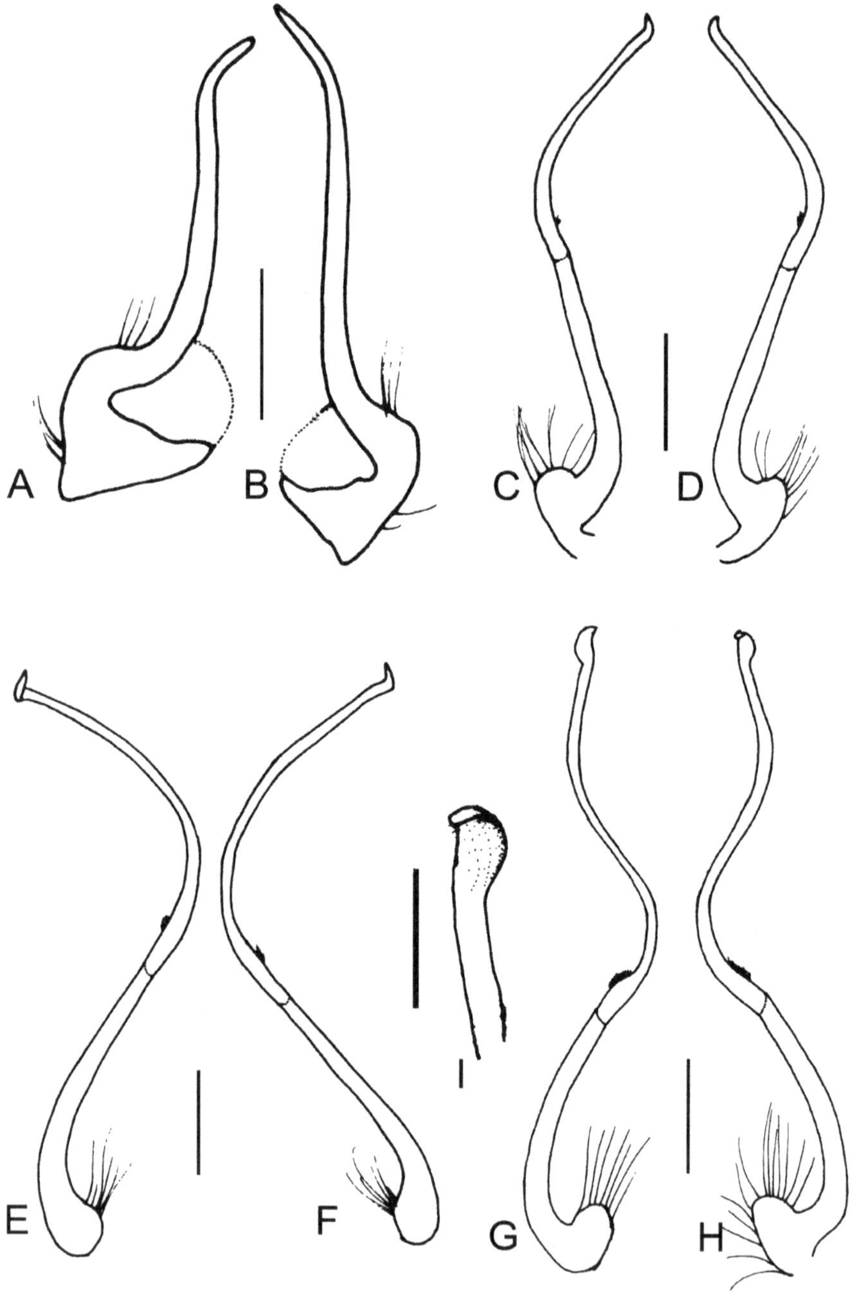

Fig. 5. *Notonyx guinotae* n. sp. Male second gonopod. A, B, paratype male (7.0 × 5.8 mm) (MZB Cru 2586); C, D, paratype male (7.8 × 6.0 mm) (MZB Cru 2585); E, F, male (11.4 × 8.2 mm) (ZRC 2009 0745); G-I, holotype male (15.5 × 11.5 mm) (MZB Cru 2583). Scales: A-F, I = 0.5 mm; G, H = 1.0 mm.

characteristic pattern of maroon lines and blotches on its carapace (Naruse & Maenosono, 2009, fig. 3) while *N. guinotae* is uniformly porcelain-white in small individuals with the carapace becoming dark brown with pinkish white ambulatory legs in larger specimens. The G1s of the two species superficially resemble each other, but that of *N. guinotae* is distinctly more slender and elongated, with the distal part relatively less dilated (fig. 4G, H) whilst in *N. kumi* it is proportionately stouter and shorter, with the distal part more dilated (cf. Naruse & Maenosono, 2009, fig. 2b, c). The G2 of *N. guinotae* is also gently curved, with the basal part only weakly expanded (fig. 5G, H) (sinuous with the basal part distinctly dilated in *N. kumi*, cf. Naruse & Maenosono, 2009, fig. 2d). The male abdomen is relatively more slender (fig. 3D) (relatively broader in *N. kumi*, cf. Naruse & Maenosono, 2009, fig. 2a), and the ambulatory propodi and dactyli are proportionately longer (fig. 3C) (relatively shorter in *N. kumi*, cf. Naruse & Maenosono, 2009, figs. 1e, 3a). As these differences are based on comparisons of males of similar sizes, they are not likely to be caused by size issues. The good series of male and female specimens of *N. guinotae* also show the third maxilliped features, ambulatory leg proportions, and male abdominal shapes do not vary substantially in the taxon, and are therefore useful taxonomic characters at the species level.

Notonyx guinotae differs from *N. latus* in its relatively less broad carapace, proportionately shorter ischium and merus of the third maxilliped (fig. 3B) (cf. Ng & Clark, 2008, fig. 2B, for *N. latus*) and proportionately shorter ambulatory legs (figs. 1A, 3C; see Ng & Clark, 2008, figs. 1A, B, 2D, for *N. latus*). More importantly, their G1s and G2s differ very markedly. The ventral subdistal part of the G1 of *N. guinotae* is simple and partially dilated (fig. 4G, H) whilst that of *N. latus* has the distal part prominently hooked and the ventral subdistal part is not dilated (Ng & Clark, 2008, fig. 3). The G2 of *N. guinotae* is also curved with the basal part small (fig. 5G, H) whilst that of *N. latus* is sinuous with the basal part relatively larger (Ng & Clark, 2008, fig. 3).

Ecology. — *Notonyx guinotae* was obtained from intertidal habitats with fine sand and gravel substrates. The area where it was collected has extensive seagrass beds. The crabs were burying in the sand and gravel, and were collected by sieving. The relatively simple and smooth carapace of *Notonyx* species is therefore very well adapted to living in this habitat.

ACKNOWLEDGEMENTS

We thank the personnel of Marine Bio-industry Implementation Unit, Research Center for Oceanography — Indonesian Institute of Sciences (LIPI) for their assistance in the field during the collection of the material for this study. The first author's research visit in Singapore was supported by a Raffles Museum Fellowship.

REFERENCES

ALCOCK, A., 1900. Materials for a carcinological fauna of India. No. 6. The Brachyura Catometopa, or Grapsoidea. Part II. Journal of the Asiatic Society of Bengal, **9** (3): 279-456.

CASTRO, P., 2007. A reappraisal of the family Goneplacidae MacLeay, 1838 (Crustacea, Decapoda, Brachyura) and revision of the subfamily Goneplacinae, with the description of ten new genera and eighteen new species. Zoosystema, **29** (4): 609-774.

CLARK, P. & P. K. L. NG, 2006. A new species of *Notonyx* A. Milne-Edwards, 1873 (Crustacea: Decapoda: Brachyuran: Goneplacidae) from the intertidal zone of Phuket, Thailand. Zoosystema, **28** (2): 539-551.

MACLEAY, W. S., 1838. On the brachyurous decapod Crustacea. Brought from the Cape by Dr. Smith. In: A. SMITH (ed.), Illustrations of the zoology of South Africa; consisting chiefly of figures and descriptions of the objects of natural history collected during an expedition into the interior of South Africa, in the years 1834, 1835, and 1836; fitted out by 'The Cape of Good Hope Association for Exploring Central Africa': together with a summary of African zoology, and an inquiry into the geographical ranges of species in that quarter of the globe, published under the Authority of the Lords Commissioners of Her Majesty's Treasury, Invertebratae. **IV** [1849]: 53-71, pls. 2, 3. (Smith, Elder & Co., London).

MILNE-EDWARDS, A., 1873. Recherches sur la faune carcinologique de la Nouvelle-Calédonie, Deuxième partie. Nouvelles Archives du Muséum d'Histoire naturelle, Paris, **9**: 155-332, pls. 4-18.

NARUSE, T. & T. MAENOSONO, 2009. *Notonyx kumi*, a new species of goneplacid crab (Decapoda; Brachyura) from the Ryukyu Islands, Japan, and Lesser Sunda Islands, Indonesia. Bulletin of the National Museum Natural Sciences, Tokyo, (Supplement) **3**: 183-189.

NG, P. K. L. & P. F. CLARK, 2008. A new species of *Notonyx* A. Milne-Edwards, 1873 (Crustacea: Decapoda: Brachyuran: Goneplacidae) from Indonesia. Zootaxa, **1987**: 20-26.

NG, P. K. L., D. GUINOT & P. J. F. DAVIE, 2008. Systema Brachyurorum, Part I. An annotated checklist of extant brachyuran crabs of the world. Raffles Bulletin of Zoology, (Supplement) **17**: 1-286.

STEPHENSEN, K., 1946. The Brachyura of the Iranian Gulf with an Appendix: The male pleopoda of the Brachyura. Danish Scientific Investigation in Iran, **4**: 57-237.

TESCH, J., 1918. The Decapoda Brachyura of the Siboga-Expedition. II. Goneplacidae and Pinnotheridae. In: Uitkomstem op Zoologisch, Botanisch, Oceanographisch en Geologisch gebied verzameld in Nederlandsch Oost-Indië 1899-1900 aan boord H.M. "Siboga" onder commando van Luitenant ter Zee 1e kl. G.F. Tydeman uitgeven door Dr. Max Weber. Siboga-Expeditie Monographie, **39c1** (84): 149-295, pls. 7-18. (Brill, Leiden).

First received 28 June 2009.
Final version accepted 14 July 2009.

PHYLOGENETIC RELATIONSHIPS OF THE PLAGUSIIDAE DANA, 1851 (BRACHYURA), WITH DESCRIPTION OF A NEW GENUS AND RECOGNITION OF PERCNIDAE ŠTEVČIĆ, 2005, AS AN INDEPENDENT FAMILY

BY

CHRISTOPH D. SCHUBART[1,3]) and JOSÉ A. CUESTA[2,4])

[1]) Biologie I, Universität Regensburg, D-93040 Regensburg, Germany

[2]) Instituto de Ciencias Marinas de Andalucía, CSIC, Avenida República Saharaui, 2, E-11519 Puerto Real, Cádiz, Spain

ABSTRACT

A molecular and morphological analysis of representatives of the family Plagusiidae, including all members of *Plagusia* Latreille, 1804, and the recently established *Davusia* Guinot, 2007, was carried out. Due to marked differences in adult and larval morphology, as well as mitochondrial and nuclear DNA, two species of *Plagusia*, viz., *P. chabrus* (Linnaeus, 1758), and *P. dentipes* De Haan, 1835, are considered sister taxa but distinct from other members of the genus. They are transferred to a new genus, *Guinusia*. A molecular phylogeny suggests that *Guinusia* is not closer related to *Plagusia* than to the plagusiid genera *Euchirograpsus* H. Milne Edwards, 1853, and *Miersiograpsus* Türkay, 1978. Furthermore, with new evidence from mitochondrial and nuclear DNA as well as a reappraisal of the larval morphology, the genus *Percnon* Gistel, 1848, is formally removed from the Plagusiidae and recognized as a separate family, Percnidae Števčić, 2005.

RÉSUMÉ

Une analyse moléculaire et morphologique des représentants de la famille des Plagusiidae comprenant tous les membres du genre *Plagusia* Latreille, 1804, et le genre récemment établi *Davusia* Guinot, 2007, a été réalisée. Pour tenir compte des nettes différences dans la morphologie adulte et larvaire ainsi que sur l'ADN nucléaire et mitochondrial, deux espèces de *Plagusia*, *P. chabrus* (Linnaeus, 1758), et *P. dentipes* De Haan, 1835, sont considérées comme espèces sœurs et transférées dans le genre *Guinusia*. Une phylogénie moléculaire suggère que ce genre n'est pas plus proche du genre *Plagusia* que des genres de Plagusiidae *Euchirograpsus*

[3]) Corresponding author; e-mail: christoph.schubart@biologie.uni-regensburg.de

[4]) e-mail: jose.cuesta@icman.csic.es

© Koninklijke Brill NV, Leiden, 2010

H. Milne Edwards, 1853 et *Miersiograpsus* Türkay, 1978 et justifie donc sa séparation. Par ailleurs, considérant les nouvelles preuves des l'ADN mitochondriaux et nucléaires et une réévaluation de la morphologie larvaire, le genre *Percnon* Gistel, 1848 est ici formellement enlevé de la famille des Plagusiidae et placé dans une famille propre, les Percnidae Števčić, 2005.

INTRODUCTION

The genus *Plagusia* Latreille, 1804, underwent a period of taxonomic confusion and re-definitions towards the end of the 19th century (Miers, 1878; De Man, 1883; Alcock, 1900; see Schubart & Ng, 2000), but this largely stabilized with the monograph by Rathbun (1918). The modern composition of the genus, and the status of species (and subspecies) were more recently summarized by Dawson (1987) and Schubart & Ng (2000). Schubart et al. (2001) addressed the issue of the subspecific status within *Plagusia depressa* (Fabricius, 1775), and recognized *Plagusia squamosa* (Herbst, 1790) (formerly referred to as *P. depressa tuberculata* Lamarck, 1818, see Schubart & Ng, 2000) as a full species. Guinot (2007) then removed *P. glabra* Dana, 1851, from *Plagusia* and placed it in a new genus, *Davusia*. This left seven species of *Plagusia* and unclear phylogenetic relationships among the genus *Davusia* and the remaining members of the family Plagusiidae.

The higher classification of *Plagusia* and its allies has also experienced recent changes. The description of the larval stages of *Euchirograpsus americanus* A. Milne-Edwards, 1880, by Wilson (1980) and *Plagusia depressa* by Wilson & Gore (1980) revealed close morphological similarities between these species, which at that time were classified in different subfamilies of the family Grapsidae, i.e., Varuninae H. Milne Edwards, 1853, and Plagusiinae, respectively. Cuesta & Schubart (1997) discussed the subfamilial arrangement based on larval morphology and confirmed that the larvae of *Euchirograpsus* were more closely related to *Plagusia* then to those of the subfamily Varuninae. They also cited unpublished evidence from mitochondrial DNA for a close phylogenetic relationship between these two genera, suggesting a "systematic position of *E. americanus* in close relationship to *P. depressa* and not linked to the main group of the Varuninae" (Cuesta & Schubart, 1997: 296). Sternberg & Cumberlidge (2000) had come to a similar conclusion in a morphological cladistic analysis, with Guinot & Bouchard (1998) having previously noted that the morphology of the male abdominal press-button locking mechanism of *Euchirograpsus* suggested that this genus was misplaced in the Varuninae. The molecular results stated in Cuesta & Schubart (1997) were published by Schubart et al. (2000), officially transferring *Euchirograpsus* to the Plagusiidae: "The Plagusiinae, next to the type genus *Plagusia*, would now also include *Euchirograpsus* (see also d'Udekem d'Acoz, 1999) and the taxonomic position of *Percnon* awaits future comparisons (i.e., incertae sedes [sic])" (Schubart et al., 2000: 184). The inclusion of *Euchirograpsus* within the Plagusiidae was reiterated by Schubart et al. (2002), and a similar fate was suggested for *Miersiograpsus*. This was confirmed by Davie (2002), who after examining adult morphology, independently transferred *Miersiograpsus* to the Plagusiidae giving the evidence of fused abdominal somites 3-6 as an autapomorphy for *Euchirograpsus*, *Miersiograpsus*, and *Plagusia*. Also around this time, two papers argued that the former grapsid subfamilies should be raised to family level (Schubart et al., 2000; Sternberg & Cumberlidge, 2000) returning to a taxonomic system that initially Guinot (1978) had put forward, but was not followed in later classifications.

The position of *Percnon* within the Plagusiinae or Plagusiidae has also been questioned repeatedly (Schubart et al., 2000, 2006). Consequently, Števčić (2005) established a new tribe for the genus, Percnini, which was later adopted as a subfamily Percninae within the Plagusiidae by Ng et al. (2008). Published molecular results have placed representatives of *Percnon* (i.e., *P. gibbesi* (H. Milne Edwards, 1853) and *P. planissimum* (Herbst, 1804)) in a polyphyletic relationship to other members of the family Plagusiidae (see Schubart et al., 2000, 2006).

In the present contribution, we use phylogenetic analyses based on adult and larval morphology, as well as mitochondrial and nuclear DNA, to examine intrageneric relationships within *Plagusia*, and intergeneric relationships among plagusiid and potentially related genera. Our results show that there is sufficient morphological and molecular diversity within the genus *Plagusia*, that a new genus can be described for two of its constituent species. In addition, *Percnon* is officially removed from the Plagusiidae and placed in its own family within the Thoracotremata. The superfamily Grapsoidea is avoided in this context following the reasoning of Schubart et al. (2006) and new evidence based on nuclear DNA markers by Landstorfer et al. (unpubl.).

MATERIAL AND METHODS

Specimens examined are deposited in museums as listed in table I. New DNA sequences were obtained between 1999 and 2009. DNA extractions and selective amplification of mitochondrial and nuclear DNA were carried out as reported in Schubart et al. (2006). PCR-amplifications of the 16S rDNA were run with four minutes denaturation at 94°C, 40 cycles with 45 s 94°C, 1 min. 48°C, 1 min. 72°C and 10 min. final denaturation at 72°C, and the primers 16L2 or 16L29 and 16H2 or 16HLeu (see Schubart, 2009). In addition, nuclear DNA of the histone H3 gene was obtained for at least one specimen of all key taxa with a cycling program of 30 s 94°C, 30 s 48°C, 30 s 72°C (otherwise as above) and the primers H3af and H3ar (Colgan et al., 1998). PCR products were purified with Microcon 100 filters (Microcon) or Quick-Clean (Bioline) and subsequently sequenced with the ABI BigDye terminator mix followed by electrophoresis in an ABI Prism 310 Genetic Analyzer (Applied Biosystems, Foster City, CA, U.S.A.). New sequence data were submitted to the European molecular database EMBL (see table I for accession numbers). In addition, the following 16S mtDNA ingroup sequences archived in molecular databases were included in our analyses: *Guinusia dentipes* (as *Plagusia dentipes AJ308421*), *Plagusia depressa* (*AJ250649, AJ311791-AJ311793*), *Plagusia squamosa* (*AJ311794-AJ311796*), *Percnon planissimum* (*AM180693*) in addition to Genbank sequences for outgroups (see table I).

The model of DNA substitution that fitted our data best was determined using the software MODELTEST 3.6 (Posada & Crandall, 1998). Two methods of phylogenetic inference were applied to our dataset: Bayesian Inference (BI) as implemented in MrBayes v. 3.1 (Huelsenbeck & Ronquist, 2001) and Minimum Evolution (ME) as implemented in MEGA 4.1 (Tamura et al., 2007) with pairwise deletions, the Maximum Composite Likelihood model, heterogeneous pattern among lineages, and gamma distributed rates among sites. Gamma shape was established and BI trees were calculated using the suggested model of evolution by MODELTEST. The Bayesian analysis was run with four MCMC (Markov chain Monte Carlo) chains for 2 000 000 generations, saving a tree every 500 generations (with a corresponding output of 4000 trees) and with *Carpilius* sp. as the outgroup. The $-\ln L$ converged on a stable value between 15 000 and 20 000 generations (burn-in phase) so that the first 25 000 generations were always excluded from the analysis to optimize the fit of the remaining trees. The posterior probabilities of the phylogeny were determined by constructing a 50% majority-rule consensus of the remaining trees. ME confidence values were established by running 2000 bootstrap pseudoreplicates.

The mitochondrial dataset consisted of 43 sequences resulting in 675 aligned characters, of which 322 were conserved, 249 parsimony-informative, and 350 variable. GTR + G + I was selected as the best-fitting evolutionary

TABLE I

List of crab specimens of which DNA (16S mtDNA or H3 nDNA) was sequenced and used for phylogenetic reconstruction or morphological comparisons [authorities of species names mentioned in the main text are not repeated here]

Species	Collection locality and date	Collection #	16S / H3 accession #
Plagusiidae			
Guinusia chabrus	Australia: Cape Peterson; 38°40′S 145°37′E; 23.1.2000	NMV J47112	FM539005 / FN434050
Guinusia chabrus	New Zealand: Kaikoura, 11.1997	SMF-24560	FM539005
Guinusia dentipes	Japan: Honshu, Chiba-ken, Banda Tateyama, 5.7.1997	SMF-24559	AJ308421 / FN434051
Davusia glabra	Australia: SE Queensland, P. Lookout, 1998	QM W23974	FM539009 / FN434052
Euchirograpsus americanus	USA: Louisiana, Sackett Bank (61-76 m), 25.3.1996	ULLZ 3626	FM539010 / FN434055
Euchirograpsus timorensis	New Caledonia: 23°37.8′S 167°42.7′E, 29.1.1993	SMF-34761	FM539011 / FN434056
Miersiograpsus australiensis	New Caledonia: 24°44.55′S 170°08.85′E, 26.10.1991	SMF-34762	FM539012 / FN434053
Miersiograpsus kingsleyi	SE Atlantic off Namibia: 27°25′S 14°50′E, 21.7.1983	ICMD2/2004	FM539013 / FN434054
Plagusia depressa	Ghana: Cape Coast, 4.7.2001	SMF-34728	FM538999 / FN434046
Plagusia depressa	São Tomé and Príncipe: Bom-Bom ∼1°N-7°E, 3.2004	SMF-34729	FM539000
Plagusia immaculata	Clipperton: 10°18.80′N 109°12.50′W, 18.1.2005	MNHN B-31976	FM539002 / FN434048
Plagusia immaculata	Maldives: Rasdhoo Atoll, Kuramathi, 12.7.1999	SMF-24941	FM539003
Plagusia immaculata	Vietnam: Hon Miau Nhatrang, 23.11.1999	MNHN B-27676	FM539004
Plagusia speciosa	Clipperton: "crête recifale à mer trés basse", 2005	MNHN B-30681	FM539008
Plagusia speciosa	Taiwan: Pingtung, Fongtzueisha, 5.9.1999	SMF-pending	FM539007 / FN434049
Plagusia squamosa	Maldives: Rasdhoo Atoll, Kuramathi, 12.7.1999	SMF-24943	FM539001
Plagusia squamosa	Hawaii: Oahu Kewalo, 2000	ZRC 2000.0412	AJ311796 / FN434047
Percnidae			
Percnon gibbesi	Spain: Ibiza, Cala Llenya, 1.4.2001	SMF-34760	FM539014 / FN434058
Percnon guinotae	Maldives: Rasdhoo Atoll, Kuramathi, 14.7.1999	SMF-24946	FM539015 / FN434059
Percnon planissimum	Kenya: Malindi: Watamu, 6.11.1999	MZUF 2735	AM180693 / FN434057

TABLE I
(Continued)

Species	Collection locality and date	Collection #	16S / H3 accession #
Outgroups			
Cardisoma carnifex (Herbst, 1794)	Thailand: Phuket: Ao Tang Khen, Dec. 1999	ZRC 1999.0787	FM539017 / FN434044
Discoplax hirtipes (Dana, 1852)	Taiwan: Pingtung: Hsiang-Chiao-Wan, 19.5.1998	ZRC 1998.398	FM863830 / FN434045
Gecarcinus lateralis (Fréminville, 1835)	Mexico: Veracruz: Pta. Delgada, 18.4.1998	ULLZ 3722	AJ130804 / FN434043
Gecarcoidea lalandii H. Milne Edwards, 1837	Taiwan: Pingtung: Hengchun, 19.5.1998	ZRC 1998.423	AM160684 / FN434042
Grapsus grapsus (Linnaeus, 1758)	Mexico: Veracruz: Pta. Delgada, 18.4.1998	ULLZ 3709	AJ250650 / FN434067
Pachygrapsus transversus (Gibbes, 1850)	Brazil: Ilha Sta. Catarina: Praia Ingleses, Nov. 2004	MZUSP-16887	FM539016 / FN434068
Chiromantes haematocheir (de Haan, 1835)	Japan: Wakayama: Samusaura, 11.6.1998	SMF-25989	AJ308414 / FN434063
Sesarma reticulatum (Say, 1817)	USA: Delaware, Woodland Beach, 17.9.1994	ULLZ 3835	AJ130799 / FN434062
Cyclograpsus integer (H. Milne Edwards, 1837)	Mexico: Veracruz: Pta. Delgada; 18.4.1998	ULLZ 3704	AJ250639 / FN434061
Varuna litterata (Fabricius, 1798)	Kenya: Mida Creek	MZUF 2503	AJ308419 / FN434060
Glyptograpsus jamaicensis (Benedict, 1892)	Jamaica: Non Such, 21.3.1997	SMF-25987	AJ308420 / FN434065
Platychirograpsus spectabilis De Man, 1896	USA: Florida: Tampa Bay, Alafia R., June 1996	SMF-24567	AJ250645 / FN434064
Xenograpsus testudinatus Ng et al., 2000	Taiwan: I-Lan, Kueishan, 16 m depth; 18.4.1999	ZRC 2009.0866	FM863827 / FN434066
Ocypode quadrata (Fabricius, 1787)	USA: Georgia: Jekyll Island, 24.8.2000	ULLZ 4261	FM539018 / FN434069
Ucides cordatus (Linnaeus, 1763)	Puerto Rico: Boquerón, July 1994	tissue only	FM539019 / FN434070
Palicus obesus (A. Milne-Edwards, 1880)	Gulf of Mexico, 1989	ULLZ 3852	AJ130808 / FN434071
Carpilius sp.	French Polynesia: Îles Australes, 17.8.1991	SMF-32771	FM208748 / FM208786

ICMD, Instituto de Ciencias Marinas, Barcelona, Spain; MNHN, Muséum national d'Histoire naturelle, Paris, France; MUZSP, Museu de Zoologia Universidade de São Paulo, Brazil; MZUF, Museo Zoologico Universitá di Firenze, Italy; NMV, National Museum Victoria, Melbourne, Australia; SMF, Forschungsinstitut und Museum Senckenberg, Frankfurt a.M., Germany; ULLZ, University of Louisiana Zoological Collection, Lafayette, USA; QM, Queensland Museum, Brisbane, Australia; ZRC, Zoological Reference Collection, Raffles Museum at the National University of Singapore, Singapore.

model by MODELTEST and implemented for subsequent Bayesian analysis. The ME bootstrap consensus tree resulted in an identical topology for all significant nodes. This topology with confidence values of both methods exceeding 50% is shown in fig. 1. The histone H3 nuclear dataset consisted of 31 taxa with a fixed length (no indels) of 328 basepairs. It was added to the 16S alignment for combined evidence, resulting in an alignment of 1003 basepairs (544 were conserved, 322 parsimony-informative, and 456 variable). Again, GTR + G + I was selected as the best-fitting evolutionary model by MODELTEST and implemented for subsequent Bayesian analysis. The resulting consensus tree with BI and ME topologies exceeding 50% is shown in fig. 2.

SYSTEMATIC ACCOUNT

Family PLAGUSIIDAE Dana, 1851

Plagusinae [sic] Dana, 1851: 288; 1852: 333, 368.
Plagusiinae — Miers, 1878: 147; Alcock, 1900: 289, 436; Guinot, 1978: 285; 1979: 106, 144, 151, 155, 206, 209; Manning & Holthuis, 1981: 238; Williams, 1984: 462; Hartnoll, 1992: 51; Ng, 1998: 1139; Guinot & Bouchard, 1998: 664. [All partim.]
Plagusiidae — Schubart et al., 2000: 183-184; Von Sternberg & Cumberlidge, 2000: 120, 133; Martin & Davis, 2001: 75; Ng et al., 2001: 47; Schubart et al., 2002: 29; Davie, 2002: 436; Števčić, 2005: 126; Schubart et al., 2006: 193, 198; Ng et al., 2008: 30, 218. [All partim.]

Type genus. — *Plagusia* Latreille, 1804 (type species *Cancer depressus* Fabricius, 1775, by monotypy and designation by Latreille, 1810: 422). Gender feminine.

Guinusia n. gen.

Plagusia — Dana, 1852: 370-371 (pro parte); Miers, 1878: 148, 152-153 (pro parte); Rathbun, 1918: 336-337 (pro parte); Barnard, 1950: 134-137 (pro parte); Griffin, 1968: 209-213; Benech, 1978: 104; Guinot, 1979: 144, 151, 209 (pro parte); Wear & Fielder, 1985: 73-76; Dawson, 1987: 37-44; Guinot & Bouchard, 1998: 664 (pro parte); González-Gordillo et al., 2000: 144-149 (pro parte); Schubart & Ng, 2000: 334 (pro parte); Guinot, 2007: 30 (pro parte).
(Not *Plagusia* Latreille, 1804.)

Type species. — *Guinusia chabrus* (Linnaeus, 1758) new combination, described as *Cancer chabrus* Linnaeus, 1758, by present designation. Gender feminine.
Other species included. — *Guinusia dentipes* (De Haan, 1835) new combination, described as *Grapsus* (*Plagusia*) *dentipes* De Haan, 1835.

Etymology. — This genus is dedicated to our esteemed and admired colleague Danièle Guinot in recognition for her decisive contributions to the understanding of the Brachyura, her Thoracotremata, the family Plagusiidae, and the genus *Plagusia*. Her recent recognition of *Davusia* as a new genus

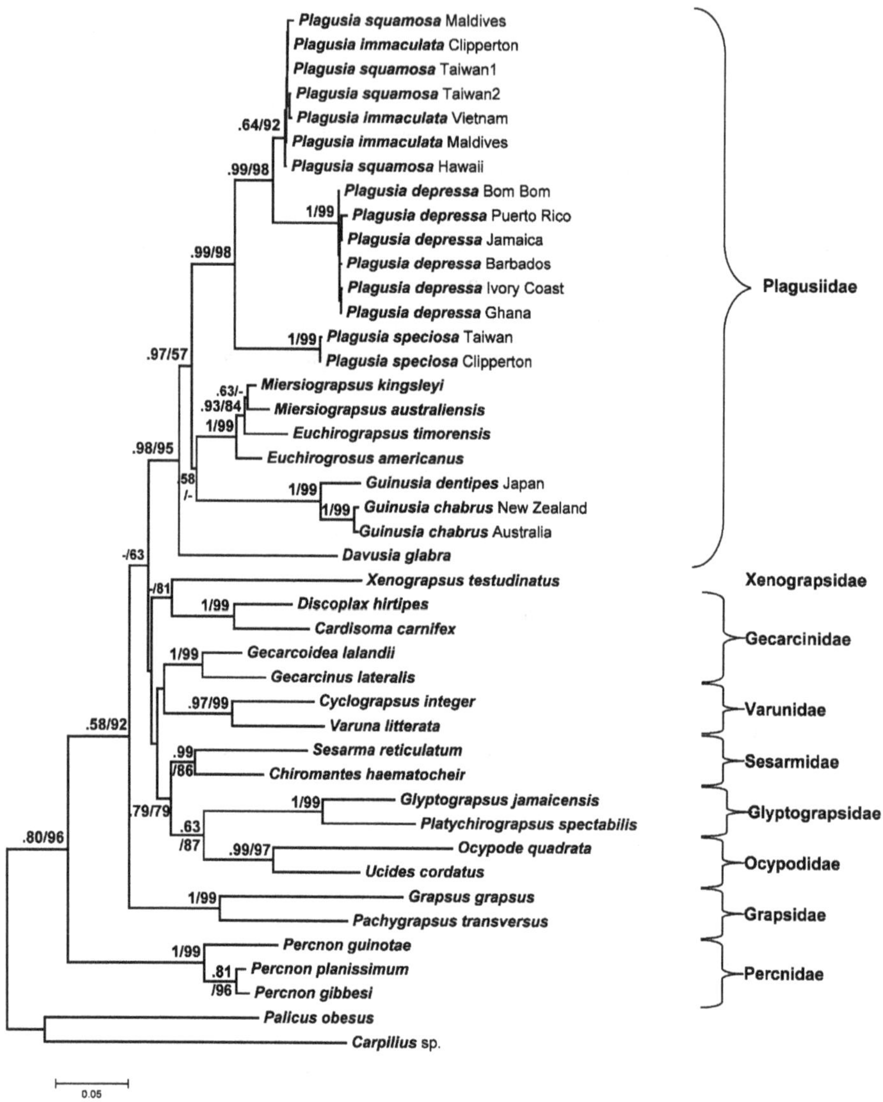

Fig. 1. Phylogenetic consensus tree of representatives of the Plagusiidae and thoracotreme outgroups based on 675 basepairs of mtDNA (16S rRNA). Bayesian Inference (BI) and Minimum Evolution (ME) confidence intervals (only ⩾50) corresponding to BI posterior probabilities/ME bootstrap values. *Carpilius* sp. was used as outgroup.

triggered the finalization of this long-planned study. The genus name is composed of the four initial letters of Danièle Guinot's last name in addition to the ending *usia*, expressing morphological similarity to the genera *Plagusia* and *Davusia*. Gender feminine.

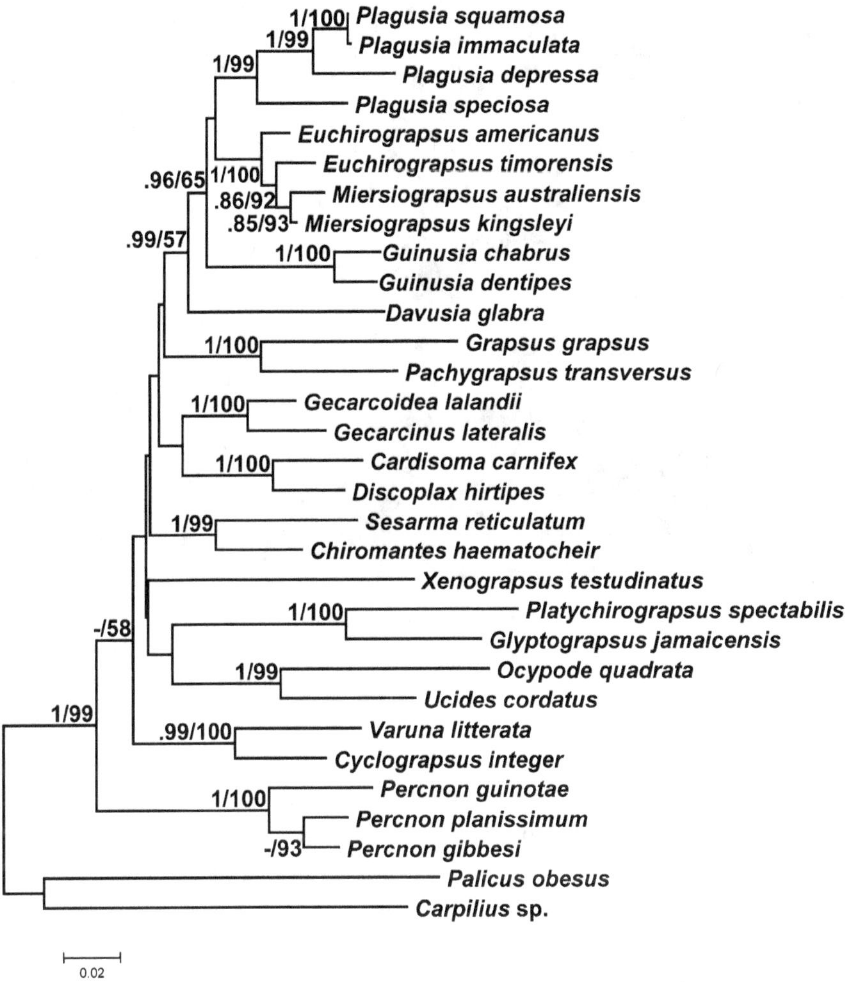

Fig. 2. Phylogenetic consensus tree of representatives of the Plagusiidae and thoracotreme outgroups based on 1003 basepairs (bp) of combined DNA (675 bp 16S rRNA, 328 bp histone H3). Bayesian Inference (BI) and Minimum Evolution (ME) confidence intervals (only $\geqslant 50$) corresponding to BI posterior probabilities/ME bootstrap values. *Carpilius* sp. was used as outgroup.

Diagnosis. — Joints of the meri of all ambulatory legs with a series of spines on their upper margin (as opposed to one subterminal and one terminal spine in *Plagusia*) (fig. 3). Carapace almost entirely destitute of tubercles (as opposed to tuberculate in *Plagusia*) (fig. 4). Larval characters as listed in table II. Other studied characters not differentiating diagnostically from *Plagusia*.

Remarks. — Guinot (2007) lists a number of characters that separate her new genus *Davusia* from the remaining species of *Plagusia*. One of these

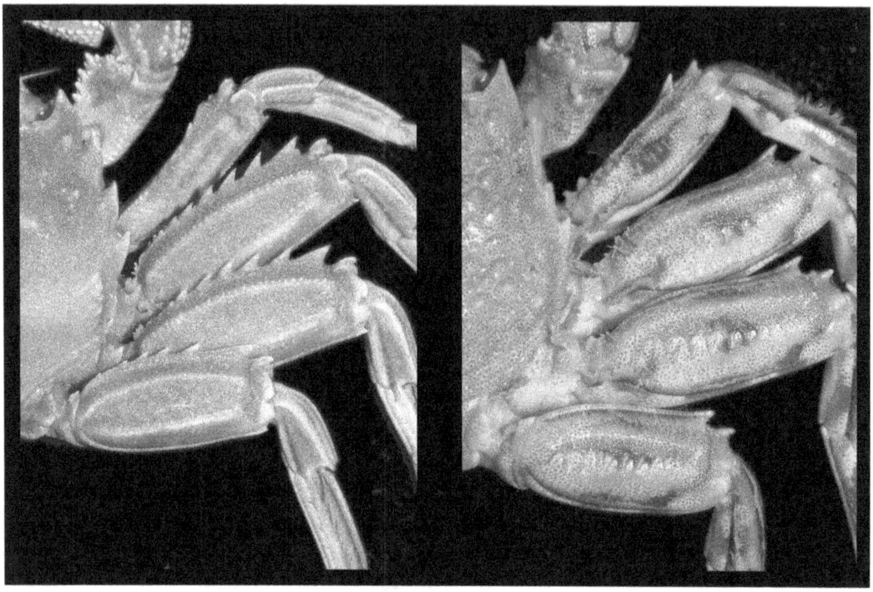

Fig. 3. Morphology of meri of walking legs in *Guinusia* gen. n. and *Plagusia*. Left, *Guinusia dentipes* (De Haan, 1835), Japan (SMF-24559); Right, *Plagusia depressa* (Fabricius, 1775) Ghana (SMF-34728).

characters is the presence of a palp on the exopods of the third maxillipeds. This palp is absent in the other species of *Plagusia*, except for *P. dentipes*, where it may be present in the form of a flagellum. She then concludes that "the generic position of *P. dentipes* should be re-examined" (Guinot, 2007: 30). On the other hand, the gonopod of *Davusia glabra* turns out to be similar to the one of *P. chabrus*. These characters are therefore not consistent in the separation of the newly recognized genus *Davusia* and suggest a mosaic of characters within the genus *Plagusia* in its former sense.

More conclusive adult morphological characters to consistently separate two species of the genus *Guinusia* n. gen., *G. chabrus* and *G. dentipes*, from the species of *Plagusia* (and *Davusia*) have in fact already been provided much earlier by Miers (1878). He separated his first group consisting of *Plagusia depressa*, *P. tuberculata* (= *P. squamosa*), *P. immaculata* Lamarck, 1818, *P. speciosa* Dana, 1852, and *P. glabra* from his second group, with *P. dentipes* and *P. chabrus*, on the basis of the absence or presence of a "series of spines" on the upper margin of the ambulatory merus and, to a lesser extent, of the presence or absence of tubercles on the carapace. It is precisely these characters (figs. 3, 4) in addition to larval morphology (fig. 5) and molecular

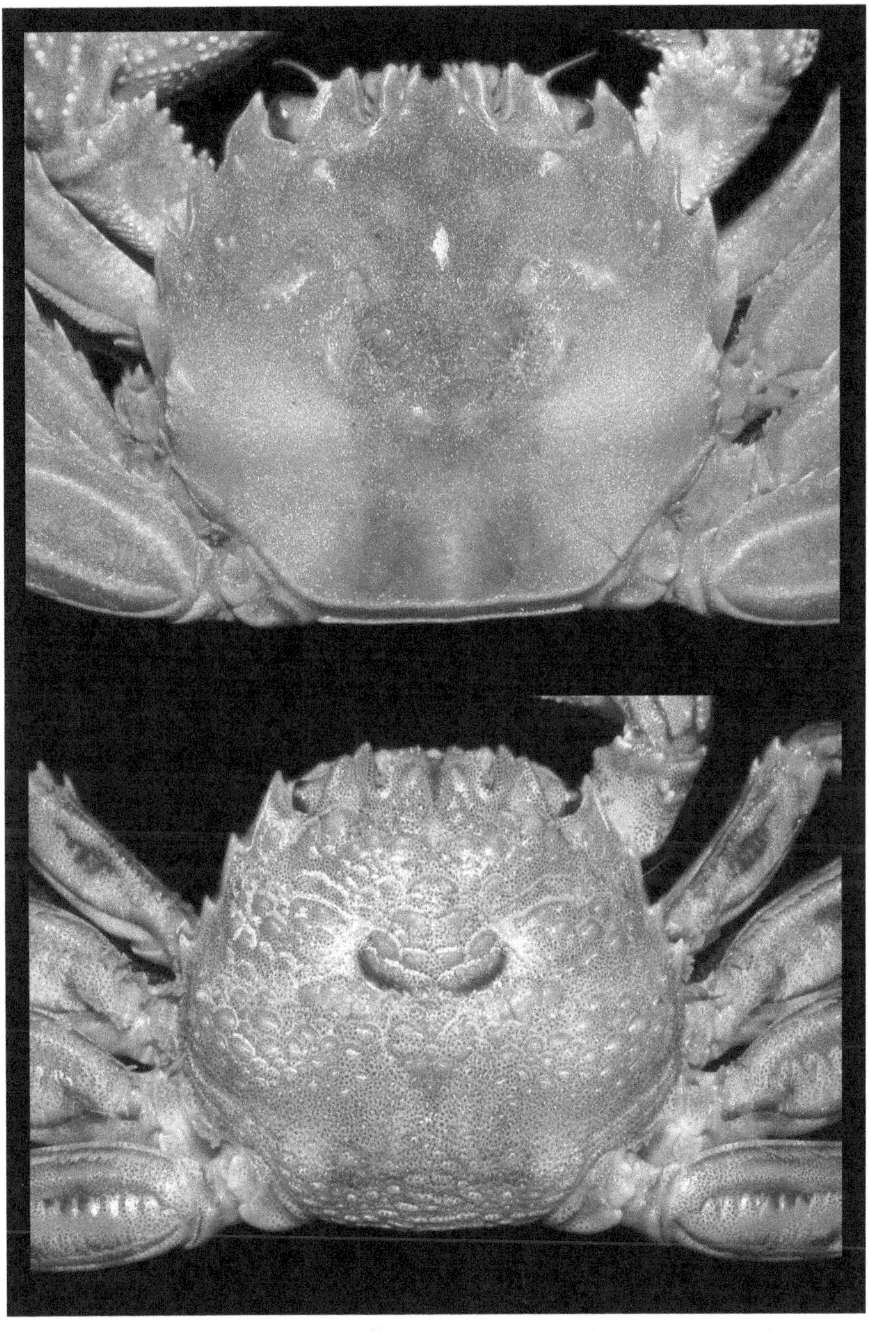

Fig. 4. Morphology of carapace structures in *Guinusia* gen. n. and *Plagusia*. Top, *Guinusia dentipes* (De Haan, 1835), Japan (SMF-24559); Bottom, *Plagusia depressa* (Fabricius, 1775), Ghana (SMF-34728).

TABLE II

Morphological and meristic differences between zoeal and megalopal stages of those species of Plagusiidae with known larval data. Abbreviations: (–) no data; (?) zoea collected in plankton samples, number of stage attributed by the author; antenna type (A) refers to exopod present as a small simple seta, (B) refers to exopod less than 1/6 of the protopod length with 1-2 terminal setae, longer one never reaching the middle of protopod length; (C) refers to exopod longer than 1/6 of the protopod length with small terminal spines and 2 terminal setae, longer one almost reaching the tip of the protopod [authorities of species names have been mentioned in the main text]

Species	No. zoeal stages	Zoeal stages		Megalopa stage	
		Antenna type	Setation of 1st maxilliped basis	No. segment on exopod of 3rd maxilliped	No. cincinuli on endopod of pleopods
Euchirograpsus americanus	5(+6)	B	2,2,2,2	–	–
Guinusia chabrus	12(?)	B	2,2,3,3	–	–
Guinusia dentipes	–	B	2,2,3,2	2	13-15
Plagusia depressa	6(?)	C	2,2,2,2	1	9
Plagusia speciosa	–	C	2,2,2,2	–	–
Plagusia squamosa	–	C	2,2,2,2	1	8-9
Percnon abbreviatum	–	–	–	–	11-15
Percnon gibbesi	6(?)	A	2,2,3,2	1(1)	13-17
Percnon guinotae	–	–	–	–	11-14
Percnon planissimum	–	–	–	1	11-14

(1) Cuesta & Schubart, unpublished data.

datasets (figs. 1, 2) that we here use as distinguishing characters for the new genus.

Supporting characters for separating the two species here included in *Guinusia* n. gen. from the remaining representatives of *Plagusia* were indeed found in the larval morphology, despite the fact that larval descriptions of Plagusiidae are incomplete and scarce (no larvae yet known for *Davusia* and *Miersiograpsus*, and only an incomplete description of the zoea of *Euchirograpsus americanus*, as opposed to published larval data for four out of seven recognized species of *Plagusia*). Zoeal data are known for *P. chabrus* (zoea I by Wear, 1970, and zoea V? and XII? by Wear & Fielder, 1985), *P. dentipes* (zoea I by Terada, 1987), *P. depressa* (zoea I-V by Wilson, 1980), and *P. squamosa* (zoea I by Schubart et al., 2001). Included in the present study are the unpublished results (by J. Cuesta, H.-C. Liu & C. D. Schubart) of the first zoea of a fifth species, *P. speciosa*.

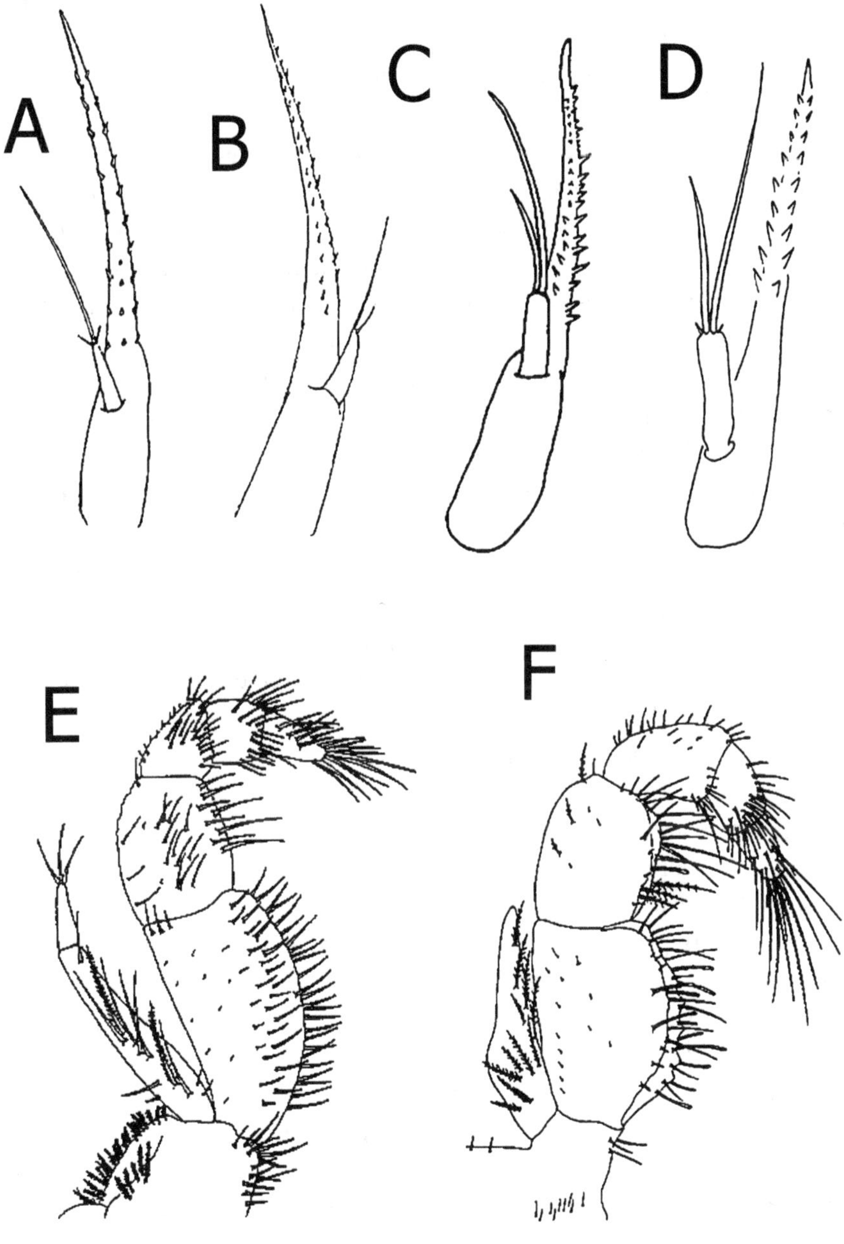

Fig. 5. Antennal morphology of zoea I of: A, *Guinusia chabrus* (Linnaeus, 1758); B, *Guinusia dentipes* (De Haan, 1835); C, *Plagusia depressa* (Fabricius, 1775); D, *Plagusia speciosa* Dana, 1852 (after Wear, 1970; Terada, 1987; Schubart et al., 2001; Cuesta, Liu & Schubart, unpubl., respectively). Third maxilliped (detail of exopod) of the megalopa of: E, *Guinusia dentipes*; F, *Plagusia depressa* (after González-Gordillo et al., 2000; Schubart et al., 2001, respectively).

All zoea I of *Plagusia* sensu stricto share a setal count of 2,2,2,2 at the basis of the first maxilliped, whereas *Guinusia dentipes* and *G. chabrus* have a 2,2,3,2 and 2,2,3,3 setal pattern, respectively. Another important difference is found in the antennal exopod, which is relatively shorter (less than 1/6 of protopod length) in *G. chabrus* and *G. dentipes*, and ends in a long seta that can reach, as maximum size, half the protopod length, with two (*G. chabrus*) or one (*G. dentipes*) additional short setae. In the species of *Plagusia* sensu stricto, the exopod is longer (more than 1/6 of protopod length) and with two long setae, the longer one reaching almost to the tip of the protopod and presenting minute terminal spines (not observed in *G. chabrus* and *G. dentipes*) (fig. 5).

The morphology of the megalopa stage also appears to be useful in distinguishing species of *Guinusia* from *Plagusia*. The megalopa has only been described for *Guinusia chabrus* (without details by Barnard, 1950; Wear, 1970; Wear & Fielder, 1985), *G. dentipes* (several descriptions, the most detailed by González-Gordillo et al., 2000), *Plagusia depressa* (Schubart et al., 2001), and *P. squamosa* (Muraoka, 1965). Three characters show important differences that allow telling apart the megalopae of *G. dentipes* (and *G. chabrus*) from those of *P. depressa* and *P. squamosa*. Firstly, the exopod of the third maxilliped, is 2-segmented in *G. dentipes*, but unsegmented in *P. depressa* and *P. squamosa* (no data available for *G. chabrus*). Secondly, the endopod of the pleopods has less than 10 (8 or 9) cincinnuli in *P. depressa* and *P. squamosa*, but more than 12 (13-17) in *G. dentipes* (no data available for *G. chabrus*). Finally, the rostrum is rather short and broadly bilobed in *G. dentipes* and *G. chabrus*, whereas it projects well forward and is bifid at the tip in *P. depressa* and *P. squamosa*.

Previous mtDNA phylogenies have included only one or two species of *Plagusia* (e.g., Schubart et al., 2001, 2006). The molecular trees generated for this study include all currently recognized species of *Plagusia* with the exception of *Plagusia integripes* Garth, 1973, from the Easter Islands, which is so far only known from the type. Morphological examination of the type specimen allowed the conclusion that the species should remain in the genus *Plagusia* based on the pattern of spines on the meri of the ambulatory legs. However, no amplifiable DNA could be obtained from it. The use of mitochondrial 16S rRNA and the nuclear histone H3 gene allowed both the recognition of intrageneric genetic distances and the construction of more robust gene phylogenies based on separate DNA sources, which should reflect the species' phylogeny. Three main clusters within *Plagusia* (sensu Schubart & Ng, 2000) could be recognized, one consisting of the type species *P. depressa*

(Atlantic) in addition to *P. squamosa*, *P. immaculata*, and *P. speciosa* (Indo-Pacific). Schubart et al. (2001) documented morphological (adult and larval) and genetic differences between the allopatric *P. depressa* and *P. squamosa*, warranting the specific status of both. In the present analysis, however, the two sympatric Indo-Pacific species, *P. squamosa*, and *P. immaculata*, could not be genetically distinguished based on the fragment used, or histone H3. This could be due to a very recent separation event, which is not expressed in the analysed DNA. However, the main morphological difference between these two species is the setation surrounding the tubercles of the carapace and in the sulcae of the pereiopods, present in *P. squamosa* but absent in *P. immaculata* (see Schubart & Ng, 2000). In most cases, this pubescence is easy to discern under a dissecting microscope, but some individuals have a faintly developed setation (C. Schubart, pers. obs.), so that we are inclined to consider these two species conspecific with a variable degree of pubescence. However a final decision on synonymy will have to await more comprehensive morphological study, and genetic comparisons with more variable markers (Schubart et al., in progress).

The genetically (and morphologically) most distinct of all former *Plagusia* species is *Davusia glabra*. The monotypic status of *Davusia* Guinot, 2007, is confirmed by the long branch with a single species at the tip (no close sister species). The molecular results indicate that *Davusia glabra* is phylogenetically basal to all other species of *Plagusia* and *Guinusia* n. gen.) and to the analysed species of *Euchirograpsus* and *Miersiograpsus*.

The third phylogenetic cluster consists of the two species, *G. chabrus*, and *G. dentipes*, which are here placed in the new genus *Guinusia*. Despite the clear morphological differences between these two species, they seem closely related genetically, forming the tip clades of a long branch with high confidence values. Also in this case, genetic distances to the remaining species of *Plagusia* are not lower than those between *Guinusia* and *Euchirograpsus*-*Miersiograpsus*, the latter two genera forming a fourth monophyletic clade within the Plagusiidae.

The final putative member of the Plagusiidae is the genus *Percnon*. The present molecular data support the results of Schubart et al. (2000, 2006), in showing that *Percnon* species do not cluster close to the Plagusiidae, but represent a basal split within the Thoracotremata. The current study included three species, viz., *Percnon gibbesi*, *P. planissimum*, and *P. guinotae* Crosnier, 1965, and evidence from mitochondrial, and for the first time also from nuclear DNA (figs. 1, 2), enforces support for an independent phylogenetic origin of

Percnon, and strengthens the insight that it should not be maintained within the Plagusiidae. This differs from the conclusion by Ng et al. (2008), who gave the Percninae subfamily status within the Plagusiidae according to the following reasoning (their p. 219): "With regards to *Percnon*, DNA (Schubart et al., 1999 [2000]) and larval evidence (J. Cuesta, pers. comm.) have suggested that *Percnon* is different and doubtfully placed in the Plagusiidae. On the basis of adult morphology, we concur that *Percnon* species are highly derived, and have many unique generic apomorphies, probably related to their unusual habit of living subtidally on rock faces exposed to heavy wave action. These, combined with unusual sternal characters and the unique grapsoid condition of the male abdomen having only segments 3-5 fused, lead us to support the recognition of a separate subfamily, Percninae, a taxon first erected as a tribe by Števčić (2005)". However, the here provided combined evidence of mtDNA, nDNA, and larvae fully support separation of the Percnidae as a full family, suggesting that apparent adult morphological similarities to the Plagusiidae are due to convergence.

Data on the larvae of *Percnon* are, unfortunately, restricted to incomplete descriptions of the development of *P. gibbesi* (see Lebour, 1944; Paula & Hartnoll, 1989) combining material reared in the laboratory and collected in the plankton, the megalopa stage of *P. planissimum* (see Muraoka, 1967), brief descriptions of the megalopa stages of *P. abbreviatum* (Dana, 1851), and *P. guinotae* and *P. planissimum* (see Hartnoll, 1992). Despite the limited data, significant differences in larval morphology also support exclusion of *Percnon* from the Plagusiidae. With regards to the zoeae, the antennal exopod has a small simple seta as in Grapsidae (sensu Schubart et al., 2002, see Cuesta & Schubart, 1999), all the cephalothoracic spines are well developed (especially dorsal and rostral ones), covered by spinules, and the tips of these spines are extended in a lanceolate fashion. The abdomen is long, thin, and with a telson having long furcal arms and a deep cleft; this is a shape not found in any other "grapsoid" family and resembles that of Dorippidae and Dotillidae (see Paula, 1991; Ng et al., in press). The setation of the endopod of the maxilla of zoea I of *P. gibbesi* has a setation of 2,2, which in the zoea III changes to 2,3, and in zoea V and VI to 2,4 according to Paula & Hartnoll (1989); this is very unusual and awaits confirmation. In all known brachyuran zoeae, this setation remains constant throughout development (Cuesta et al., 2006). The megalopae are also distinct, with 10 segments on the antennular flagellum (eight in Plagusiidae) and a trifid rostrum (bilobate or bifid at the tip in Plagusiidae).

In an mtDNA phylogeny of selected Thoracotremata, Schubart et al. (2006) found two representatives of the Plagusiidae (*Plagusia depressa* and *Euchi-*

rograpsus americanus) clustering with low support among the Gecarcinidae, suggesting an apparent paraphyletic relationship of the latter family. Due to the low confidence values and limited number of taxa, this was not discussed in detail (except perhaps by Ng et al., 2008: 214), and Schubart et al. (2006: 196) concluded that "Further studies with all gecarcinid taxa and additional molecular markers are necessary to confirm their phylogenetic relationship and their position within the Thoracotremata". The molecular trees in the present study support monophyly of the Plagusiidae (now including all the genera) with confidence values always exceeding 0.95 posterior probability, and show no affinity to gecarcinid representatives. In contrast, the phylogenetic status of the Gecarcinidae as well as the basal relationships among the families of Thoracotremata remain unresolved (see also N. K. Ng et al., 2007). Very remarkable, however, is the close relationship of *Ocypode* and *Ucides* so that, in contrast to Ng et al. (2008), for the time being, we continue recognizing *Ucides* as member of the family Ocypodidae (and the Ucididae Števčić, 2005 a junior synonym of the Ocypodidae). Further studies are currently underway to resolve basal relationships among the Thoracotremata, including more conserved nuclear markers (28S and nuclear coding genes) (Schubart et al., in prep.; Landstorfer et al., in prep.).

The present study represents a comprehensive phylogenetic study of the Plagusiidae and has resulted in recognition of the new genus *Guinusia* and the establishment of a new phylogenetic understanding of relationships among four genera and 11 species of the family Plagusiidae. In order to make the Plagusiidae monophyletic we formally exclude *Percnon* and the subfamily Percninae from the Plagusiidae, and hereby recognize the independent family Percnidae Števčić, 2005.

Family PERCNIDAE Števčić, 2005 [new status]

Subfamily PERCNINAE Števčić, 2005 (see Ng et al., 2008)

Tribe PERCNINI Števčić, 2005

Type genus. — *Percnon* Gistel, 1848 (only genus in family).

Type species. — *Acanthopus planissimus* Herbst, 1804.

Other species. — *Percnon abbreviatum* (Dana, 1851), *P. affine* (H. Milne Edwards, 1853), *P. gibbesi* (H. Milne Edwards, 1853), *P. guinotae* Crosnier, 1965; *P. sinense* Chen, 1977.

Diagnosis. — Carapace conspicuously flattened; front with two antennal clefts, third maxillipeds with reduced exopod and lacking palp (shared with Plagusiidae); meri of walking legs always with several spines on dorsal crest, male abdomen with somites 3-5 fused (3-6 in Plagusiidae).

ACKNOWLEDGEMENTS

We would like to thank Peter Castro, Bertrand Richer de Forges, Peter Davie, and Peter Ng for their invitation to contribute to this special volume of Crustaceana (and for comments on the first version), thus allowing us to honour Danièle Guinot. We also would like to express our gratitude to a number of colleagues who have contributed to previous studies of *Plagusia* and this way prompted our attention to this genus, among them predominantly Peter K. L. Ng, Hung-Chang Liu, Nacho González-Gordillo, and Shinyi Tsuchida. We are also greatly indebted to Alain Crosnier, Régis Cleva and the MNHN Paris, Pere Abelló and the Instituto de Ciencias Marinas, Barcelona, Peter J. F. Davie and the Queensland Museum Brisbane, Jody Martin and the Natural History Museum of Los Angeles County, Peter K. L. Ng and the Zoological Reference Collection of the Raffles Museum, Gary Poore and the Victoria Museum, Melbourne, Michael Türkay, Kristin Pietratus and the Senckenberg Museum Frankfurt, Darryl L. Felder, Rafael Robles, Emma Palacios Theil, Rudolf Diesel, Sara Fratini, Carsten H. G. Müller, and Peter Wirtz, for kindly photographing, donating or making available further specimens from their collections. Darryl Felder and Joe Neigel (University of Louisiana at Lafayette) and Jürgen Heinze (Universität Regensburg) are acknowledged for providing the laboratory facilities to generate some of the sequences. Most especially, we would like to thank Danièle Guinot for her important contributions to the field, for so many interesting discussions, and for her constant encouragement.

REFERENCES

ALCOCK, A., 1900. Materials for a carcinological fauna of India. No. 6. The Brachyura Catometopa, or Grapsoidea. Part II. Journal of Asiatic Society of Bengal, **69** (3): 279-456.

BARNARD, K. H., 1950. Descriptive catalogue of South African decapod Crustacea. Annals of the South African Museum, **38**: 1-837.

BENECH, S. V., 1978. Ocean transport of a community of the grapsid crab *Plagusia dentipes* (De Haan, 1833). Crustaceana, **35**: 104.

COLGAN, D. J., C. MCLAUGHLIA, G. D. F. WILSON, S. P. LIVINGSTON, G. D. EDGECOMBE, J. MACARANAS, G. CASSIS & M. R. GRAY, 1998. Histone H3 and U2 snRNA DNA sequences and arthropod molecular evolution. Australian Journal of Zoology, **46**: 419-437.

CUESTA, J. A., G. GUERAO, H.-C. LIU & C. D. SCHUBART, 2006. Morphology of the first zoeal stages of eleven Sesarmidae (Crustacea, Brachyura, Thoracotremata) from the Indo-West Pacific, with a revision of larval characters of the family. Invertebrate Reproduction and Development, **49** (3): 151-173.

CUESTA, J.A. & C. D. SCHUBART, 1998. The first zoeal stage of *Glyptograpsus impressus*, with comments on the subfamilial arrangement of Grapsidae (Crustacea: Brachyura). Cahiers de Biologie Marine, **38**: 291-299 [as 1997].

— — & — —, 1999. First zoeal stages of *Geograpsus lividus* and *Goniopsis cruentata* from Panama confirm consistent larval characters within the subfamily Grapsinae (Crustacea: Brachyura: Grapsidae). Ophelia, **51** (3): 163-176.

DANA, J. D., 1851. On the classification of the Crustacea Grapsoidea. American Journal of Science and Arts, (2) **12**: 283-290.

— —, 1852. Crustacea. United States Exploring Expedition during the years 1838, 1839, 1840, 1841, 1842 under the command of Charles Wilkes, U.S.N., **13** (1): i-viii, 1-685. (C. Sherman, Philadelphia).

DAVIE, P. J. F., 2002. Crustacea: Malacostraca: Eucarida (Part 2): Decapoda-Anomura, Brachyura. In: A. WELLS & W. W. K. HOUSTON (eds.), Zoological Catalogue of Australia, **19.3B**: i-xiv, 1-641. (CSIRO Publications, Melbourne).

DAWSON, E. W., 1987. A key to the world species of *Plagusia* (Crustacea: Brachyura), with a new record of *P. depressa tuberculata* Lamarck from New Zealand. National Museum of New Zealand Records, **3** (4): 37-45.

GONZÁLEZ-GORDILLO, J. I., S. TSUCHIDA & C. D. SCHUBART, 2000. Redescription of the megalopa of *Plagusia dentipes* (Brachyura: Plagusiidae) from Japan. Crustacean Research, **29**: 143-151.

GRIFFIN, D. J. G., 1968. The designation of a neotype for *Cancer chabrus* Linnaeus and the status of *Plagusia capensis* De Haan (Decapoda, Grapsidae). Crustaceana, **15** (2): 209-213.

GUINOT, D., 1978. Principes d'une classification évolutive des Crustacés Décapodes Brachyoures. Bulletin Biologique de la France et de la Belgique, **112** (3): 211-292.

— —, 1979. Données nouvelles sur la morphologie, la phylogenèse et la taxonomie des Crustacés Décapodes Brachyoures. Mémoires du Muséum national d'Histoire naturelle, (A, Zoologie) **112**: 1-354.

— —, 2007. A new genus of the family Plagusiidae Dana, 1851, close to *Plagusia* Latreille, 1804 (Crustacea, Decapoda, Brachyura). Zootaxa, **1498**: 27-33.

GUINOT, D. & J.-M. BOUCHARD, 1998. Evolution of the abdominal holding systems of brachyuran crabs (Crustacea, Decapoda, Brachyura). Zoosystema, **20** (4): 613-694.

HARTNOLL, R. G., 1992. Megalopae and early postlarval stages of East African *Percnon* (Decapoda: Brachyura: Grapsidae). Journal of Zoology, London, **228**: 51-67.

HUELSENBECK, J. P. & F. RONQUIST, 2001. MrBayes: Bayesian inference of phylogenetic trees. Bioinformatics, **17**: 754-755.

LATREILLE, P. A., 1804. Crustacés (1). Crustacea. Classe Septième. Tableau méthodique des Crustacés. In: Nouveau Dictionnaire d'Histoire Naturelle, **24**: 123-127.

— —, 1810. Considérations générales sur l'ordre naturel des animaux composant les classes des Crustacés, des Arachnides et des Insectes; avec un tableau méthodique de leurs genres, disposés en familles: 1-144. (F. Schoell, Paris).

LEBOUR, M. V., 1944. Larval crabs from Bermuda. Zoologica, **29**: 113-128.

MAN, J. G. DE, 1883. Carcinological studies in the Leyden Museum. No. 3. Notes of the Leyden Museum, **5** (15): 150-169.

MANNING, R. B. & L. B. HOLTHUIS, 1981. West African brachyuran crabs (Crustacea: Decapoda). Smithsonian Contributions to Zoology, **306**: i-xii, 1-379.

MARTIN, J. W. & G. E. DAVIS, 2001. An updated classification of the recent Crustacea. Natural History Museum of Los Angeles County, Science Series, **39**: 1-124.

MIERS, E. J., 1878. Revision of the Plagusiinae. Annals and Magazine of Natural History, (5) **1**: 147-154.

MURAOKA, K., 1965. On the post-larval stage of *Plagusia depressa tuberculata* Lamarck (Grapsidae). Research on Crustacea, **2**: 83-90.

— —, 1967. On the post-larval stages of *Percnon planissimum* (Herbst), Grapsidae. Research on Crustacea, **3**: 61-67.

NG, N. K., P. J. F. DAVIE, C. D. SCHUBART & P. K. L. NG, 2007. Xenograpsidae, a new family of grapsoid crabs (Crustacea: Brachyura) associated with shallow water hydrothermal vents. Raffles Bulletin of Zoology, (Supplement) **16**: 233-256.

NG, P. K. L., 1998. Crabs. In: K. E. CARPENTER & V. H. NIEM (eds.), FAO Species identification guide for fishery purposes. The living marine resources of the western central Pacific, **2**. Cephalopods, crustaceans, holothurians and sharks: 1045-1155. (Food and Agriculture Organisation, Rome).

NG, P. K. L., P. F. CLARK & J. A. CUESTA, in press. Establishment of a new subfamily for *Shenius anomalus* (Shen, 1953) (Decapoda, Brachyura, Dotillidae). Journal of Natural History, London.

NG, P. K. L., D. GUINOT & P. J. F. DAVIE, 2008. Systema Brachyurorum: Part I. An annotated checklist of extant brachyuran crabs of the world. Raffles Bulletin of Zoology, (Supplement) **17**: 1-286.

NG, P. K. L., C. H. WANG, P. H. HO & H. T. SHIH, 2001. An annotated checklist of brachyuran crabs from Taiwan (Crustacea: Decapoda). National Taiwan Museum Special Publications Series, Taipei, **11**: 1-186.

PAULA, J., 1991. The zoeal stages of the crab *Medorippe lanata* (Linnaeus, 1767) (Brachyura, Dorippidae) reared in the laboratory, and the larval characters of the Dorippidae. Journal of Natural History, London, **25**: 75-89.

PAULA, J. & R. G. HARTNOLL, 1989. The larval and post-larval development of *Percnon gibbesi* (Crustacea, Brachyura, Grapsidae) and the identity of the larval genus *Pluteocaris*. Journal of Zoology, London, **218**: 17-37.

POSADA, D. & K. A. CRANDALL, 1998. MODELTEST: testing the model of DNA substitution. Bioinformatics, **14**: 817-818.

RATHBUN, M. J., 1918. The grapsoid crabs of America. U.S. National Museum Bulletin, **97**: 1-461, pls. 1-161.

SCHUBART, C. D., 2009. Mitochondrial DNA and decapod phylogenies: the importance of pseudogenes and primer optimization. In: J. W. MARTIN, K. A. CRANDALL & D. L. FELDER (eds.), Decapod crustacean phylogenetics. Crustacean Issues, **18**: 47-66. (Taylor & Francis/CRC Press, Boca Raton, Florida).

SCHUBART, C. D., S. CANNICCI, M. VANNINI & S. FRATINI, 2006. Molecular phylogeny of grapsoid crabs (Decapoda, Brachyura) and allies based on two mitochondrial genes and a proposal for refraining from current superfamily classification. Journal of Zoological Systematics and Evolutionary Research, **44** (3): 193-199.

SCHUBART, C. D., J. A. CUESTA, R. DIESEL & D. L. FELDER, 2000. Molecular phylogeny, taxonomy, and evolution of nonmarine lineages within the American grapsoid crabs (Crustacea: Brachyura). Molecular Phylogenetics and Evolution, **15** (2): 179-190.

SCHUBART, C. D., J. A. CUESTA & D. L. FELDER, 2002. Glyptograpsidae, a new brachyuran family from Central America: larval and adult morphology, and a molecular phylogeny of the Grapsoidea. Journal of Crustacean Biology, **22** (1): 28-44.

SCHUBART, C. D., J. I. GONZÁLEZ-GORDILLO, N. REYNS, H.-C. LIU & J. A. CUESTA, 2001. Are Atlantic and Indo-Pacific populations of the rafting crab, *Plagusia depressa*, distinct? New evidence from larval morphology and mtDNA. Raffles Bulletin of Zoology, **49** (2): 301-310.

SCHUBART, C. D. & P. K. L. NG, 2000. On the identities of the rafting crabs *Cancer depressus* Fabricius, 1775, *Cancer squamosus* Herbst, 1790, *Plagusia immaculata* Lamarck, 1818, and *Plagusia tuberculata* Lamarck, 1818 (Crustacea: Decapoda: Brachyura: Plagusiidae). Raffles Bulletin of Zoology, **48** (2): 327-336.

STERNBERG, R. VON & N. CUMBERLIDGE, 2000. Taxic relationships within the Grapsidae MacLeay, 1838 (Crustacea: Decapoda: Eubrachyura). Journal of Comparative Biology, **3** (2): 115-136 [as 1998].

ŠTEVČIĆ, Z., 2005. The reclassification of brachyuran crabs (Crustacea, Decapoda: Brachyura). Natura Croatica, **14** (supplement 1): 1-159.

TAMURA K., J. DUDLEY, M. NEI & S. KUMAR, 2007. MEGA4: Molecular Evolutionary Genetics Analysis (MEGA) software version 4.0. Molecular Biology and Evolution, **24**: 1596-1599.

TERADA, M., 1987. Zoeal forms of 14 species of crabs from Enshunada. Research on Crustacea, **16**: 93-120.

UDEKEM D'ACOZ, C. D', 1999. Inventaire et distribution des crustacés décapodes de l'Atlantique nord-oriental, de la Méditerranée et des eaux continentales adjacentes au nord de 25°N. Collection "Patrimoines Naturels", **40**: 1-383. (Muséum national d'Histoire naturelle, Paris).

WEAR, R. G., 1970. Life-history studies on New Zealand Brachyura. 4. Zoea larvae hatched from crabs of the family Grapsidae. New Zealand Journal of Marine and Freshwater Research, **4** (1): 3-35.

WEAR, R. G. & D. R. FIELDER, 1985. The marine fauna of New Zealand: Larvae of the Brachyura (Crustacea, Decapoda). New Zealand Oceanographic Institute Memoir, **92**: 1-90.

WILLIAMS, A. B., 1984. Shrimps, lobsters, and crabs of the Atlantic coast of the eastern United States, Maine to Florida: 1-550. (Smithsonian Institution Press, Washington, D.C.).

WILSON, K. A., 1980. Studies on decapod Crustacea from the Indian River region of Florida. XV. The larval development under laboratory conditions of *Euchirograpsus americanus* A. Milne-Edwards, 1880 (Crustacea: Decapoda: Grapsidae) with notes on grapsid subfamilial larval characters. Bulletin of Marine Science, Miami, **30** (4): 756-775.

WILSON, K. A. & R. H. GORE, 1980. Studies on decapod Crustacea from the Indian River region of Florida. XVII. Larval stages of *Plagusia depressa* (Fabricius, 1775) cultured under laboratory conditions (Brachyura: Grapsidae). Bulletin of Marine Science, Miami, **30** (4): 776-789.

First received 28 June 2009.
Final version accepted 13 August 2009.

POPULATION DYNAMICS AND DISTRIBUTION OF *OZIUS TRUNCATUS* H. MILNE EDWARDS, 1834 (BRACHYURA, OZIIDAE) ON ECHINODERM REEF, LEIGH MARINE RESERVE, NEW ZEALAND

BY

KALAYARASY SIVAGURU[1,3]) and COLIN L. MCLAY[2,4])

[1]) Department of Biological Sciences, University of Auckland, Auckland, New Zealand
[2]) School of Biological Sciences, Canterbury University, Christchurch, New Zealand

ABSTRACT

Knowledge about the population ecology of predatory crabs from the Pacific Ocean is very limited. *Ozius truncatus* H. Milne Edwards, 1834 is a relatively large brachyuran crab found on rocky intertidal shores in northern New Zealand. Densities of this crab at Leigh Marine Reserve were as high as about 2 m^{-2} in summer, but the distribution was very patchy. Juvenile crabs are recruited in late summer and grow to about 15 mm in carapace length (CL) in one year when they begin to reach sexual maturity. In their second and third years they reach about 25 mm and 35 mm CL, respectively. Males and females have a similar average size (26.9 mm and 28.1 mm CL) and the sex ratio is 1 : 1. A female living to the maximum age, estimated at 3+ yrs, could produce three to four annual broods. Crabs were ovigerous from October to February during the two years of the study. Monitoring of tagged crabs shows that they have a high level of site fidelity and limited spatial range. This suggests that any impact of these predators will only be localized.

RÉSUMÉ

La connaissance de l'écologie des populations de crabes prédateurs de l'océan Pacifique est très limitée. *Ozius truncatus* H. Milne Edwards, 1834 est une relativement grande espèce de crabe vivant sur les fonds rocheux de la zone intertidale au nord de la Nouvelle-Zélande. Les densités de cette espèce dans la réserve marine de Leigh atteignent 2 par m^2 en été, mais leur distribution est très irrégulière. Les juvéniles sont recrutés à la fin de l'été et grandissent jusqu'à 15 mm de longueur (CL) en un an, lorsqu'ils atteignent la maturité sexuelle. Dans leur seconde et troisième année ils atteignent 25 puis 35 mm (CL). Les males et les femelles sont de tailles similaires (26,9 et 28,1 mm) avec un sexa ratio de 1. Une femelle vivant jusqu'à l'âge maximal de trois ans pourra produire trois à quatre pontes, une par an. Les crabes ont

[3]) Corresponding author; e-mail: ksivaguru@doc.govt.nz
[4]) e-mail: colin.mclay@canterbury.ac.nz

© Koninklijke Brill NV, Leiden, 2010

été ovigères d'octobre à février pendant les deux années de cette étude. Le suivit des crabes marqués montre qu'ils sont fidèles au même site et font des déplacements réduits. Cela suggère que l'impact de ces prédateurs sera seulement localisé.

INTRODUCTION

Several studies have suggested that crabs have a significant impact on rocky intertidal reefs (Kitching & Lockwood, 1974; Bertness & Cunningham, 1981; Menge, 1983; Lubchenco et al., 1984; Lowell, 1986; Hughes, 1989), but this depends upon their numbers and distribution on the shore. Being larger and often more aggressive the prime requirement of these crabs is a refuge where they can hide from each other when the tide is out (Bertness & Cunningham, 1981). Often very few of these crabs show up in randomly placed quadrats, traditionally used for shore surveys, because their distribution is patchy or they hide in places hard to sample, so it is difficult to assess their importance in rocky shore communities. The impact of predators on the shore depends upon their numbers and prey consumption: i.e., their "numerical" and "functional" responses (Holling, 1966). Each of these responses has several components. The components of the numerical response are recruitment, mortality, and growth, all of which determine mean size and numbers of crabs; the components of the functional response are per capita prey consumption and mobility (foraging area). Taken together, the product numbers x consumption, gives an estimate of impact. Here we provide estimates of the population size of an oziid crab in northern New Zealand, *Ozius truncatus* H. Milne Edwards, 1834, its dynamics and mobility, providing a numerical component of predatory impact. The functional component and assessment of the overall impact of *O. truncatus* in the ecosystem will be presented separately (Sivaguru & McLay, in prep.) (see Sivaguru, 1997 for details of the entire study).

Ozius truncatus is normally found under intertidal boulders and cobbles. It is a generalist predator with dimorphic chelipeds that is capable of eating a wide variety of gastropods (Chilton & Bull, 1984, 1986; Skilleter & Anderson, 1986; and Hughes, 1993). Chilton & Bull (1984) identified a strong, negative correlation between the position of *O. truncatus* and the distribution of gastropods on shores in southern Australia: low abundance of small snails low on the shore corresponded to high densities of reef crabs. They suggested that this pattern could be largely caused by these crabs because they selectively consume small snails (Chilton & Bull, 1984) and larger snails have a size refuge from predation. The large cheliped is used to open snails and shells of the same gastropod species inhabited by hermit crabs (C. L. McLay, unpubl. data) by chipping fragments from around the aperture or, with small shells, by crushing the entire shell. Given their relatively large size, *O. truncatus* could have a significant impact on gastropod and hermit crab populations on rocky New Zealand shores. At our study site in the Leigh Marine Reserve the distribution of *O. truncatus* overlaps with that of several species of gastropods: *Nerita melanotragus* E. A. Smith, 1884, *Diloma aethiops* (Gmelin, 1791), *Haustrum scobinum* (Quoy & Gaimard, 1833), *Lunella smaragda* (Gmelin, 1791), *Diloma nigerrima* (Gmelin, 1791), *Haustrum haustorium* (Gmelin, 1791), and the hermit crabs (*Pagurus novizealandiae* (Dana, 1852) and *P. traversi* (Filhol, 1885), Paguridae)) that occupy the empty shells of these gastropods. Both snails and hermit crabs are potential prey items.

No estimates of the density of *O. truncatus* in New Zealand have been attempted, probably because they would only be rarely detected using strict random sampling. Our study is unusual because it involves repeated sampling of the same marked crabs (sampling with replacement). During the course of the study, the same crabs were counted several times from a defined area of the rocky intertidal. Little is known about the abundance of this crab in New Zealand and nothing is known about its population dynamics other than that ovigerous females have been found from October to January, planktonic development includes 4 zoeal stages (Wear, 1968; Wear & Fielder, 1985), and settlement occurs about 4-6 weeks after hatching (McLay, 1988).

Some aspects of the population dynamics of *O. truncatus* on Echinoderm Reef in the Leigh Marine Reserve, Auckland (fig. 1a) are here described: (1) distribution and abundance on the shore; (2) population size structure, sex ratio, period of recruitment, and (3) the patterns of crab movements.

METHODS

The study area was a rocky intertidal reef of gently sloping sandstone and mudstone in the Leigh Marine Reserve (36°16′S 174°48′E). Intertidal rock makes up about 38.2 ha (7% of the Reserve) (Ayling, 1978), but habitat suitable for *Ozius truncatus* is perhaps only 3-4 ha. The tidal range is approximately 2.2 m and the climate is warm temperate with mean sea surface temperatures ranging from 13.5°C in August to 21°C in February (Evans, 1992). The predominant assemblages on the rocky reef have been described by Creese (1988). Initial sampling by random quadrats (4 m^2) of Echinoderm Reef resulted in only small numbers of *O. truncatus* being found. A more intensive search established patches of boulders where crabs were normally found and so the study site was divided into labelled areas with the results recorded separately for each clump (fig. 1b, c). These were separated by natural discontinuities of unsuitable habitat. From each area the numbers, sizes (measured as carapace length to the nearest mm), and sex of every crab found were recorded. Thus our study is a meta-population analysis of fixed sites. Sampling commenced in May 1996, initially using the four clumps that had the highest densities of crabs in the preliminary survey. The numbers of crabs in these clumps declined steadily over the next two months, and in August 1996 the sampling programme was expanded to include all 15 major clumps in the upper-mid intertidal area of Echinoderm Reef. Sampling on this basis continued monthly until July 1998. Inability to move some boulders meant that small crabs (<10 mm CL) tend to be under-represented. Newly recruited juveniles are pale pink in colour. Patterns on the carapace start to develop when crabs are approximately 13 mm CL, and by 20 mm they are of the same coloration, deep chocolate brown with black fingers, as adult crabs. The number of ovigerous females was recorded at each sample date in order to determine the reproductive period.

To investigate the extent of movements between boulders, crabs were tagged by gluing numbered plastic labels (Hallmark Ltd., Adelaide, Australia). This tagging method was tested with control crabs held in seawater tanks at the Leigh Laboratory and no mortality resulted. Marked rocks were used as reference points to locate each crab and to measure the distance moved between monthly samples. As new crabs were encountered they were tagged in the same fashion at each monthly census from May 1996 to June 1997. During their first year crabs less than 20 mm CL may have been tagged more than once because of moulting, but this could only influence the picture of apparent movements because population size was estimated by total count.

RESULTS

Population dynamics

Our study of *Ozius truncatus* covered the period May 1996 to July 1998 (table I, figs. 2, 3). Amongst crabs CL > 15 mm there were no signs of a cohort structure at the beginning of the study in May 1996. With larger sample sizes from July 1996, two groups become evident: one group with CL of about 25 mm and another group with CL of about 35 mm. These two groups were

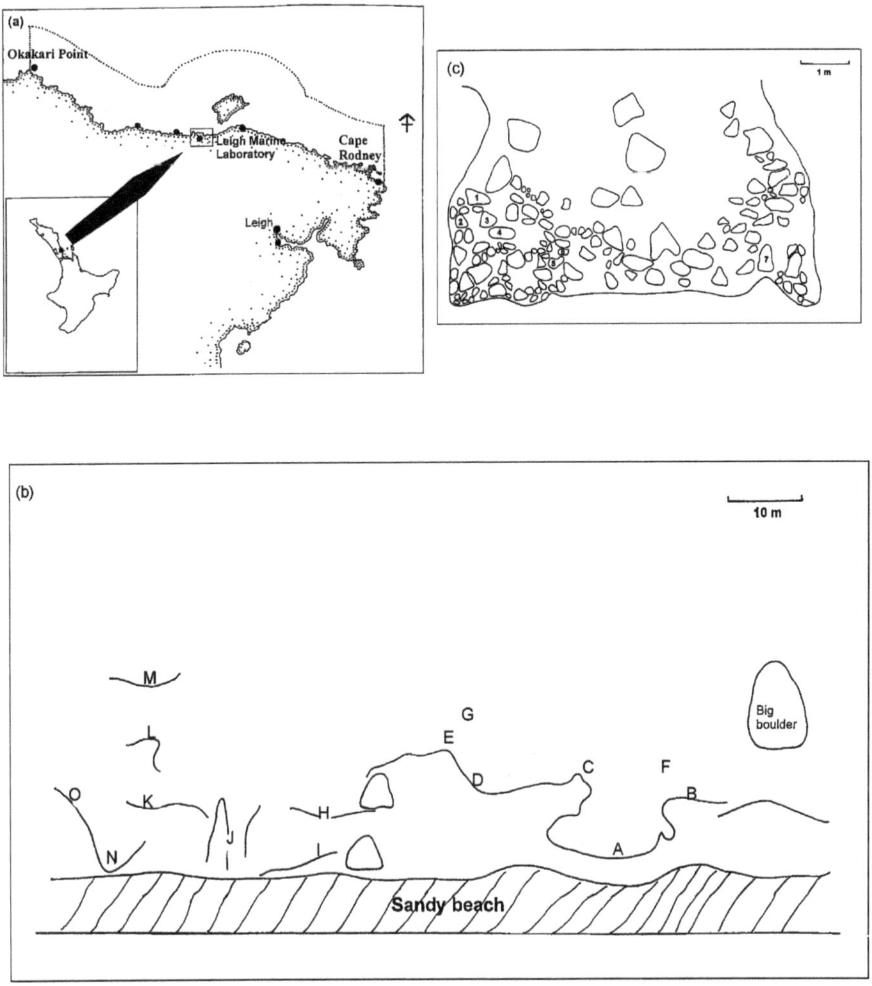

Fig. 1. a, Map of north-eastern New Zealand showing the location of Leigh Marine Reserve and its boundaries (dotted line) with study area indicated by a small box; distribution: 1b, location of marked crab clump sites A-O on Echinoderm Reef (scale bar = 10 m); 1c, detailed map of Clump A showing micro-spatial pattern, traced from a photograph, showing location of crabs under stones (1-7) in April 1996 (scale bar = 1 m).

evident in August 1996 so the smaller group may be a cohort recruited in the preceding summer, 1996-97, while the larger group may be a cohort from the summer previous to that, 1995-96. These two cohorts remained evident in some months following August 1996. By September 1997 the 1996-97 cohort became the 35 mm CL group and these continued to increase in size through the 1997-98 summer but dwindled in numbers by the end of the study. The largest crab (CL = 55 mm) was seen in September 1996 and this may well

TABLE I

Numbers of juveniles of *Ozius truncatus* H. Milne Edwards (CL < 15 mm) males and females in monthly samples. Mean size and standard error of the mean are given in mm carapace length. Numbers of females include ovigerous crabs; *, significant difference from expected 1 : 1 ratio at 0.05 confidence level

Time	No. of juveniles (Average size ± SE)	No. of males (Average size ± SE)	No. of females (Average size ± SE)	No. of ovigerous females	Male female ratio	χ^2
May 1996	3 (13.7 ± 0.88)	20 (25.8 ± 0.81)	21 (30.3 ± 1.22)	0	0.95	0.02
June 1996	1 (12)	11 (28.5 ± 1.34)	8 (31.75 ± 1.6)	0	1.37	0.47
July 1996	0	11 (29 ± 1.89)	10 (27.7 ± 2.3)	0	1.1	0.05
Aug. 1996	0	36 (30.39 ± 0.9)	28 (29.9 ± 1.1)	0	1.28	0.89
Sept. 1996	0	33 (27.5 ± 0.84)	30 (29.8 ± 0.78)	0	1.1	0.14
Oct. 1996	1 (12)	26 (28.96 ± 0.8)	48 (29.9 ± 0.8)	21 (29.5 ± 0.8)	0.54	6.54*
Nov. 1996	1 (8)	48 (27.5 ± 0.68)	48 (29.1 ± 0.75)	32 (28.0 ± 0.7)	1	0
Dec. 1996	1 (9)	41 (27.8 ± 0.82)	35 (27.5 ± 0.9)	13 (26.5 ± 0.8)	1.17	0.47
Jan. 1997	16 (10.5 ± 0.35)	60 (28.1 ± 0.53)	56 (29.1 ± 0.7)	6 (32.7 ± 0.9)	1.07	0.14
Feb. 1997	19 (10.3 ± 0.59)	54 (28.1 ± 0.72)	60 (29.4 ± 0.65)	2 (32.0)	0.9	0.32
Mar. 1997	45 (11.3 ± 0.35)	53 (28.7 ± 0.86)	46 (29 ± 0.7)	0	1.15	0.50
Apr. 1997	37 (11.3 ± 0.38)	25 (29.5 ± 1.26)	51 (30.9 ± 0.9)	0	0.49	8.9*
May 1997	25 (11.2 ± 0.31)	37 (25.9 ± 0.92)	40 (31.58 ± 1.2)	0	0.925	0.12
June 1997	19 (12.6 ± 0.49)	30 (28.1 ± 0.97)	36 (28.1 ± 1.2)	0	0.83	0.55
July 1997	12 (9.7 ± 0.54)	22 (26.7 ± 1.83)	20 (25.3 ± 1.47)	0	1.1	
Aug. 1997	9 (10.1.1 ± 0.8)	22 (28.45 ± 1.3)	14 (25.43 ± 2.0)	0	1.57	
Sept. 1997	18 (10.2 ± 0.5)	29 (28.72 ± 1.1)	47 (29.1 ± 0.99)	0	0.62	
Oct. 1997	15 (12.4 ± 0.49)	26 (25.12 ± 1.1)	26 (23.1 ± 1.16)	2 (23.0)	1.0	
Nov. 1997	18 (9.1 ± 0.43)	38 (28.24 ± 1.1)	38 (27.1 ± 1.08)	29 (29.8 ± 0.9)	1.0	

TABLE I
(Continued)

Time	No. of juveniles (Average size ± SE)	No. of males (Average size ± SE)	No. of females (Average size ± SE)	No. of ovigerous females	Male female ratio	χ^2
Dec. 1997	17 (10.5 ± 0.34)	49 (26.6 ± 1.06)	60 (28.4 ± 0.85)	16 (31.7 ± 0.7)	0.82	
Jan. 1998	15 (13.2 ± 0.34)	76 (25.6 ± 0.72)	76 (27.5 ± 0.78)	33 (31.8 ± 0.7)	1.0	
Feb. 1998	10 (12.9 ± 0.4)	42 (24.9 ± 1.12)	49 (28.5 ± 1.08)	3 (37.7)	0.86	
Mar. 1998	3 (13.1 ± 0.34)	37 (22.54 ± 0.9)	31 (25.32 ± 1.4)	0	1.19	
Apr. 1998	4 (13.5 ± 0.59)	41 (24.34 ± 1.1)	32 (26.5 ± 1.32)	0	1.28	
May 1998	3 (13.7 ± 0.58)	35 (23.5 ± 0.95)	24 (24.42 ± 1.1)	0	1.46	
June 1998	1 (14)	38 (23.68 ± 0.9)	35 (27.88 ± 1.3)	0	1.1	
July 1998	2 (14.0)	37 (25.54 ± 1.1)	44 (26.98 ± 0.9)	0	0.84	

be a survivor from an even earlier cohort, perhaps 1994-95 or even 1993-94. This suggests that the life span of these crabs is of at least 3 years. The modal CL of the 1+, 2+ and 3+ cohorts is approximately 15 mm, 25 mm, and 35 mm. It is possible to follow the growth of the 1996-97 cohort recruited around December 1996. This group first becomes clearly evident amongst the rocks of Echinoderm Reef from January 1997, when CL was around 10 mm, and can be traced through to February 1998 when most of them have joined the CL > 15 mm crabs. Overall the mean male CL = 27.0 mm and female CL = 28.1 mm.

Ovigerous females were recorded from October 1996 to February 1997 and again during the same months in the following summer (fig. 4). Their size ranged from 20 to 47 mm CL with mean size tending to increase towards the end of the egg-bearing season (see table I). Peak numbers of ovigerous females occurred during the November-January period. These data suggest that upon reaching 20 mm CL, females produce a single brood of eggs and that egg and larval development take about 3 months (approx. 2 + 1 months, respectively). While it can be argued that significant cohorts entered the population in the summers of 1995-96 and 1996-97, there does not appear to have been any significant recruitment in the summer of 1997-98. There was no 10 mm

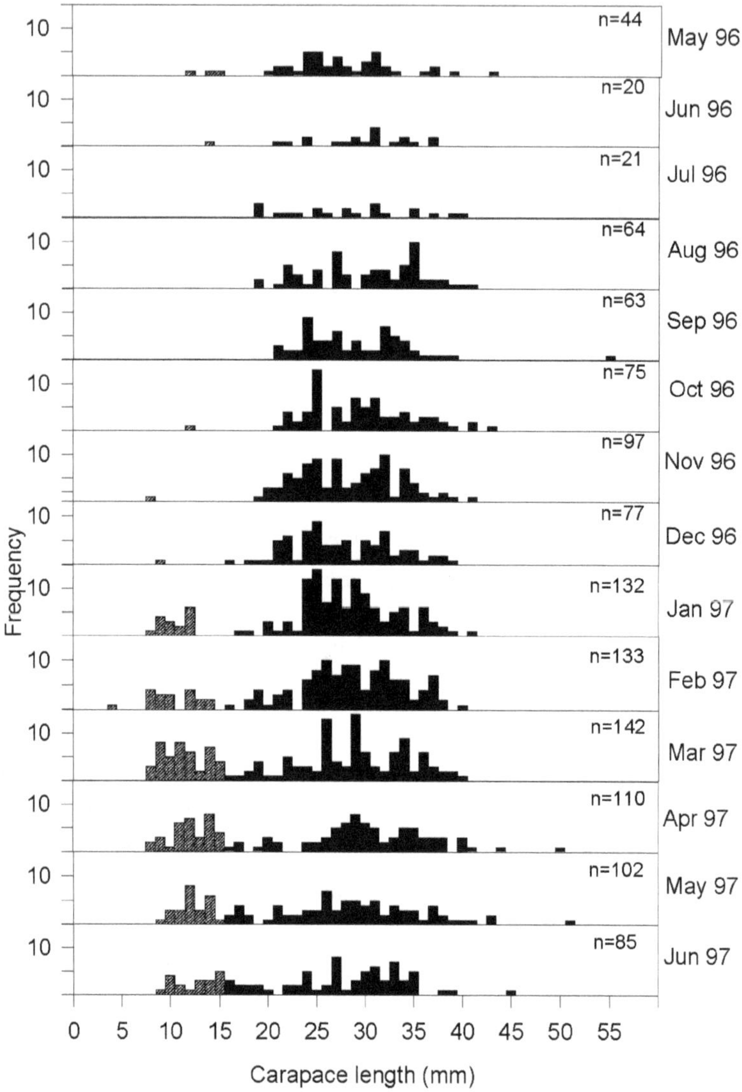

Fig. 2. Monthly population structure of *Ozius truncatus* H. Milne Edwards, 1834, males and females combined, from May 1996 to June 1997. Sample sizes shown in right-hand corner of each monthly histogram. Crabs greater than 15 mm carapace length shown as solid bars; smaller crabs shown as cross-hatched bars.

CL group in January 1998 similar to that seen in January 1997. Females produce their first clutch of eggs in the summer following their recruitment (age of approximately 1 year) and then annually for 2 to 3 further years, should they survive, giving a maximum of 3 to 4 annual broods during their lifetime. Recruitment to the shore population occurs from December to April.

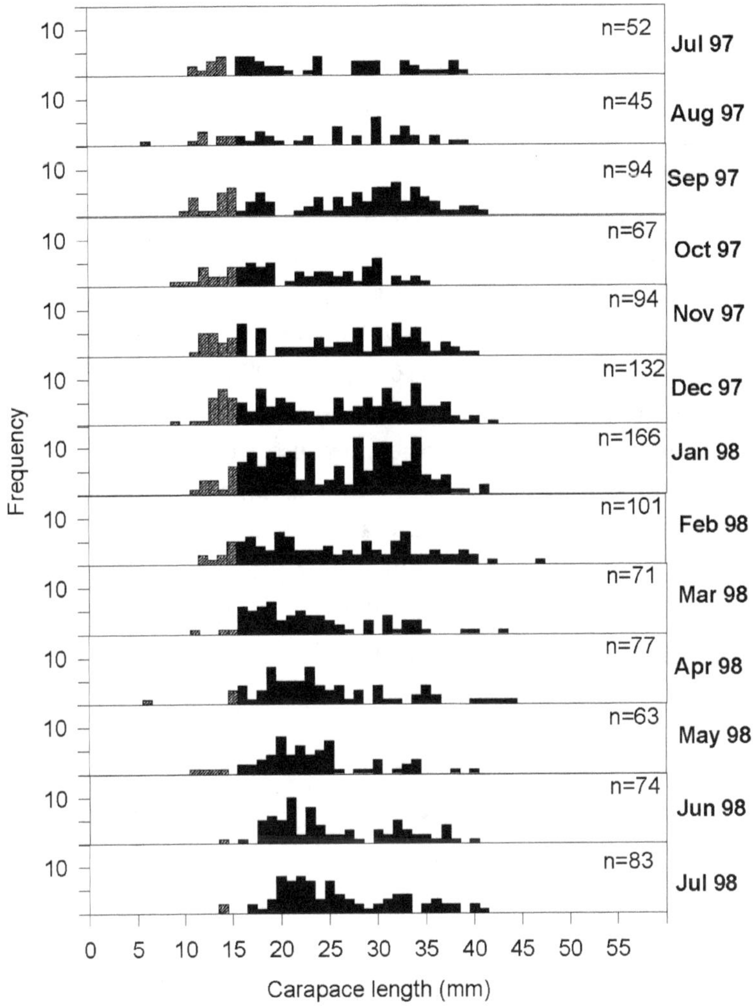

Fig. 3. Combined population structure of *Ozius truncatus* H. Milne Edwards, 1834, males and females combined, from July 1997 to July 1998. Sample sizes shown in right-hand corner of each monthly histogram. Crabs greater than 15 mm carapace length shown as solid bars; smaller crabs shown as cross-hatched bars.

The smallest crab found during sampling was 4 mm, observed in February 1997.

The sex ratio of *Ozius truncatus* throughout the sampling period was approximately 1 : 1, with significant deviations only in October 1996 and April 1997 when females were approximately twice as common as males, and August 1997 and May 1998 when males out-numbered females (table I). Overall 1013 (51.7%) crabs were female while 948 (48.3%) were male. In most months females were larger than males: mean CL of females = 28.1 mm and mean CL = 26.9 mm for males, but there was no seasonal pattern of CL differences.

During the 27-month study, the population size reached a peak of 166 in summer (January) and a minimum of 45 in late winter (August) (fig. 4). The pattern of seasonal fluctuations was remarkably similar between 1996 to 1998. The seasonal change is a reflection of summer recruitment by juveniles and winter mortality of larger crabs. The stable pattern between years is a reflection of the lack of physical disturbance on this relatively sheltered shore. Mean population size between August 1996 to July 1998 was 91.9 crabs (SE = 6.3). The sampling methods used in this study do not allow us to transform number of crabs counted to density estimates, because we do not know how much of the area searched was habitable by *O. truncatus*. However, in April 1996 when areas A, C, J, and K were quadrat sampled it was found that the mean density was 0.6 crabs m^{-2}. During the following months, total crab numbers increased so that by the summer of 1996/97 the density had risen to 2.16 crabs m^{-2}. The actual density during the year will closely mirror the pattern seen in fig. 4.

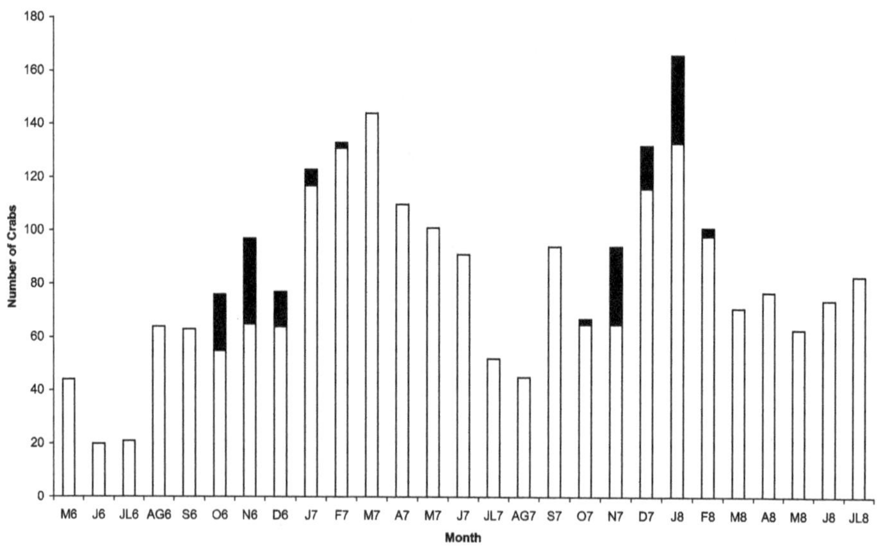

Fig. 4. Total population size of *Ozius truncatus* H. Milne Edwards, 1834 for clumps A-O from May 1996-July 1998. Ovigerous female numbers are shown by a filled bar. The first 3 months do not include clumps B, D, E, F, G, H, I, L, M, N, and O. The complete data set for all 15 clumps is from August 1996 to July 1998.

Variation in crab numbers in each clump

Most specimens of *O. truncatus* are solitary, each found under a particular stone. Aggressive behaviour can be high when crabs have nowhere to hide (Sivaguru & McLay, in prep.). Thus the maximum number of crabs in each area is set by the numbers of stones that provide habitable hiding spaces. Since this remains fairly constant over time one would expect the number of crabs in each clump to remain roughly the same, as long as the area is saturated with crabs. Fig. 5 shows the variation in numbers during the period May 1996 to June 1997. The total number of crabs recorded at each site over the 11-14 month period ranged from 10 (clump F; see fig. 1b) to 215 (clump A), mean = 77.2 (±15.0). For most of the less favoured sites (mean number of crabs per month <5) the variance/mean ratio was approximately 1.0, indicating that changes from month to month were random. For sites where more crabs were found (mean > 7) the ratio ranged from 2.0-5.3, indicating more extreme changes. The large increases in numbers within some

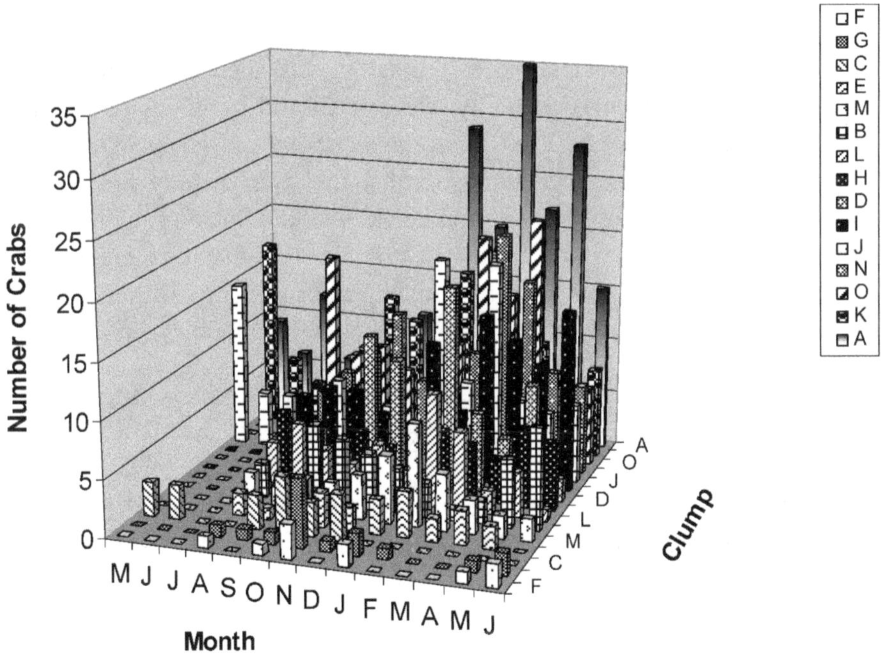

Fig. 5. Numbers of *Ozius truncatus* H. Milne Edwards, 1834 in each clump from May 1996 to June 1997. Note: prior to August 1996 clumps B, D, E, F, G, H, I, L, M, N, and O were not included. Data are complete for all 15 clumps for 11 months from August 1996 to June 1997. Clumps are arranged in abundance rank order from smallest (F) at the front to largest (A) at the back (see legend on graph).

clumps (notably D, I, J, N, O, K, and A, see fig. 1b) was due to the recruitment of juveniles in the summer. Thus it seems that recruitment is greater in areas where there are already larger numbers of adults. These constituted about half of the sites. If these favoured sites are over-saturated, it does not seem that the overflow disperses (see below) to less favoured sites, which remain small sub-populations.

Crab movements

Specimens of *Ozius truncatus* were found underneath medium sized (0.003 to 0.15 m^2) flat boulders of various shapes, surrounded by small pebbles or cobbles, and providing enough space to conceal the entire crab (see fig. 1c for example). Other species such as the filter-feeding half crab (*Petrolisthes elongatus* (H. Milne Edwards, 1837), Porcellanidae) and the detritus-feeding pebble crab (*Heterozius rotundifrons* A. Milne-Edwards, 1867; Belliidae) were also found in large numbers within the habitat sampled.

Most (93.5%) tagged adult *O. truncatus* remained within the clumps where they were originally tagged, moving on average only 1.12 m (fig. 6). Only occasionally (six individuals) did they move outside the original clump to a distance of greater than 10 m. Only one crab moved to an adjacent clump, moving a total distance of 8.2 m. The average distance moved by all recovered crabs was 2.29 m. These data suggest that the crabs mainly fed on prey species within the clumps in their immediate vicinity. Some of the tags were unreadable because they were damaged by the movement of crabs under boulders. This tagging method was not successful with juveniles because of frequent moulting, and so is more of an estimate of adult movements.

DISCUSSION

Ozius truncatus inhabits the upper middle intertidal levels on rocky shores living beneath larger cobbles and medium-sized boulders. Within these areas the crab has a patchy distribution. The population dynamics of *O. truncatus* at the Leigh Marine Reserve provided evidence of at least two cohorts that were recruited in the summers of 1996-97 and 1995-96 and possibly remnants of the 1995-1995 cohort. The progression of these cohorts through the population suggested that crabs reach CL ~15 mm by age 1+ yr, ~25 mm by age 2+ and ~35 mm by age 3+. Males and females grow to a similar maximum size (57 mm and 55 mm CL, respectively; see McLay, 1988) and the sex ratio at

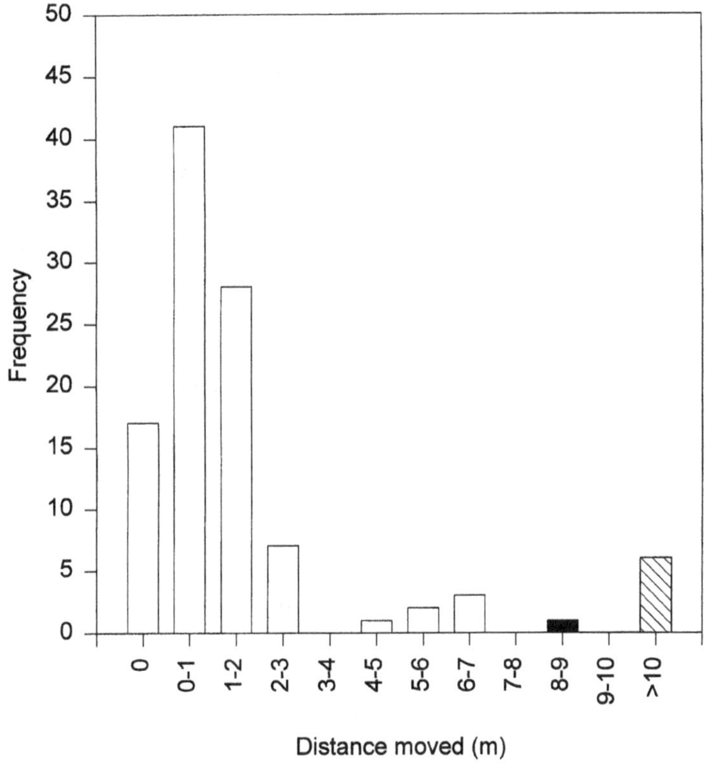

Fig. 6. Frequency distribution of the movements of tagged *Ozius truncatus* H. Milne Edwards, 1834 on Echinoderm Reef, New Zealand. Movements are pooled over the May-December 1996 period. Movements of less than 7 m (empty bars) were all within a clump; the single move of 8 m (filled bar) was between adjacent clumps (I and J); moves greater than 10 m (cross-hatched bar) were among widely-spaced clumps.

this location does not differ from a 1 : 1 ratio. Females reach sexual maturity during their first year (min. CL for ovigerous crabs was 20 mm) and if they survive to maximum age they could produce 3 or 4 annual broods. Ovigerous females were found in the population from October to February, similar to that reported by McLay (1988). During the breeding season the mean CL of ovigerous females tends to increase as larger females become ovigerous later in the summer. Eggs are brooded by females for approximately 2 months, with the planktonic phase lasting about 1 month, so that recruitment begins in December. Juvenile crabs may settle in crevices and other inaccessible areas so that the exact timing of recruitment is uncertain. We were not able to detect juveniles entering the population until they became ~5-10 mm CL. There was no evidence that recruitment was successful during 1997-98 (see fig. 3). The *O. truncatus* population is nevertheless maintained by surviving adults.

The distribution of tagged crabs on the shore demonstrated that *O. truncatus* adults are largely sedentary. They show a high level of site fidelity and were only rarely recorded outside the clump where they were originally tagged. The mean distance moved between monthly samples was only 2.3 m. This implies that the predatory impact of these crabs should be quite localized, but it is possible that they may move further afield when covered by the tide, returning to their original site at low tide. Bertness & Cunningham (1981) found that *Ozius verrauxii* Saussure, 1853 in the Bay of Panama forages in the splash zone during both day and night time low tides. They suggest that close sympatry of the crabs and their gastropod prey may be enforced by fish predation. Unlike *O. truncatus*, *O. verrauxii* retires to a burrow, which may imply a high level of site fidelity. By contrast, *Menippe mercenaria* (Say, 1818) living in oyster clumps, made frequent short range (mean = 4.6 m) movements and so only had a transient presence in a clump (Wilber, 1986).

Another predator in the Leigh Reserve that has been more intensively studied is the rock lobster, *Jasus edwardsii* (Hutton, 1875) (Palinuridae). It is a large predator that occupies the same subtidal sites for long periods, but ventures out of its shelter for local foraging on shellfish. Intensive studies of the movement of tagged lobsters show a high level of site fidelity. MacDiarmid et al. (1991) found that about 40% returned to the same shelter on consecutive days, and Kelly & MacDiarmid (2003) found that 55% of lobsters, tagged over two years, were found at the same site in 1996. In Tasmania, Barrett et al. (2009) recaptured 90% of tagged rock lobsters within 200 m of the release site on weekly to annual time scales. *Jasus edwardsii* forage chiefly at night when the median foraging range was only 24 m (MacDiarmid et al., 1991). Thus the impact on bivalve prey populations is restricted to near the daytime shelter.

Mobility and diet of crustacean predators may be related. Stachowicz & Hay (1999) found that there was a relationship between diet and mobility of three seaweed eating crabs from North Carolina: the majid *Mithraculus forceps* A. Milne-Edwards, 1875 (as *Mithrax*) exhibited the lowest mobility, highest site fidelity, and least selective diet of the 3 species, whereas another majid *Libinia dubia* H. Milne Edwards, 1834 was intermediate in both mobility and selectivity, and the panopeid *Panopeus herbstii* H. Milne Edwards, 1834 had the greatest mobility and narrowest diet. We might thus anticipate that *O. truncatus* could have a broad diet and eat any prey within the vicinity of its site. It is better to eat whatever is at hand rather than be exposed to danger and eaten.

Our estimates of the population densities of *O. truncatus* can only be approximate because we did not use random sampling. Had strict random sampling been employed, it is very unlikely that individuals would even have been detected as being present. If any shore sampling can be said to be random, it is only random within areas likely to yield positive samples (see McLay & McQueen, 1995). In areas of Echinoderm Reef where it was known that there were relatively high numbers of crabs present (areas A, C, J, and K) we estimated that in April 1996 the mean density was 0.6 crabs m^{-2} and by the summer of 1996/97 the density had risen to 2.16 crabs m^{-2}. This should be regarded as the maximum density likely to be found at the site, and that the annual change in density should follow that of total numbers (fig. 4). The only other estimates of the density of *O. truncatus* were provided by Chilton & Bull (1984) for a South Australia rocky platform. They used transect sampling methods, making direct comparison with our results difficult (only total numbers within 5 m intervals along transects are given), but based on their stated method we estimate that crab numbers ranged from 0.13 to 0.66 per m^2. It thus seems that density on the South Australia rocky platform and the shore at Leigh is similar, but it is not clear whether we are comparing crabs of the same size range. Chilton & Bull (1984) found evidence that the numbers of *O. truncatus* had a significant impact on the numbers of small gastropods (*Nerita atramentosa* Reeve, 1855, *Bembicium nanum* (Lamarck, 1822), and *Diloma concamerata* (Wood, 1828) (as *Austrocochlea*)) along the shore. Crabs selectively ate smaller snails.

Ozius truncatus is sparsely distributed on rocky shores, being restricted to patches of cobbles in mid to upper regions. Results suggest that *O. truncatus* is a slow-growing, long-lived, generally sedentary species that is unlikely to have more than a localized effect on shore communities, but as a result of selective grazing they could still be responsible for the size structure of their prey populations.

ACKNOWLEDGEMENTS

We would like to thank Bob Creese, Bob Lewis, and Richard Taylor for their guidance and help during this study, which was part of an M.Sc. thesis completed by the first author. Thanks are also due to all the technical staff at the Leigh Marine Laboratory. Assistance with field work was generously given by many contemporary graduate students. It is a privilege for both of us to acknowledge and recognize the important contribution that Daniele Guinot has made to knowledge and understanding of the Brachyura. Her influence will be enduring.

REFERENCES

AYLING, A. M., 1978. Cape Rodney to Okakari Point Marine Reserve Survey. Leigh Laboratory Bulletin, **1**: 1-98.

BARRETT, N., C. BUSXTON & C. GARDNER, 2009. Rock lobster movement patterns and population structure within a Tasmanian Marine Protected Area inform fishery and conservation management. Marine and Freshwater Research, **60**: 417-425.

BERTNESS, M. D. & C. CUNNINGHAM, 1981. Crab shell-crushing predation and gastropod architectural defense. Journal of Experimental Marine Biology and Ecology, **50**: 213-230.

CHILTON, N. B. & C. M. BULL, 1984. Influence of predation by a crab on the distribution of the size groups of three intertidal gastropods in Australia. Marine Biology, Berlin, **83**: 163-169.

— — & — —, 1986. Size-related selection of two intertidal gastropods by the reef crab *Ozius truncatus*. Marine Biology, Berlin, **93**: 475-480.

CREESE, R. G., 1988. Ecology of molluscan grazers and their interactions with marine algae in north-eastern New Zealand: a review. New Zealand Journal of Marine and Freshwater Research, **22**: 427-444.

EVANS, J., 1992. Leigh climate report. Leigh Laboratory Bulletin, **27**: 1-56.

HOLLING, C. S., 1966. The functional response of invertebrate predators to prey density. Memoirs of the Entomological Society of Canada, **48**: 1-86.

HUGHES, R. N., 1989. Foraging behaviour of a tropical crab: *Ozius verreauxii*. Proceedings of the Royal Society of London, (B) **237**: 201-212.

— —, 1993. Consumption of gastropods by the reef crab *Ozius truncatus* H. Milne Edwards; the role of opportunistic foraging behaviour. In: F. E. WELLS, D. I. WALKER, H. KIRKMAN & R. LEETHBRIDGE (eds.), The marine flora and fauna of Rottnest Island, Western Australia: 443-453. (Western Australian Museum, Perth).

KELLY, S. & A. B. MACDIARMID, 2003. Movement patterns of mature spiny lobsters, *Jasus edwardsii*, from a marine reserve. New Zealand Journal of Marine and Freshwater Research, **37**: 149-158.

KITCHING, J. A. & J. LOCKWOOD, 1974. Observations on shell form and its ecological significance in thaid gastropods of the genus *Lepsiella* in New Zealand. Marine Biology, Berlin, **28**: 131-141.

LOWELL, R. B., 1986. Crab predation on limpets: predator behavior and defensive features of the shell morphology of the prey. Biological Bulletin, Woods Hole, **171**: 577-596.

LUBCHENCO, J., B. A. MENGE, S. D. GARRITY, P. J. LUBCHENCO, L. R. ASHKENAS, S. D. GAINES, R. EMLET, J. LUCAS & S. STRAUSS, 1984. Structure, persistence, and role of consumers in a tropical rocky intertidal community (Taboguilla Island, Bay of Panama). Journal of Experimental Marine Biology and Ecology, **78**: 23-73.

MACDIARMID, A. B., B. HICKEY & R. A. MAILER, 1991. Daily movement patterns of the spiny lobster *Jasus edwardsii* (Hutton) on a shallow reef in northern New Zealand. Journal of Experimental Marine Biology and Ecology, **147**: 185-205.

MCLAY, C. L., 1988. Crabs of New Zealand. Leigh Laboratory Bulletin, **22**: 1-463.

MCLAY, C. L. & D. J. MCQUEEN, 1995. Intertidal zonation of *Cyclograpsus lavauxi* H. Milne Edwards, 1853 (Brachyura: Grapsidae) along the coast of the South Island of New Zealand. Crustacean Research, **24**: 49-64.

MENGE, B. A., 1983. Components of predation intensity in the low zone of the New England rocky intertidal region. Oecologia, **58**: 141-155.

MILNE EDWARDS, H., 1834-1837. Histoire naturelle des Crustacés comprenant l'anatomie, la physiologie et la classification de ces animaux. Paris, Librairie Encyclopédique de Roret, **1**: i-xxxv, 1-468; **2**: 1-531; Atlas, 1837: 1-32, pls. 1-42; **3**, 1840: 1-638.

SIVAGURU, K., 1997. The feeding ecology and behaviour of a predatory shore crab *Ozius truncatus* (Decapoda: Xanthidae): 1-145. (M.Sc. Thesis, University of Auckland).

SIVAGURU, K. & C. L. MCLAY, in prep. Predation of intertidal gastropods on Echinoderm Reef, Leigh Marine Reserve, New Zealand by *Ozius truncatus* H. Milne Edwards, 1834 (Brachyura: Oziidae).

SKILLETER, G. A. & D. T. ANDERSON, 1986. Functional morphology of the chelipeds, mouthparts and gastric mill of *Ozius truncatus* (Milne Edwards) (Xanthidae) and *Leptograpsus variegatus* (Fabricius) (Grapsidae) (Brachyura). Australian Journal of Marine & Freshwater Research, **37**: 67-79.

STACHOWICZ, J. J. & M. E. HAY, 1999. Reduced mobility is associated with compensatory feeding and increased diet breadth of marine crabs. Marine Ecology Progress Series, **188**: 169-178.

WEAR, R. G., 1968. Life-history studies on New Zealand Brachyura. 2. Family Xanthidae. Larvae of *Heterozius rotundifrons* A. Milne-Edwards, 1867, *Ozius truncatus* H. Milne Edwards, 1834, and *Heteropanope (Pilumnopeus) serratifrons* (Kinahan, 1856). New Zealand Journal of Marine and Freshwater Research, **2**: 293-332.

WEAR, R. G. & D. R. FIELDER, 1985. The Marine Fauna of New Zealand: Larvae of the Brachyura (Crustacea, Decapoda). New Zealand Oceanographic Institute Memoirs, **92**: 1-90.

WILBER, D. H., 1986. The distribution and daily movement of stone crabs (*Menippe mercenaria*) in an intertidal oyster habitat on the Northwest coast of Florida. Marine Behaviour and Physiology, **12**: 279-291.

First received 17 June 2009.
Final version accepted 13 July 2009.

TWO NEW SPECIES OF THE FAMILY GONEPLACIDAE (DECAPODA, BRACHYURA) FROM THE WEST PACIFIC

BY

MASATSUNE TAKEDA[1,3]) and HIRONORI KOMATSU[2])

[1]) Faculty of Modern Life, Teikyo Heisei University, 2-51-4 Higashi-Ikebukuro, Toshima-ku, Tokyo, 170-8445 Japan

[2]) Department of Zoology, National Museum of Nature and Science, 3-23-1 Hyakunincho, Shinjuku-ku, Tokyo, 169-0073 Japan

ABSTRACT

Two new species of the family Goneplacidae are described from Japan and Taiwan. *Singhaplax danielae* n. sp. can be distinguished from its congeners by a wider abdomen and the shape of the male gonopods. *Microgoneplax guinotae* n. sp. can be distinguished from its congeners by its robust male first gonopod.

RÉSUMÉ

Deux nouvelles espèces de la famille Goneplacidae sont décrites du Japon et Taiwan. *Singhaplax danielae* n. sp. peuvent être distinguées des congénères par l'abdomen plus large et de la forme des gonopodes masculins. *Microgoneplax guinotae* n. sp. peuvent être distinguées des congénères par le premier gonopode de mâle robuste.

INTRODUCTION

While re-examining the crabs collected by the late Mr. S. Nagai and donated to the Wakayama Prefectural Museum of Natural History, a small crab of the family Goneplacidae from Japan was noticed with a well-developed first male gonopod in spite of its small size. The specimen has long eyestalks, and it was, thus, referred to the genus *Ommatocarcinus* White, 1852 by Marumura & Kosaka (2003). A detailed examination of the specimen following the up-to-date revision by Castro (2007) revealed that it is actually a new species

[3]) Corresponding author; e-mail: takeda-m@thu.ac.jp

© Koninklijke Brill NV, Leiden, 2010

of *Singhaplax* Serène & Soh, 1976. On the way to describe the species, we received important information from Drs. P. Castro, P. K. L. Ng, and T. Naruse that a male specimen from Taiwan appears to be conspecific with the new species from Japan. In the present paper a new species, *S. danielae*, is to be described based on these two specimens.

As part of our study, we also re-examined the specimens referred to *Singhaplax* in the collections of the National Museum of Nature and Science, Tokyo. Those specimens are mostly represented by *S. styrax* Castro, 2007, but there is a small specimen from Japan having a short second gonopod in contrast to a well-developed, stout first gonopod. These characters suggest that it belongs to *Microgoneplax* Castro, 2007. Comparing with the known species listed by Castro (2007), the specimen represents a new species, *M. guinotae*, which is also described herein.

Descriptive terminology and measuring methods follow Castro (2007). The material examined is deposited in the National Museum of Nature and Science (NSMT), the Wakayama Prefectural Museum of Natural History (WMNH), and the Zoological Reference Collection, Raffles Museum of Biodiversity Research, National University of Singapore (ZRC). Other abbreviations used in the text are: cl, carapace length; cw, carapace width; G1, G2, first and second male gonopods, respectively; P1, first pair of pereopods (chelipeds); P2-P5, second to fifth pairs of pereopods (first to fourth ambulatory legs).

We take great pleasure in dedicating these two new species of crabs to Prof. Danièle Guinot. The first author, M. Takeda, especially wishes to offer his cordial thanks for her continuous guidance and encouragement over a span of 40 years.

TAXONOMY

Family GONEPLACIDAE MacLeay, 1838

Singhaplax danielae n. sp. (figs. 1-3)

Ommatocarcinus sp. — Marumura & Kosaka, 2003: 47 (in list).

Material examined. — Holotype, male (cl 4.4 × cw 9.5 mm), WMNH-Na-Cr 676, off Oshima I., Kushimoto, Kii Peninsula, Japan, 90 m deep, coll. S. Nagai, 15 Mar. 1980. Paratype, male (cl 4.7 × cw 10.3 mm), ZRC 2009.0424, Taiwan.

Description. — Carapace (figs. 1A, 3) transversely rectangular, much wider than long (2.2 times as wide as long in holotype). Carapace convex, without clear indication of regions. Front almost straight. Base of front distinctly

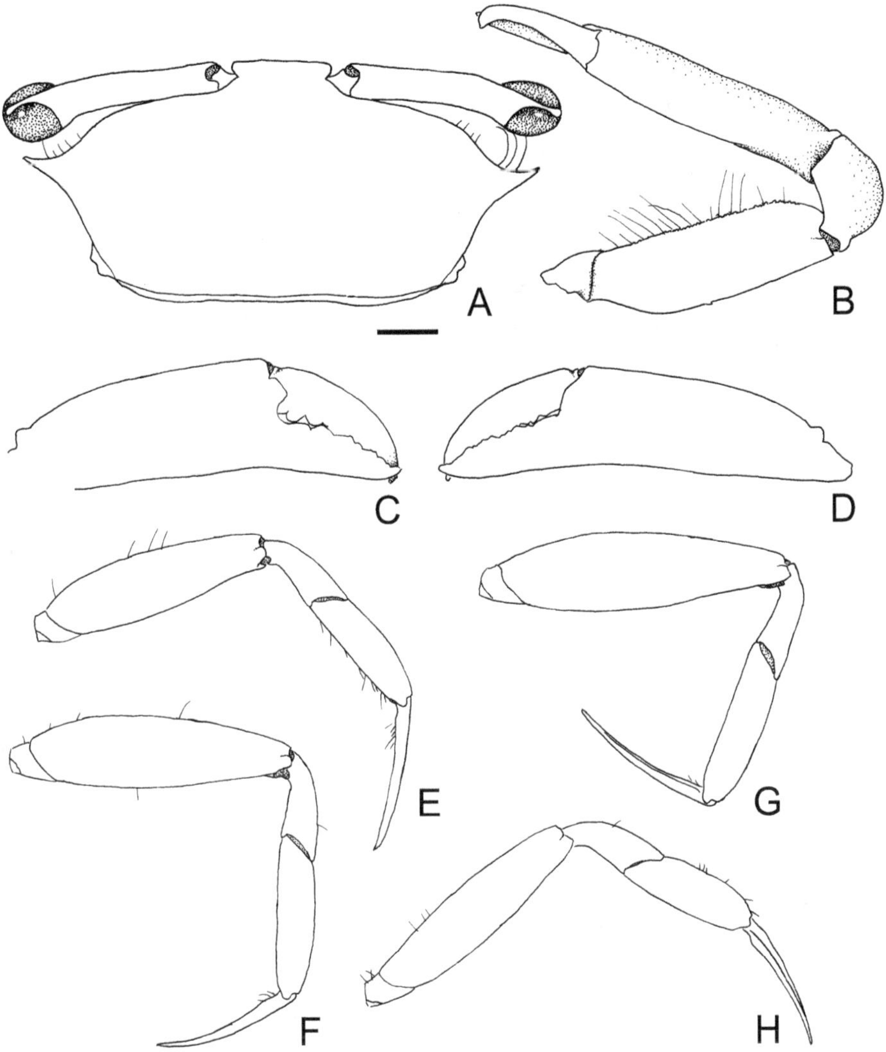

Fig. 1. *Singhaplax danielae* n. sp. Holotype, male (cl 4.4 × cw 9.5 mm), WMNH-Na-Cr 676. A, carapace, dorsal view; B, right cheliped, dorsal view; C, right chela, outer view; D, left chela, outer view; E-H, right P2-P5, respectively, dorsal view. Scale bar: 1 mm.

constricted, with notch between front and inner edge of supraorbital border. Supraorbital borders broad, conspicuously sinuous, smooth, with simple setae, setae on lateral part long. Suborbital borders sinuous, fringed with fine granules; short, wide inner tooth not visible dorsally. Outer orbital angle long, slender, slightly directed anteriorly; tooth placed on middle of carapace. Lateral borders strongly convergent posteriorly, slightly sinuous, no anterolateral

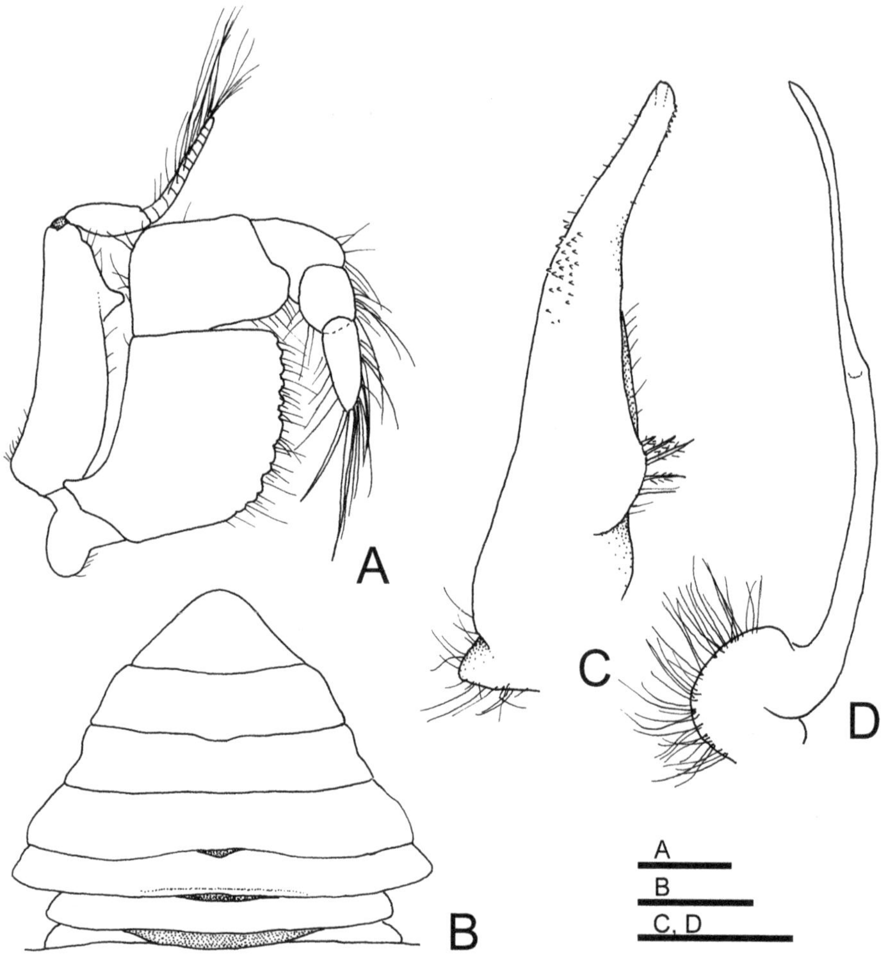

Fig. 2. *Singhaplax danielae* n. sp. Holotype, male (cl 4.4 × cw 9.5 mm), WMNH-Na-Cr 676. A, right third maxilliped, ventral view; B, abdomen, ventral view; C, right G1, ventral view; D, right G2, ventral view. Scales bars: 0.5 mm for A, C, D; 1 mm for B.

teeth. Subhepatic and pterygostomial regions smooth; pterygostomial ridge with plumose setae on lateral half.

Eye peduncles (fig. 1A) very long (1.5 front width in holotype), cornea elongated, spherical distal margin. Third maxillipeds (fig. 2A) flattened; ischium divided from basis by suture, with denticulate mesial margin; merus trapezoidal.

Chelipeds (P1) (fig. 1B-D) unequal (right slightly larger than left in holotype), 3.1 cl in right in holotype; merus triangular in cross section, flattened on dorsal surface, with fine granules and long setae on inner margin, subacute granule on midlength of outer margin. Fingers 0.6 propodus length in holotype; dactylus curved, 2 rounded teeth on proximal part of cutting edge.

Ambulatory legs (P2-P5) (fig. 1E-H) long, slender, unarmed, varying number of setae, setae on meri of P2-P5 and propodi of P5 plumose; propodi of P5 slightly broadened; dactyli long, slender, each with 2 carinae along upper and lower sides; relative length P4 > P3 > P2 > P5, length of P5 merus 0.9 cl.

Male abdomen (fig. 2B) wide, triangular, with 6 freely movable somites plus telson; telson triangular, 2.1 times as wide as long. Somite 1 very short, medially concealed beneath carapace (not concealed in paratype). Somite 3 widest in abdominal somites, about 0.7 width of thoracic sternite 8. Somites 3 to 6 gradually decreasing in width.

G1 (fig. 2C) long, basal part wide, distal part curved mesially, small denticles on distal 0.4 of ventro-lateral surface and distal 0.3 of dorsal surface. G2 (fig. 2D) long, slender, slightly longer than G1, flagellum slightly shorter than proximal part (peduncle), tip pointed.

Notes on paratype. — The paratype specimen (fig. 3B) looks more mature than the holotype. The cheliped is elongate (3.9 cl in right) and the propodus of the cheliped is also elongate (fingers 0.5 propodus length). The cheliped merus does not have a distinct granule on the inner margin, and the first somite of the abdomen is not concealed beneath the carapace. The shape of the male gonopods agrees well with that of the holotype.

Remarks. — *Singhaplax danielae* n. sp. can be distinguished from its congeners by the wider male abdomen and the shape of the male gonopods. The wide carapace (>2.0 cl) associates the new species with *S. ockelmanni* (Serène, 1971) from the Andaman Sea coast of Thailand and the Philippines, and with *S. wolffi* (Serène, 1964) from New Caledonia. *Singhaplax danielae* n. sp. is distinguished from *S. ockelmanni* by the following characters: the width of somite 3 of the male abdomen is 1.4 times as long as the combination length of somite 3 to telson (fig. 2B) (as long as that in *S. ockelmanni*); the tip of the G1 is simple (fig. 2C) (bilobed and beak-like in *S. ockelmanni*) (cf. Serène & Umali, 1972; Serène & Soh, 1976). *Singhaplax danielae* n. sp. is also distinguished from *S. wolffi* by the following characters: none of the anterolateral borders of the carapace has a tubercle (figs. 1A, 3) (0-4 tubercles or prominences in *S. wolffi*); the tip of the G2 is simply acute (fig. 2D) (with a terminal spinule in *S. wolffi*) (cf. Serène, 1964; Castro, 2007).

The number of species of *Singhaplax* now known stands at eight with the addition of this new species (see Castro, 2007; Ng et al., 2008). Additional species are currently being described by T. Naruse and P. Castro (work in progress).

Fig. 3. *Singhaplax danielae* n. sp. A, holotype, male (cl 4.4 × cw 9.5 mm), WMNH-Na-Cr 676; B, paratype, male (cl 4.7 × cw 10.3 mm), ZRC 2009.0424.

Microgoneplax guinotae n. sp. (figs. 4-6)

Material examined. — Holotype, male (cl 5.0 × cw 9.0 mm), NSMT-Cr 10017, KT-84-12 cruise, stn 12-1, off Susami, Kii Peninsula, Japan, 33°29.9′N 135°35.9′E – 33°30.0′N 135°30.7′E, 100-101 m, dredge, shell and sandy mud bottom, coll. E. Tsuchida, 31 Aug. 1984.

Description. — Carapace (figs. 4A, 6) transversely rectangular, much wider than long (1.8 times as wide as long in holotype). Carapace convex, without clear indication of regions. Front almost straight. Without distinct notch between front, inner edge of supraorbital border. Supraorbital borders broad, conspicuously sinuous, smooth, with simple setae. Suborbital borders sinuous,

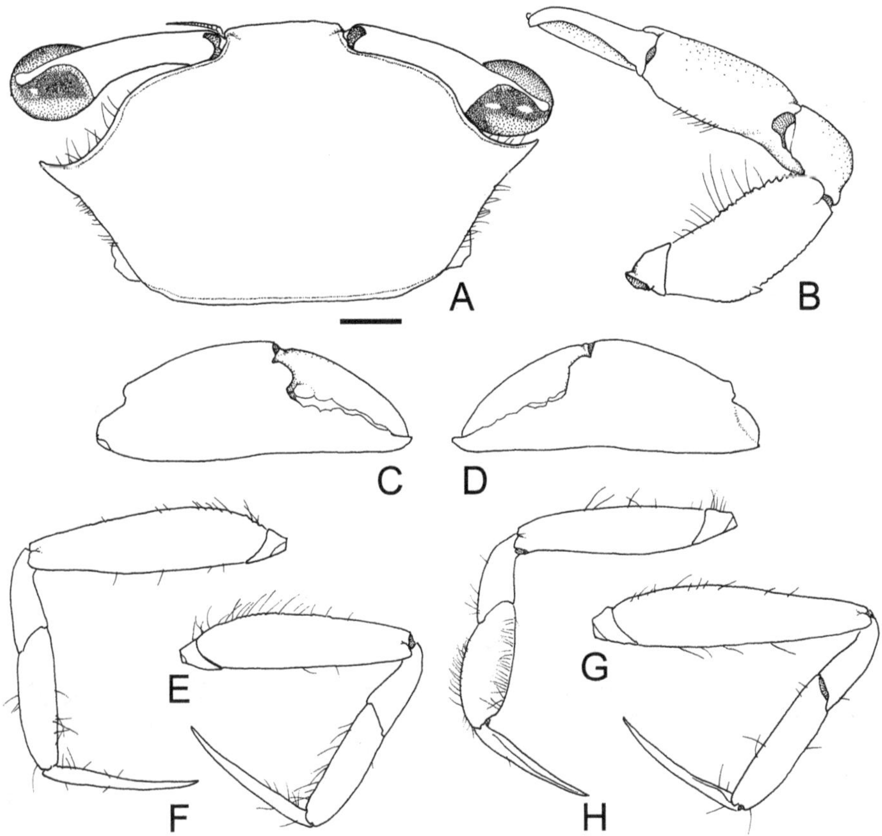

Fig. 4. *Microgoneplax guinotae* n. sp. Holotype, male (cl 5.0 × cw 9.0 mm), NSMT-Cr 10017. A, carapace, dorsal view; B, right cheliped, dorsal view; C, right chela, outer view; D, left chela, outer view; E, right P2; F, left P3; G, right P4; H, left P5, dorsal view. Scale bar: 1 mm.

fringed with fine granules; large, wide inner tooth not visible dorsally. Outer orbital angle long, slender, slightly directed anteriorly; tooth medially placed on carapace. Lateral borders beyond outer orbital teeth slightly concave, no anterolateral teeth. Subhepatic, pterygostomial regions with low, round tubercles; pterygostomial ridge granulate.

Eye peduncles (fig. 4A) long (equal to front width), cornea elongate, spherical distal margin. Third maxillipeds (fig. 5A) flattened; ischium divided from basis by suture, with denticulate mesial margin; merus trapezoidal.

Chelipeds (P1) (fig. 4B-D) unequal (right slightly larger than left in holotype), 2.1 cl in right; merus triangular in cross section, flattened on dorsal surface, with low tubercles and long setae on inner margin, subacute granule on midlength of outer margin. Fingers as long as propodus; no dark colour on

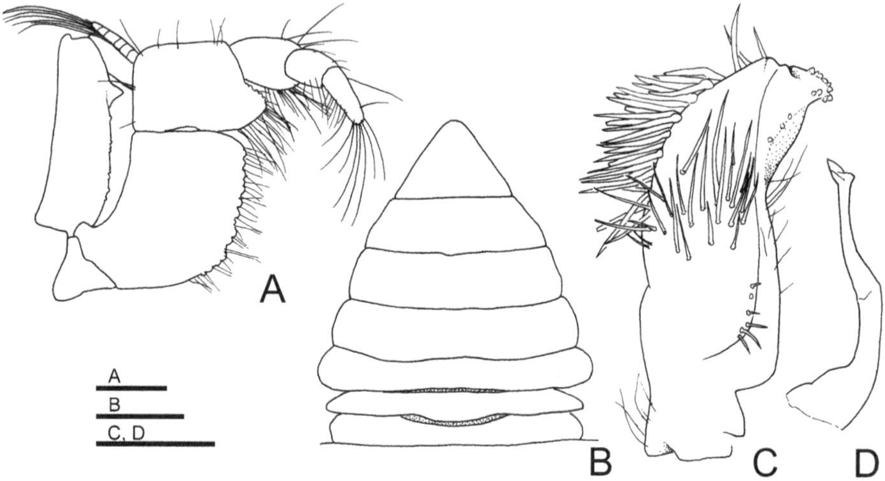

Fig. 5. *Microgoneplax guinotae* n. sp. Holotype, male (cl 5.0 × cw 9.0 mm), NSMT-Cr 10017. A, right third maxilliped, ventral view; B, abdomen, ventral view; C, right G1, ventral view; D, right G2, ventral view. Scales bars: 0.5 mm for A, C, D; 1 mm for B.

Fig. 6. *Microgoneplax guinotae* n. sp. Holotype, male (cl 5.0 × cw 9.0 mm), NSMT-Cr 10017.

fingers in preserved specimen; dactylus curved, two rounded teeth on proximal part of cutting edge in major chela.

Ambulatory legs (P2-P5) (fig. 4E-H) long, slender, unarmed, varying number of setae, setae on meri of P2-P5 and propodi of P5 plumose; propodi of P5 slightly broadened; dactyli long, slender, each with 2 carinae along upper and lower sides; relative length P4 > P3 > P2 > P5, length of P5 merus 0.7 cl.

Male abdomen (fig. 5B) wide, triangular, with 6 freely movable somites plus telson; telson triangular, 1.4 times as wide as long. Somites 1, 2 very short. Somite 3 widest of abdominal somites, about 0.6 width of thoracic sternite 8. Somites 3 to 6 gradually decreasing in width.

G1 (fig. 5C) robust, with dense, plumose, stout setae on distal 0.5 of lateral surface, plumose setae on mid-portion of ventral surface; tip directed mesially, denticulate. G2 (fig. 5D) short, slender, 0.7 times as long as G1, flagellum about same length as proximal part (peduncle), tip with terminal spinule.

Remarks. — *Microgoneplax guinotae* n. sp. is readily distinguished from its congeners by the characteristic, robust G1 having dense and stout setae on the distal half of its lateral surface. The holotype specimen was collected with a male specimen of *Singhaplax styrax* Castro, 2007 from the same station. They are similar in general appearance, but can be easily distinguished by the shape of the male gonopods.

ACKNOWLEDGEMENTS

The descriptions of two new species were made possible through the permission of the authorities of the Wakayama Prefectural Museum of Natural History, and the kind arrangement of Drs. Peter Castro (California State Polytechnic University, Pomona), Peter K. L. Ng (National University of Singapore), and Tohru Naruse (University of the Ryukyus). The manuscript was brushed up by P. Castro, P. K. L. Ng, and an anonymous reviewer, to whom we have to tender our sincere thanks.

REFERENCES

CASTRO, P., 2007. A reappraisal of the family Goneplacidae MacLeay, 1838 (Crustacea, Decapoda, Brachyura) and revision of the subfamily Goneplacinae, with the description of 10 new genera and 18 new species. Zoosystema, **29**: 609-774.

MACLEAY, W. S., 1838. On the brachyurous decapod Crustacea brought from the Cape by Dr. Smith. In: A. SMITH (ed.), Illustrations of the zoology of South Africa; consisting chiefly of figures and descriptions of the objects of natural history collected during an expedition into the interior of South Africa, in the years 1834, 1835, and 1836; fitted out by 'The Cape of Good Hope Association for Exploring Central Africa': together with a summary of African zoology, and an inquiry into the geographical ranges of species in that quarter of the globe. Invertebratae, **4** [1849]: 53-71, pls. 2, 3. (Smith, Elder & Co., London).

MARUMURA, M. & A. KOSAKA, 2003. Catalogue of brachyuran and anomuran crabs collection donated by the late Mr. Seiji Nagai to the Wakayama Prefectural Museum of Natural History: 1-74. (Wakayama Prefectural Museum of Natural History, Kainan).

SERÈNE, R., 1964. Goneplacidae et Pinnotheridae récoltés par le Dr. Mortensen. Papers from Dr. Th. Mortensen's Pacific Expedition 1914-1916, part 80. Videnskabelige Meddelelser fra Dansk Naturhistorisk Forening i København, **126**: 181-282, pls. 16-24.

— —, 1971 [1970]. Observations préliminaires sur des brachyoures nouveaux ou mal connus du sud-est asiatique (Crustacea Decapoda). Bulletin du Muséum national d'Histoire naturelle, Paris, (2) **42**: 903-918.

SERÈNE, R. & C. L. SOH, 1976. Brachyura collected during the Thai-Danish Expedition (1966). Research Bulletin Phuket Marine Biological Center, Thailand, **12**: 1-37, figs. 1-28, pls. 1-7.

SERÈNE, R. & A. F. UMALI, 1972 [1970]. The family Raninidae and other new and rare species of brachyuran decapods from the Philippines and adjacent regions. Philippine Journal of Science, **99**: 21-105, pls. 1-9.

WHITE, A., 1852. Descriptions of some apparently new species of Annulosa (collected by Mr. Macgillivray during the voyage of H.M.S. Rattlesnake). Appendix no. 6. In: J. MACGILLIVRAY (ed.), Narrative of the voyage of H.M.S. Rattlesnake, commanded by the late Captain Owen Stanley, R.N., F.R.S. etc. during the years 1846-1850. Including discoveries and surveys in New Guinea, the Louisiade Archipelago, etc. to which is added the account of Mr. E. B. Kennedy's expedition for the exploration of the Cape York Peninsula, **2**: 387-395, pls. 4, 5. (T.&W. Boone, London).

First received 15 June 2009.
Final version accepted 20 July 2009.

A NEW SPECIES OF *PSEUDOLAMBRUS* (BRACHYURA, PARTHENOPIDAE) FROM NEW CALEDONIA

BY

S. H. TAN[1])

Raffles Museum of Biodiversity Research, Department of Biological Science, National University of Singapore, 6 Science Drive 2, Singapore 117546, Republic of Singapore

ABSTRACT

A new species of parthenopid crab, *Pseudolambrus guinotae* n. sp., is described. The single male specimen, which was collected from New Caledonia, most closely resembles *Pseudolambrus planus* (Rathbun, 1911) but can be easily differentiated from the latter by possessing a much longer rostrum.

RÉSUMÉ

Une nouvelle espèce de crabe Parthenopidae, *Pseudolambrus guinotae* n. sp. est décrite. Le seul spécimen mâle, récolté en Nouvelle-Calédonie, ressemble beaucoup à *Pseudolambrus planus* (Rathbun, 1911) mais se différencie aisément de ce dernier par la possession d'un rostre beaucoup plus long.

INTRODUCTION

The taxonomic status of the parthenopid genus *Pseudolambrus* Paul'son, 1875, has long been confusing. This is partly due to the general belief that sexual dimorphism accounts for much of the external morphological variation among species, and is infrequently reported by authors, possibly due to lack of comparative material (see Tan, 2004). Tan (2004) revised the Parthenopidae and in a subsequent paper restricted *Pseudolambrus* to almost all western Indo-Pacific Ocean species, while the remaining species were placed in two other

[1]) e-mail: dbstansh@nus.edu.sg

© Koninklijke Brill NV, Leiden, 2010

genera, *Ochtholambrus* Tan & Ng, 2007, and *Velolambrus* Tan & Ng, 2007 (see Tan & Ng, 2007: 99).

Ng et al. (2008) recognized 15 species, with one additional species, *Pseudolambrus bato* Tan, 2008, recently described. As part of the revision of the Parthenopidae by Tan (2004), several *Pseudolambrus* species were described in that work that were not formerly published. I take this occasion to dedicate one of these species to one of the most respected figures in modern carcinology, Danièle Guinot.

MATERIAL AND METHODS

The material used for this study mainly originated from the Muséum national d'Histoire naturelle, Paris (MNHN), and the U.S. National Museum of Natural History, Smithsonian Institution, Washington, D.C. (USNM). The terminology follows McLay & Tan (2009). Measurements presented here, in millimetres (mm), are those of the carapace width and length, respectively, and in that very order.

TAXONOMIC ACCOUNT

Family PARTHENOPODAE MacLeay, 1838

Pseudolambrus Paul'son, 1875

Pseudolambrus guinotae n. sp. (fig. 1)

Material examined. — Holotype, male, 9.9 by 9.5 mm (MNHN), New Caledonia, Grotte Merlet, 20-30 m, J.-L. Menou coll., 19-21 January 1993.

Comparative material. — *Pseudolambrus planus* (Rathbun, 1911), male holotype, 18.1 by 16.1 mm (USNM 41446), Seychelles, Amirante I., HMS "Sealark", stn E21, 30 fms (55 m), 17 October 1905; 1 female, 8.1 by 7.4 mm (MNHN), CHALCAL 1, Chesterfield and Bellona Plateau, stn D19, 19°06.73′S 158°41.75′E, 60 m, 12-31 July 1984; 1 female, 16.0 by 13.9 mm (MNHN), CORAIL 2, Chesterfield lagoon, stn DW144, 19°27.73′S 158°23.28′E, 50 m, 30 August 1988.

Description. — Carapace (fig. 1) broadly pentagonal, slightly wider than long, surface rugose with fine reticulated patterning; rostrum much longer than broad, ventral surface with 2 rows of 6 oval depressions each, tip truncated. Epistome with lattice-like patterning, without any sub-circular structures. Posterior, metabranchial margins not continuous, notch at junction of 2 margins. Cheliped merus outer margin dentate, distalmost tooth largest,

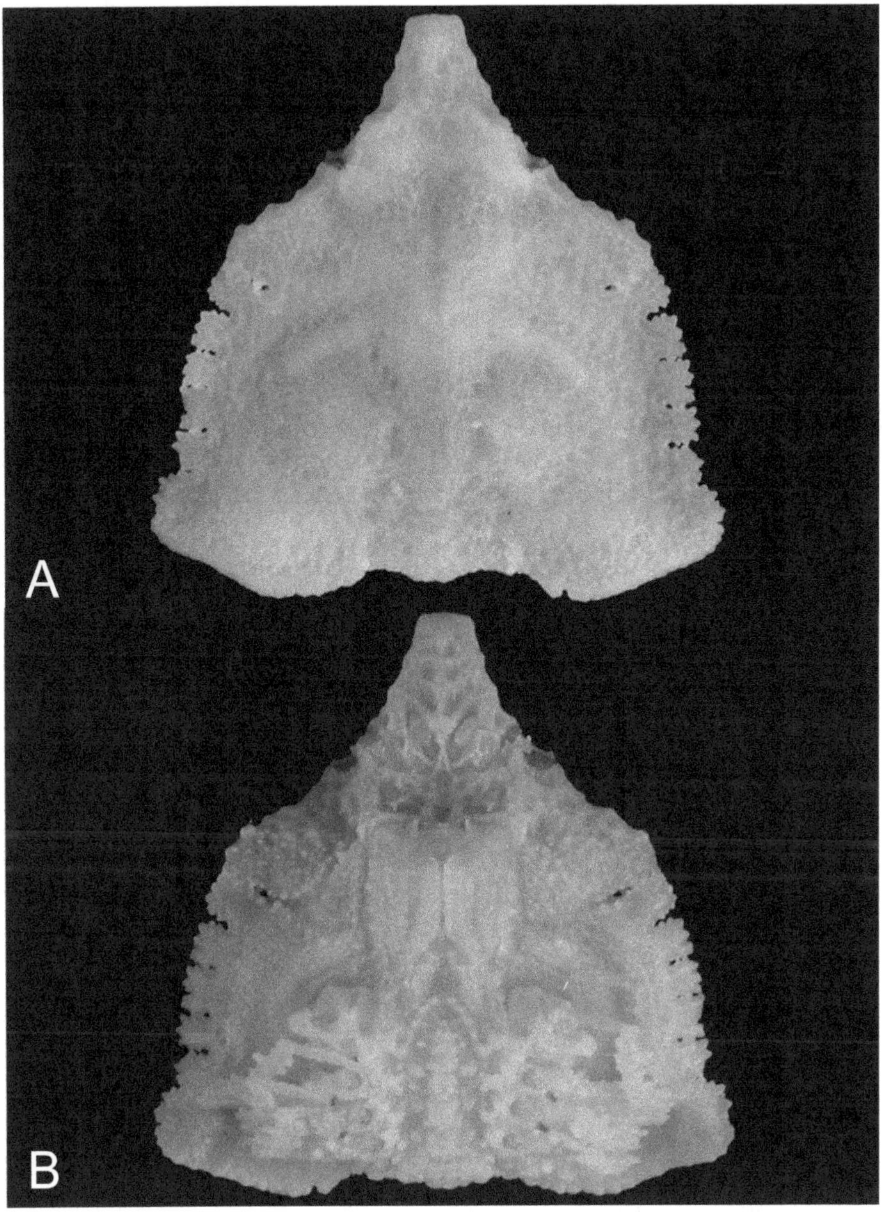

Fig. 1. *Pseudolambrus guinotae* n. sp.: A, B, holotype, male 9.9 × 9.5 mm (MNHN), New Caledonia, Grotte Merlet, 20-30 m, J.-L. Menou coll., 19-21 January 1993. A, carapace, dorsal view; B, carapace, ventral view.

lamelliform, tooth tip rounded; teeth not fused with each other. Cheliped dactylus upper margin dentate, teeth prominent.

Etymology. — I have great pleasure in naming this unique parthenopid species after the grande doyenne of carcinology, Danièle Guinot, a great friend and colleague whom I have the pleasure of working with.

Distribution. — Known only from the type locality, New Caledonia, from a depth between 20 and 30 m.

Remarks. — The new species is so distinctive from all the other *Pseudolambrus* species that a diagnosis is sufficient to characterize it. Amongst all 15 *Pseudolambrus* species currently known (Ng et al., 2008), this species most closely resembles *P. planus* (Rathbun, 1911) in that the carapaces of both species are flattened dorso-ventrally. *Pseudolambrus guinotae*, however, is easily distinguished from *P. planus* in that it possesses a very long rostrum as well as two parallel rows of circular depressions on the ventral surface of the rostrum. In addition, the posterior margin is not continuous with the metabranchial margin in *P. guinotae*, but it is continuous in *P. planus*.

Tan (2004) diagnosed several other *Pseudolambrus* species similar to both *P. planus* and *P. guinotae*, but those will be described in a separate, comprehensive revision of the Parthenopidae.

ACKNOWLEDGEMENTS

I thank Peter Ng and Peter Castro for the opportunity to participate in this volume. I would also like to thank the staff, past and present, of the Muséum national d'Histoire naturelle for hosting me while studying this group of crabs for my dissertation. I thank three referees for their comments, which improved the quality of this paper. Part of this project is supported by the National University of Singapore grant ARF R-154-000-334-112.

REFERENCES

MACLEAY, W. S., 1838. On the brachyurous decapod Crustacea. Brought from the Cape by Dr. Smith. In: A. SMITH (ed.), Illustrations of the zoology of South Africa; consisting chiefly of figures and descriptions of the objects of natural history collected during an expedition into the interior of South Africa, in the years 1834, 1835, and 1836; fitted out by 'The Cape of Good Hope Association for Exploring Central Africa': together with a summary of African zoology, and an inquiry into the geographical ranges of species in that quarter of the globe, published under the Authority of the Lords Commissioners of Her Majesty's Treasury, Invertebratae, **4** [1849]: 53-71, pls. 2, 3. (Smith, Elder & Co., London).

MCLAY, C. L. & S. H. TAN, 2009. Revision of the crab genus *Garthambrus* Ng, 1996, with the description of two new genera and discussion of the status of *Tutankhamen* Rathbun, 1925 (Crustacea: Brachyura: Parthenopidae). Zootaxa, **2122**: 1-50.

NG, P. K. L., D. GUINOT & P. J. F. DAVIE, 2008. Systema Brachyurorum: Part I. An annotated checklist of extant brachyuran crabs of the world. Raffles Bulletin of Zoology, (Supplement) **17**: 1-286.

PAUL'SON, O., 1875. Studies on Crustacea of the Red Sea with notes regarding other areas. Part 1. Podophthalmata and Edriophthalmata (Cumacea): 1-164, pls. 1-21. (Israel Programme for Scientific Translations, Jerusalem). [English translation of original Russian edition.]

RATHBUN, M. J., 1911. No. XI. Marine Brachyura. In: The Percy Sladen Trust Expedition to the Indian Ocean in 1905, under the leadership of J. Stanley Gardiner, **3**. Transactions of the Linnean Society of London, (Zoology) (2) **14**: 191-261, figs. 1, 2, pls. 15-20.

TAN, S. H., 2004. A systematic revision of the Parthenopidae (Crustacea: Decapoda: Brachyura): 1-730. (Unpubl. Ph.D. Thesis, Department of Biological Sciences, National University of Singapore).

— —, 2008. A new species of *Pseudolambrus* (Crustacea: Brachyura: Parthenopidae: Parthenopinae) from Panglao, Bohol, the Philippines. Zootaxa, **1731**: 63-68.

TAN, S. H. & P. K. L. NG, 2007. Descriptions of new genera from the subfamily Parthenopinae (Crustacea: Decapoda: Brachyura: Parthenopidae). Raffles Bulletin of Zoology, (Supplement) **16**: 95-119.

First received 15 July 2009.
Final version accepted 29 July 2009.

FIRST SPECIES OF *PARAGALENE* KOSSMANN, 1878 (DECAPODA, BRACHYURA, PROGERYONIDAE) FROM THE WESTERN ATLANTIC

BY

MARCOS TAVARES[1,2] and GUSTAVO A. S. DE MELO[1,3]

[1]) Museu de Zoologia, Universidade de São Paulo, Ipiranga 04263-000, São Paulo, SP, Brazil

ABSTRACT

A new species of deep-water crab, *Paragalene danieleae* (Progeryonidae), from the western Atlantic is described and illustrated. The new species can be easily separated from its eastern Atlantic and Mediterranean counterpart, *P. longicrura* (Nardo, 1868), by a suite of carapace and appendage characters.

RÉSUMÉ

Une nouvelle espèce de crabe d'eau profonde, *Paragalene danieleae* (Progeryonidae) de l'Atlantique occidental est décrite et illustrée. La nouvelle espèce se sépare aisement de sa analogue de l'Atlantique oriental et Mediterranée, *P. longicrura* (Nardo, 1868), par divers caractères de la carapace et des appendices.

INTRODUCTION

For more than thirteen decades *Paragalene longicrura* (Nardo, 1868) has been the only valid species assigned to the genus *Paragalene* Kossmann, 1878. *Paragalene longicrura* is rare and has been known only from a small number of localities in the Mediterranean Sea and two localities in the eastern Atlantic (see d'Udekem d'Acoz, 1999). Its taxonomic position has long been controversial. The species was originally described in the genus *Eriphia* Latreille, 1817, and accordingly attributed to the Xanthidae (Nardo, 1868). It was later included in the Geryonidae (e.g., Guinot, 1969: 692, 696; 1971:

[2]) Corresponding author: e-mail: mdst@usp.br

[3]) e-mail: gasmelo@usp.br

© Koninklijke Brill NV, Leiden, 2010

1077-1078; Türkay, 1976a, b), until Manning & Holthuis (1989) restricted the Geryonidae to *Geryon* Krøyer, 1897, *Chaceon* Manning & Holthuis, 1989, and *Zariquieyon* Manning & Holthuis, 1989. Manning & Holthuis (1989: 51) also suggested that *Paragalene* could be included in a family of its own, along with *Progeryon* Bouvier, 1922. Instead of dealing with the uncertain taxonomic position of *Paragalene* as well as *Progeryon*, Števčić (2005: 64) erected two intermediate categories, the redundant monotypic tribes Paragalenini Števčić, 2005, and Progeryonini Števčić, 2005, both included in the Geryoninae (Geryonidae). Karasawa & Schweitzer (2006) attributed full family status to the Progeryonini (as Progeryonidae), to which *Paragalene* was eventually transferred (Ng & Manuel-Santos, 2007; Ng et al., 2008).

Trapping operations recently conducted between 100 and 1000 m depth off the coast of Guadeloupe, West Indies, revealed a wealth of diversity of decapod crustaceans (Poupin, 1994; Tavares, 1994; Tavares & Guinot, 1996), and among them a western Atlantic counterpart of *P. longicrura*, herein described as new, *Paragalene danieleae* n. sp.

The following abbreviations are used: CBR (Colleccions Biològiques de Referència, Institut de Ciències del Mar, Barcelona); MZUSP (Museu de Zoologia, Universidade de São Paulo); cl, carapace length; cw, carapace width; P1, cheliped; P2-P5, pereiopods 2 to 5.

SYSTEMATIC ACCOUNT

Family PROGERYONIDAE Števčić, 2005

Paragalene Kossmann, 1878

Paragalene Kossmann, 1878: 253.

Type species. — *Paragalene neapolitana* Kossmann, 1878, by monotypy [= *Eriphia longicrura* Nardo, 1868]. Gender feminine.

Species included. — *Paragalene longicrura* (Nardo, 1868), eastern Atlantic and Mediterranean Sea (d'Udekem d'Acoz, 1999: 241, and references therein); *Paragalene danieleae* n. sp., western Atlantic.

Paragalene danieleae n. sp. (figs. 1, 3A, C)

Type material. — Holotype: female (cl 37 mm, cw 48 mm) (MZUSP 19938), Caribbean Sea, West Indies, Guadeloupe, west coast of Basse Terre, Mission ORSTOM-IRPM-SMCB, stn A210, 16°11.80′N 61°48.72′W, G. Leblond and J. Poupin leg., April 1993, 250 m.

Comparative material. — *Paragalene longicrura* (Nardo, 1868), western Mediterranean Sea, Balearic Is., Cape Freu, northeastern Mallorca, 39°45′N 03°26′E, La Salada cave, 1984, approximately 10 m, 1 female (cl 40 mm, cw 55 mm) (CBR).

Etymology. — The new species is named after our colleague and friend Danièle Guinot (Muséum national d'Histoire naturelle, Paris), whose carcinological oeuvre changed our understanding of brachyuran systematics worldwide.

Description of holotype female. — Carapace subpentagonal, slightly wider than long, cl/cw ratio 1.29 (figs. 1C, 3A). Front cut into 4 lamellar teeth, 2 mesial, 2 lateral, inner orbital teeth excluded. Median notch between mesial rostral teeth distinctly U-shaped; distance between mesial rostral teeth markedly greater than distance between mesial rostral tooth and lateral rostral

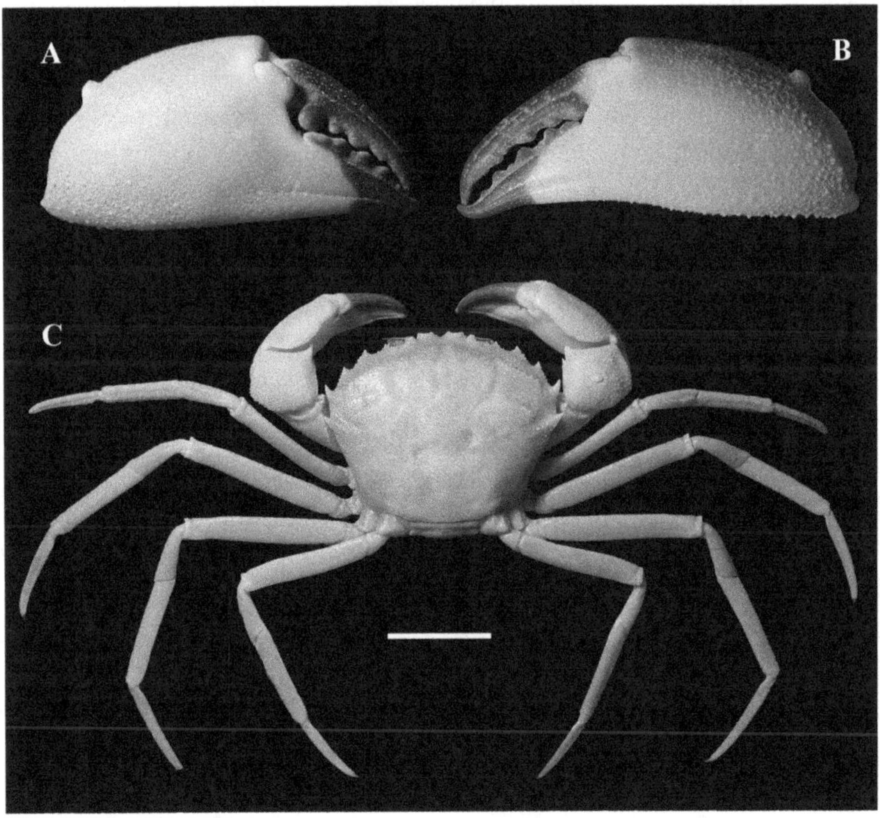

Fig. 1. *Paragalene danieleae* n. sp. female holotype cl 37 mm, cw 48 mm (MZUSP 19938). A, B, right and left cheliped propodus and dactylus, respectively; C, habitus, dorsal view. Scale bar: 20 mm.

tooth. Orbit circular, deep. Supraorbital border entire, narrow, semicircular, with large granules (figs. 1C, 3C). Inner supraorbital tooth triangular, stronger than outer tooth; inner suborbital tooth acute, much stronger than outer suborbital tooth. Infraorbital border incomplete. First antennal article short, mobile; article 2 + 3 closing infraorbital border; fourth longer than 2 + 3, reaching supraorbital border; fifth slender, oblique. Ocular peduncle short, cornea well formed, dark brown pigmented. Carapace anterolateral borders curved, each armed with 4 anterolateral teeth, outer supraorbital excluded. First, second teeth sharp, well defined; third anterolateral tooth strongest, acute, curved forward; fourth straight, outwardly directed, smooth acute apex. Posterolateral borders straight. Carapace dorsal surface strongly convex anteriorly, granules much larger, dense anteriorly. Granular ridge across hepatic region almost straight, well marked; granular ridge across mesobranchial region shorter. Protogastric region with 2 weak, granular elevations, shallowly grooved medially. Urogastric region sided by distinct depression. Cardiac region slightly elevated, finely granular. Anterior border of buccal cavern smooth, prominent, clearly demarcating epi- and endostome; endostomial ridge strong, smooth. Third maxillipeds completely closing buccal frame; palp long, overreaching ischio-meral suture; anterolateral angle of merus expanded, wing-like; ischium much longer than merus.

Chelipeds (P1) moderately heterochelic, heterodont. Right cheliped strongest, massive. Dactylus slender, less than half propodus maximal length, dark brown coloured; cutting edge with strong, rounded proximal tooth, remaining teeth low (fig. 1A-C). Dorsal, lateral, ventral faces of propodus finely granular, mesial face smooth; carpus granular, with 2 acute teeth on inner margin, superior tooth much stronger than inferior. Left P1 distinctly less massive; dactylus, pollex slender, proximal tooth on cutting edge of dactylus lacking, left P1 otherwise similar to right cheliped. P2-P5 markedly long, slender, similar in shape (fig. 1C). Dactyli similar in shape, slightly curved, acute; upper, ventral surfaces densely setose; mesial, lateral surfaces, and tip devoid of setae. Dactyli of P2, P5 similar in size, distinctly shorter than those of P3, P4. Meri, carpi, propodi long, slender, similar in shape. Meri of P2, P5 similar in size, distinctly shorter than those of P3, P4; distal end of merus of P5 extending far beyond tip of the last anterolateral carapace tooth.

Female thoracic sternum wide, sterno-abdominal cavity shallow, formed by sternites 8-4. Sternal suture 2/3 distinct, uninterrupted; 3/4 faint, recognizable laterally only; 4/5, 5/6 deep, medially interrupted; 6/7, 7/8 deep, complete.

Female abdomen with 6 freely-movable somites plus telson. Somites 1, 2 completely covering thoracic sternum between P5 coxae, thoracic sternite 8 not visible; somite 3 very close to margin of sternite 7, small portion of thoracic sternite 7 visible only laterally. Abdominal somite 6 longest. Telson large, little wider than long, extending from thoracic sternite 5 to sternal sutures 2/3. Vulvae small, completely covered by abdominal somite 6, opening near thoracic sternal suture 5/6.

Male unknown.

Remarks. — *Paragalene* was until now known from one species, *P. longicrura* from the eastern Atlantic (Madeira and Canary Islands: Türkay, 1976a, b; González Pérez, 1995) and the Mediterranean Sea (Bouvier, 1940; Dieuzeide & Goëau-Brissonnière, 1955; Guinot, 1969; Gili & Macpherson, 1987; d'Udekem d'Acoz, 1999; Pallaoro, 2005; Castro & Ng, 2008; Ng et al., 2008). *Paragalene longicrura* occurs in dark caves and on hard bottoms with algal growth and shells, at depths between 30 and 160 m, and is considered to be rare (see d'Udekem d'Acoz, 1999: 241). The discovery of a congener, *P. danieleae* n. sp., after thirteen decades is quite a surprise. Although *Paragalene danieleae* n. sp. is represented by only one adult female, it can be easily distinguished from *P. longicrura* (figs. 2, 3B, D) by: (1) distal end of P5 merus extending far beyond the tip of the last (fourth) anterolateral carapace tooth (almost reaching to the tip of the tooth in *P. longicrura*); (2) distal end of merus of P4 extending far beyond the imaginary line connecting the tips of the third anterolateral carapace teeth (slightly overreaching that line in *P. longicrura*); (3) median notch between mesial rostral teeth clearly U-shaped (distinctly V-shaped in *P. longicrura*); (4) distance between the two mesial rostral teeth markedly greater than the distance between one mesial rostral tooth and its adjacent lateral rostral tooth (distance between mesial and adjacent lateral rostral teeth about the same in *P. longicrura*); (5) second anterolateral tooth sharp, well defined (broadly triangular in *P. longicrura*); (6) granular ridge across the hepatic region almost straight (distinctly convex in *P. longicrura*); (7) convexity of gastric regions of carapace more pronounced (gastric regions only slightly convex in *P. longicrura*); and (8) carapace subpentagonal, cl/cw 1.29 (carapace hexagonal, proportionally wider in relation to its length in *P. longicrura*, cl/cw 1.37). The above differences refer to the females only. These characters, however, are not known to vary between sexes (see also Castro & Ng, 2008) so the differences clearly suggest the West Atlantic species is new. The discovery of the male of

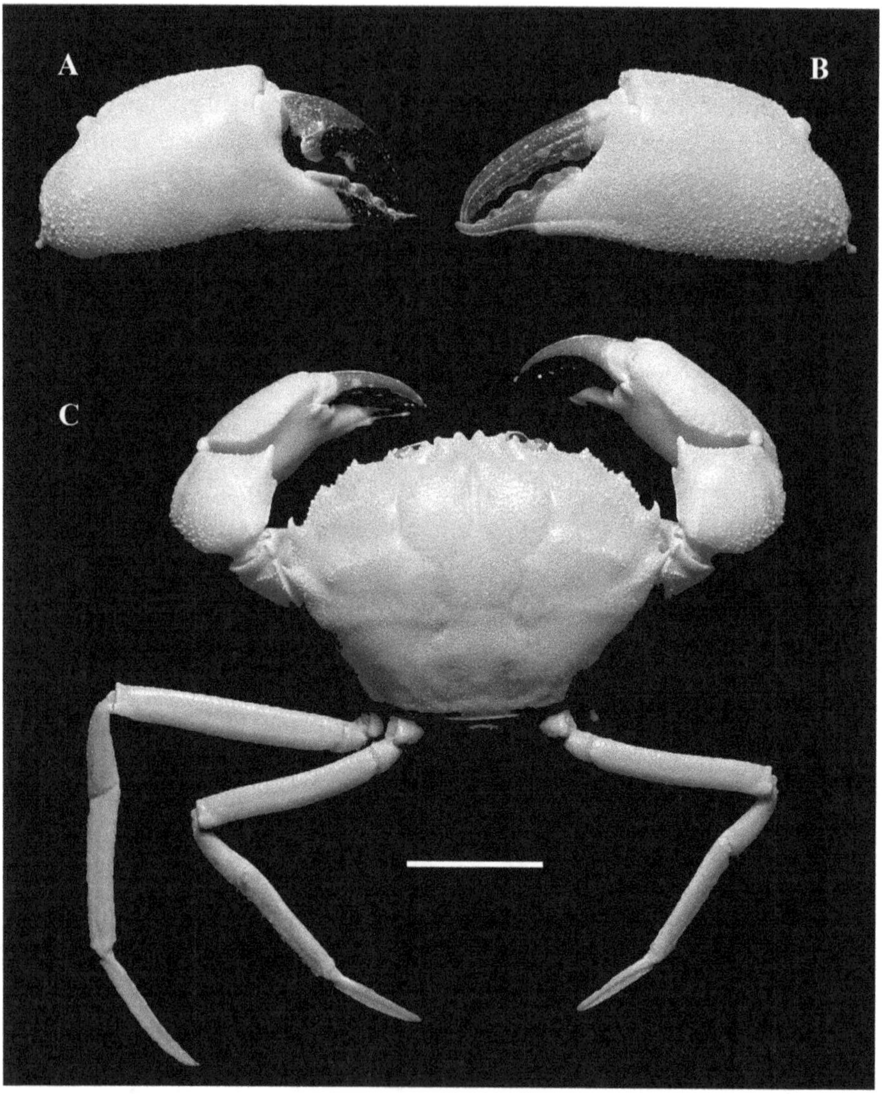

Fig. 2. *Paragalene longicrura* (Nardo, 1868) female cl 40 mm, cw 55 mm (CBR unnumbered). A, B, right and left cheliped propodus and dactylus, respectively; C, habitus, dorsal view. Scale bar: 20 mm.

P. danieleae n. sp. will permit additional taxonomic details to be added to the original description and probably show additional differences between the two species.

Distribution. — Known so far from the type locality, Guadeloupe in the Caribbean Sea.

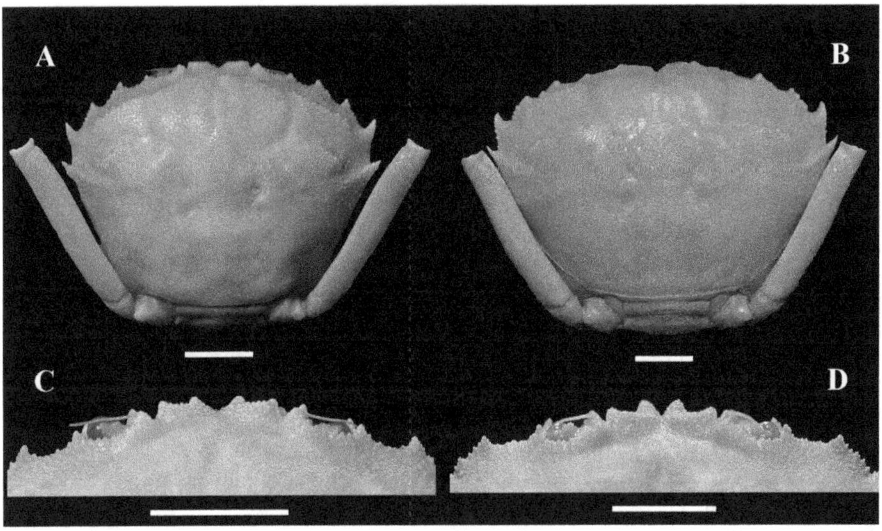

Fig. 3. A, C, *Paragalene danieleae* n. sp. female holotype cl 37 mm, cw 48 mm (MZUSP 19938). B, D, *Paragalene longicrura* (Nardo, 1868) female cl 40 mm, cw 55 mm (CBR unnumbered). A, B, dorsal view of carapace and merus of pereiopod 5; part of pereiopod 5 and remaining appendages removed electronically from figure; C, D, anterior border of carapace. Scale bars: 10 mm.

ACKNOWLEDGEMENTS

We thank the editors for the opportunity to contribute to this special volume. This contribution would not have been possible without the help of our colleagues and friends Enrique Macpherson (Consejo Superior de Investigaciones Cientificas and Centre d'Estudis Avancats de Blanes, Girona, Spain), who arranged the loan of *Paragalene longicrura*, and Joseph Poupin (Institut de Recherche de l'École Navale, BCRM-Brest, France), who made the material of *P. danieleae* n. sp. available for study. Thanks also to Allyson Pinheiro and William Santana (both from MZUSP) for preparing the photographs and plates, respectively. We also thank the CNPq (Conselho Nacional de Desenvolvimento Científico e Tecnológico, Brasília) for supporting studies on the systematics of decapod crustaceans in the form of ongoing grants 302065/2007-5 and 303224/87-8, respectively. MT thanks PETROBRAS for financial support (process number 4600224970).

REFERENCES

BOUVIER, E.-L., 1940. Décapodes marcheurs. Faune de France, **37**: 1-404. (Paul Le Chevalier, Paris).

CASTRO, P. & P. K. L. NG, 2008. *Rhadinoplax*, a new genus of Progeryonidae Števčić, 2005 for *Carcinoplax microphthalmus* Guinot & Richer de Forges, 1981, and a redescription of *Paragalene longicrura* (Nardo, 1868) (Crustacea: Decapoda: Brachyura: Goneplacoidea). Zootaxa, **1777**: 53-68.

DIEUZÉIDE, R. & W. GOËAU-BRISSONNIÈRE, 1955. *Paragalene longicrura* Nardo (Décapode Brachyoure) aux environs d'Alger. Bulletin des Travaux de la Station d'Aquiculture et Pêche de Castiglione, (n. ser.) **7**: 275-279.

GILI, J. M. & E. MACPHERSON, 1987. Crustáceos decápodos capturados em cuevas submarinas del litoral Balear. Investigacion Pesquera, Bercelona, **51** (Supplement 1): 285-291.

GONZÁLEZ PÉREZ, J. A., 1995. Catálogo de los crustáceos decápodos de las Islas Canarias. Gambas, langostas, cangrejos: 1-281. (Publicaciones Turquesa, Santa Cruz de Tenerife).

GUINOT, D., 1969. Les Goneplacidae (suite et fin). Recherches préliminaires sur les groupements naturels chez les Crustacés Décapodes Brachyoures. VII. Bulletin du Muséum national d'Histoire naturelle, Paris, (2) **41** (3): 688-724.

— —, 1971. Synthèse et bibliographie. Recherches préliminaires sur les groupements naturels chez les Crustacés Décapodes Brachyoures. VIII. Bulletin du Muséum national d'Histoire naturelle, Paris, (2) **42** (5): 1063-1090.

KARASAWA, H. & M. SCHWEITZER, 2006. A new classification of the Xanthoidea sensu lato (Crustacea: Decapoda: Brachyura) based on phylogenetic analysis and traditional systematics and evaluation of all fossil Xanthoidea sensu lato. Contributions to Zoology, **75** (1/2): 23-73.

KOSSMANN, R., 1878. Kurze Notizen über einige neue Crustaceen sowie über neue Fundorte einiger bereits beschriebenen. Archiv für Naturgeschichte, **44** (1): 251-258.

MANNING, R. B. & L. B. HOLTHUIS, 1989. Two new genera and nine new species of geryonid crabs (Crustacea: Decapoda: Geryonidae). Proceedings of the Biological Society of Washington, **102** (1): 50-77.

NARDO, G. D., 1868. Annotazioni illustranti cinquantaquattro specie di Crostacei podottalmi, endottalmi e succhiatori del mare Adriatico, alcune delle quali nuove o male conosciute, accompagnate da trentare figure litografate, e precedute dalla storia della carcinologia Adriatica antica e rencente. Memorie del Reale Istituto Veneto di Scienze Lettere e Arti, Venezia, **14** (2): 217-343.

NG, P. K. L., D. GUINOT & P. J. F. DAVIE, 2008. Systema Brachyurorum: Part 1. An annotated checklist of extant brachyuran crabs of the world. Raffles Bulletin of Zoology, (Supplement) **17**: 1-286.

NG, P. K. L. & M. R. MANUEL-SANTOS, 2007. Establishment of the Vultocinidae, a new family for an unusual new genus and new species of Indo-West Pacific crab (Crustacea: Decapoda: Brachyura: Goneplacoidea), with comments on the taxonomy of the Goneplacidae. Zootaxa, **1558**: 39-68.

PALLAORO, A., 2005. The discovery of the crab, *Paragalene longicrura* (Nardo, 1868) (Decapoda, Brachyura, Xanthidae) in the Adriatic Sea. Crustaceana, **78** (6): 749-753.

POUPIN, J., 1994. Faune marine profonde des Antilles françaises. Récoltes du navire Polka faites en 1993: 1-79. (ORSTOM Éditions, Collection Études et Thèses, Paris).

ŠTEVČIĆ, Z., 2005. The reclassification of brachyuran crabs (Crustacea: Decapoda: Brachyura). Fauna Croatica, **14** (1): 1-159.

TAVARES, M., 1994. Description de *Cymonomus leblondi* sp. nov. de la Guadeloupe, avec une clef des *Cymonomus* américains (Decapoda, Brachyura, Cymonomidae). Bulletin du Muséum national d'Histoire naturelle, Paris, (4) **16** (1): 203-208.

TAVARES, M. & GUINOT, D., 1996. Description de *Neopilumnoplax gervaini* sp. nov. des eaux profondes des Caraïbes (Decapoda, Brachyura, Goneplacidae). Bulletin du Muséum national d'Histoire naturelle, Paris, (4) **18** (1-2): 225-232.

TÜRKAY, M., 1976a. Die Madeirensischen Brachyuren des Museu Municipal do Funchal und des Forschungs-Instituts Senckenberg. I. Familien: Dromiidae, Homolidae, Calappidae, Leucosiidae, Cancridae, Portunidae, Xanthidae, Geryonidae, Goneplacidae und Palicidae (Crustacea: Decapoda). Boletin du Museu Municipal do Funchal, **30** (133): 57-74.

— —, 1976b. Ein Fund von *Paragalene longicrura* (Nardo, 1868) in der Ägäis (Decapoda: Brachyura). Crustaceana, **30** (1): 108.

UDEKEM D'ACOZ, C. D', 1999. Inventaire des Crustacés Décapodes de l'Atlantique nord-oriental, de la Méditerranée et des eaux continentales adjacentes au nord de 25°N. Muséum national d'Histoire naturelle/Service du Patrimoine Naturel, Collection Patrimoines Naturels, Paris, **40**: 1-383.

First received 26 June 2009.
Final version accepted 16 July 2009.

A NEW SPECIES OF *PILOSAMON* NG, 1996 (BRACHYURA, POTAMIDAE) FROM THAILAND

BY

DARREN C. J. YEO[1,3]) and PHAIBUL NAIYANETR[2,4])

[1]) Department of Biological Sciences, National University of Singapore, 14 Science Drive 4, Singapore 117543, Republic of Singapore

[2]) Department of Biology, Faculty of Science, Chulalongkorn University, Bangkok 10330, Thailand

ABSTRACT

A new species of potamid freshwater crab, *Pilosamon guinotae* sp. nov., is described from northeastern Thailand. It is morphologically closest to *Pilosamon palustre* (Rathbun, 1904), but distinguished by differences in the carapace, male abdomen and male first pleopod. Taxonomic notes on the poorly known *P. palustre* are also provided.

RÉSUMÉ

Une nouvelle espèce de crabe d'eau douce potamidé rattaché au genre *Pilosamon* Ng, 1996 est décrite du nord est de la Thailande. Elle est morphologiquement proche de *Pilosamon palustre* (Rathbun, 1904) mais s'en distingue par des differencesdifférences sur la carapace, l'abdomen male et le premier pleopod male. Des informations sur l'espèce mal connue *P. palustre* sont également fournies.

INTRODUCTION

Potamid freshwater crabs from the northeastern region of Thailand, examined as part of ongoing taxonomic studies of the fauna of that region, were found to include an undescribed species from the genus *Pilosamon* Ng, 1996. Ng (1996) originally established *Pilosamon* for one species, *Potamon* (*Potamon*) *laosense* Rathbun, 1904, but in a recent re-appraisal, Yeo and Ng (2007:

[3]) e-mail: darrenyeo@nus.edu.sg

[4]) e-mail: sirati2001@yahoo.com

© Koninklijke Brill NV, Leiden, 2010

289) also assigned the poorly known species, *Potamon* (*Potamon*) *palustre* Rathbun, 1904, to the genus. They also mentioned that they had on hand specimens of a new species from this genus. The new species possesses the characteristic coarse setae on the margins of the ambulatory legs and a G1 form that are diagnostic of *Pilosamon*.

The present paper serves to describe the new species, herein named *Pilosamon guinotae* sp. nov., which is distinguished by external and male first pleopod characters from *Pilosamon palustre*, its morphologically closest congener. The opportunity is taken here to also present a diagnosis and additional notes on the adult morphology of *P. palustre*, which was hitherto only known from a single juvenile specimen (Ng & Naiyanetr, 1993).

The following abbreviations are used: G1 and G2 for the male first and second pleopods, respectively. Measurements, in millimetres, are of carapace width and length, respectively. Terminology used in the descriptive accounts essentially follows Ng (1988). Material examined is deposited in the Chulalongkorn University Natural History Museum (CUMZ), Bangkok, Thailand; Swedish Museum of Natural History, Stockholm, Sweden (NRM); Nationaal Natuurhistorisch Museum [formerly Rijksmuseum van Natuurlijke Historie (RMNH)], Leiden, The Netherlands; Senckenbergischen Naturforschenden Gesellschaft (SMF), Frankfurt, Germany; and the Zoological Reference Collection (ZRC) of the Raffles Museum of Biodiversity Research, National University of Singapore.

TAXONOMY

Family POTAMIDAE Ortmann, 1896

Subfamily POTAMISCINAE Bott, 1970 (sensu Yeo & Ng, 2003)

Pilosamon guinotae sp. nov. (figs. 1A, C, E, 2A, B, E-H)

Material examined. — Holotype: male, 73.8 by 58.1 mm, ZRC 2002.0111, Yotdom sanctuary, Nam Yun District, Ubon Ratchathani Province, Thailand, coll. Komon Boonchai, 16 November 1983. Paratypes: 1 male, 75.4 by 59.8 mm, CUMZ, same data as holotype; 1 male, 1 juvenile female, male 52.5 by 42.1 mm, ZRC 2002.0112, 1 male, 1 female, 11 juvenile males, largest male 47.5 by 38.8 mm, CUMZ, 1 male, 1 juvenile male, RMNH, Huai Luang, Phibun Mangsahan District, Ubon Ratchathani Province, Thailand, coll. Prachuap & P. Naiyanetr, 5 November 1974, 24 October 1982; 1 male, 1 female, male 54.0 by 44.1 mm, SMF, 1 female, 59.8 by 46.6 mm, CUMZ, mountains of Buntharik District, Ubon Ratchathani Province, Thailand, coll. Somlak Kanthapuk, 1 June 1991; 3 males, 2 females, 1 juvenile male, largest male 54.1 by 43.2 mm, ZRC 2002.0113, 9 males, 15 females, largest male 55.1 by 44.6 mm,

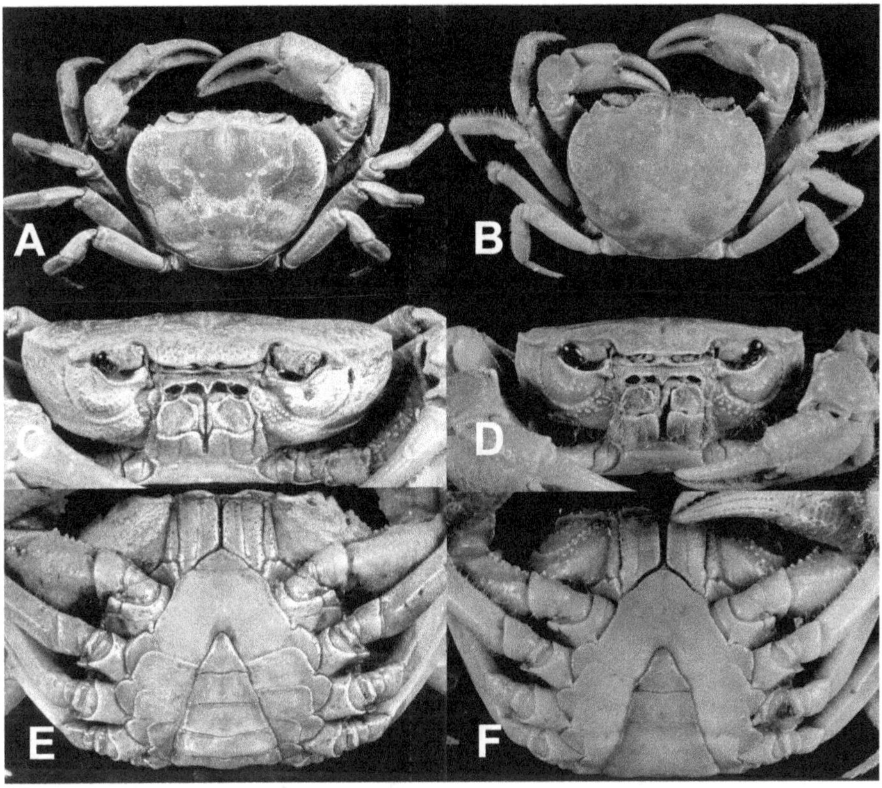

Fig. 1. A, C, E, *Pilosamon guinotae* sp. nov., holotype male, 73.8 by 58.1 mm, ZRC 2002.0111; B, D, F, *Pilosamon palustre* (Rathbun, 1904), male, 45.6 by 36.9 mm, NRM. A, B, dorsal view; C, D, frontal view; E, F, ventral view.

CUMZ, Huai Wer, Phibun Mangsahan District, Ubon Ratchathani Province, Thailand, coll. P. Naiyanetr & Siri, 24 October 1982, 18 April 1984.

Diagnosis. — External orbital angle acutely triangular; anterolateral margin cristate, distinctly serrated; branchial region not inflated. Third maxilliped highly setose; exopod long, with flagellum longer than width of merus. Ambulatory legs lined with coarse setae, decreasing in length, density from first pair to fourth pair. Suture between anterior thoracic somites 2, 3 convex. Male abdomen lateral margins straight. G1 of adults strongly bent outwards; terminal segment stout, curved, distal part straight, appearing pointed; tip acuminate, without distinct dorsal, subdistal opening, dorsal flap broad, extending along almost entire length of terminal segment, high, semicircular, with rounded apex, which is skewed towards distal portion, proximal part gradually tapering, distal part tapering abruptly, with shallow subdistal notch.

Etymology. — This species is named after Danièle Guinot, for her kind help over the years, and in tribute to her substantial contributions to brachyuran systematics.

Remarks. — The structure of the G1 of *Pilosamon guinotae* is most similar to that of *P. palustre* (Rathbun, 1904), but it can be distinguished from the latter by the terminal segment of the G1 possessing a relatively narrower and higher dorsal flap, with the rounded apex skewed slightly towards the distal portion (relatively broader and lower dorsal flap of the terminal segment of the G1, with rounded apex skewed slightly towards the proximal portion in *P. palustre*) (fig. 2A-F). In addition, *P. guinotae* differs externally in having a more transverse carapace (more quadrate carapace in *P. palustre*); less distinctly cristate and relatively more distinctly serrated anterolateral margins (more distinctly cristate and more weakly serrated in *P. palustre*); acutely triangular external orbital angle (relatively more broadly triangular in *P. palustre*), and straight lateral margins of the male abdomen (gently but distinctly concave in *P. palustre*) (fig. 1A, B).

The material examined for this species included juvenile, subadult and adult males of various sizes. Variation in the G1 structures of this series of specimens can easily be attributed to allometric growth. The G1 changes from generally being more slender and sinuous in smaller (younger) males to broader and less sinuous in larger (older) males. The G1 in smaller males is less strongly bent outwards, and the terminal segments are less curved and appear to have relatively broader and slightly lower dorsal flaps. The G1 tends to become distinctly more bent outwards and terminal segments more curved in larger specimens, with the dorsal flaps also narrowing slightly and increasing in height (see fig. 2A, B, E-H). Similar allometric growth of the G1 has been noted for other potamid species (Türkay & Naiyanetr, 1987: 391; Yeo & Ng, 1998: 640; unpublished data). Despite this variation, the differences mentioned above to separate *P. guinotae* from *P. palustre* are still valid.

Pilosamon guinotae is easily distinguished from the type species, *P. laosense*, by its relatively stouter G1 terminal segment with a semicircular to convex dorsal flap (G1 terminal segment slender with bluntly triangular dorsal flap in *P. laosense*) (fig. 2A, B, E-H; cf. Ng, 1996, fig. 2B-E). Externally, the coarse setae lining the ambulatory legs of *P. guinotae* decrease in density and length from the first pair to the fourth pair, while in *P. laosense*, the setae is equally dense and long on all the ambulatory legs (fig. 1A; cf. Rathbun, 1904, pl. 11 fig. 3; Bott, 1970, pl. 49 fig. 39).

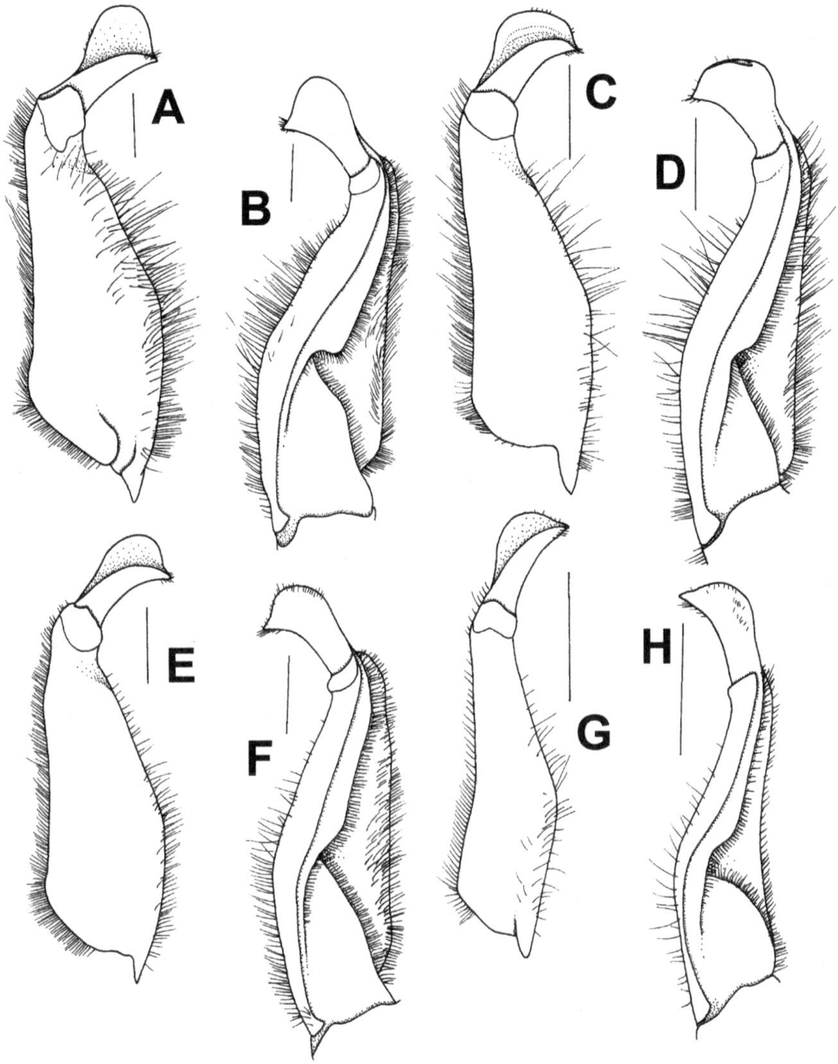

Fig. 2. A, B, E-H, *Pilosamon guinotae* sp. nov; A, B, holotype male, 73.8 by 58.1 mm, ZRC 2002.0111; E, F, male, 54.1 by 43.2 mm, ZRC 2002.0113; G, H, male, 34.8 by 28.8 mm, ZRC 2002.0113. C, D, *Pilosamon palustre* (Rathbun, 1904), male, 45.6 by 36.9 mm, NRM. A, C, E, G, dorsal view of right G1; B, D, F, H, ventral view of right G1. Scale bars: 2.0 mm.

Pilosamon palustre (Rathbun, 1904) (figs. 1B, D, F, 2C, D)

Potamon (*Potamon*) *palustris* Rathbun, 1904: 273, fig. 13, pl. 10 fig. 3.
Ranguna (*Ranguna*) *luangprabangensis luangprabangensis* — ? Bott, 1970: 169 (part) (not *Potamon* (*Potamon*) *luangprabangensis* Rathbun, 1904).
Potamon palustre — Ng & Naiyanetr, 1993: 14, fig. 42; Yeo & Ng, 1999: 639.

Pilosamon palustre — Yeo & Ng, 2007: 289; Ng et al., 2008: 165.

Material examined. — Holotype: male, 28.0 by 22.5 mm, MNHN-B 5265, Montagnes de Laos, Siam, coll. Harmand, 1877. Male, 45.6 by 36.9 mm, NRM, Xe Set River, Saravan Province, southern Laos, coll. unknown, September 1988.

Diagnosis. — External orbital angle triangular; anterolateral margin distinctly cristate, weakly serrated; branchial region not inflated. Third maxilliped highly setose; exopod long, with flagellum longer than width of merus. Ambulatory legs lined with coarse setae, decreasing in length and density from first pair to fourth pair. Suture between anterior thoracic somites 2 and 3 almost straight. Male abdomen lateral margins very gently but distinctly concave. G1 strongly bent outwards; terminal segment stout, curved, distal part straight, appearing pointed, tip acuminate, without distinct dorsal subdistal opening, dorsal flap broad, extending along almost entire length of terminal segment, high, semicircular, with rounded apex slightly skewed towards proximal portion, gradually tapering proximally, distal part tapering abruptly, with shallow subdistal notch.

Remarks. — Bott (1970) had previously synonymised this species under *Ranguna* (*Ranguna*) *luangprabangensis luangprabangensis* (Rathbun, 1904) [now in the genus *Iomon* Yeo & Ng, 2007] with a question mark, but without explanation. Ng & Naiyanetr (1993), however, re-examined the types of both species and found their external carapace and G1 structures to be quite different. *Potamon palustre* has since been recognised as a valid species and reassigned to *Pilosamon* (Yeo & Ng, 1999, 2007; Yeo et al., 2008).

Rathbun's (1904) description of *Potamon* (*Potamon*) *palustre* was based on a single juvenile male specimen (28.0 by 22.5 mm, MNHN-B 5265) from "Montagnes de Laos, Siam", which Ng & Naiyanetr (1993) referred to present-day northeastern Thailand. The material examined herein includes a second, considerably larger, adult male (45.6 by 36.9 mm, NRM), from Xe Set River in southern Laos. The holotype is clearly immature, as its G1 structure is not as well developed, with the G1 differences between the two specimens being most probably size-related. The G1 of the holotype differs from that of second specimen by the following characters: (i) more slender, sinuous, gently bent outwards (broader, less sinuous, strongly bent outwards in the second specimen); (ii) terminal segment slightly curved, with tip bluntly acute (terminal segment more distinctly curved, with tip acuminate in the second specimen); (iii) dorsal flap of terminal segment appears relatively lower, with broadly convex apex in median portion, slight trace of subdistal notch (dorsal flap broad, high, with rounded apex slightly skewed towards proximal portion and shallow subdistal notch in the second specimen) (fig. 2C, D; Ng & Naiyanetr, 1993: 14, fig. 42). Similar variations were observed in the large series of specimens of *P. guinotae* sp. nov. (see Remarks for the species). There are also some slight external differences between the two specimens, the holotype appearing to have a carapace that is slightly less quadrate in shape; the third maxillipeds are not setose; and only the first pair of ambulatory legs are lined with coarse setae (see Rathbun, 1904, pl. 10 fig. 3; Ng & Naiyanetr, 1993: 14). These differences, however, are not specifically significant as they are also most probably size-related, reflecting the considerably smaller size and maturity level of the holotype compared to the second specimen. It is also not uncommon for old specimens like the holotype to lose some setae through handling over time. The diagnosis given above therefore is based on the second, non-type adult specimen.

Distribution. — "Montagnes de Laos, Siam" (present-day northeastern Thailand or Laos).

ACKNOWLEDGEMENTS

We thank Peter Ng (National University of Singapore (NUS)) and Peter Castro (California State Polytechnic University) for their helpful suggestions and comments on the manuscript. We are also grateful to the following for their generous help in facilitating access to material from their respective museums: Danièle Guinot (MNHN); Charles Fransen (RMNH); Michael Türkay (SMF); Karin Sindemark (NRM); and Kelvin Lim (ZRC). The black and white photographs were developed by Yip Hoi Kee (NUS). The study was partially supported by research grant R-154-000-222-112 from NUS.

LITERATURE CITED

BOTT, R., 1970. Die Süsswasserkrabben von Europa, Asien, Australien und ihre Stammesgeschichte. Eine Revision der Potamoidea und Parathelphusoidea (Crustacea, Decapoda). Abhandlungen der Senckenbergischen Naturforschenden Gesellschaft, Frankfurt, **526**: 1-338, pls. 1-58.

NG, P. K. L., 1988. The freshwater crabs of Peninsular Malaysia and Singapore: i-viii, 1-156. (Department of Zoology, National University of Singapore, Shinglee Press, Singapore).

——, 1996. Establishment of two new genera for *Potamon lacunifer* Rathbun, 1904, and *Potamon laosensis* Rathbun, 1904 (Decapoda, Brachyura, Potamidae) from Laos. Crustaceana, **69** (7): 898-906.

NG, P. K. L., D. GUINOT & P. J. F. DAVIE, 2008. Systema Brachyurorum: Part I. An annotated checklist of extant brachyuran crabs of the world. Raffles Bulletin of Zoology, (Supplement) **17**: 1-286.

NG, P. K. L. & P. NAIYANETR, 1993. New and recently described freshwater crabs (Crustacea: Decapoda: Brachyura: Potamidae, Gecarcinucidae and Parathelphusidae) from Thailand. Zoologische Verhandelingen, Leiden, **284**: 1-117.

ORTMANN, A., 1896. Das system der Decapoden-Krebse. Zoologische Jahrbücher, (Systematik) **9**: 409-453.

RATHBUN, M. J., 1904. Les crabes d'eau douce. Nouvelles Archives du Muséum d'Histoire Naturelle, Paris, (4) **6**: 225-312, pls. 9-18.

TÜRKAY, M. & P. NAIYANETR, 1987. The identity of *Potamon rangoonense* Rathbun, 1904 and *Thelphusa larnaudii* A. Milne-Edwards, 1869, with introduction of *Neolarnaudia botti* n. g. n. sp. (Crustacea: Decapoda: Potamidae). Senckenbergiana Biologica, Frankfurt, **67** (4/6): 389-396.

YEO, D. C. J. & P. K. L. NG, 1998. Freshwater crabs of the *Potamon tannanti* species group (Crustacea, Decapoda, Brachyura, Potamidae) of northern Indochina. Raffles Bulletin of Zoology, **46** (2): 627-650.

— — & — —, 1999. The state of freshwater crab taxonomy in Indochina (Decapoda, Brachyura). In: F. R. SCHRAM & J. C. VON VAUPEL KLEIN (eds.), Crustaceans and the Biodiversity Crisis, Proceedings of the Fourth International Crustacean Congress, Amsterdam, 1998, **1**: 637-646.

— — & — —, 2003. Recognition of two subfamilies in the Potamidae Ortmann, 1896 (Brachyura, Potamidae) with a note on the genus *Potamon* Savigny, 1816. Crustaceana, **76** (10): 1219-1235.

— — & — —, 2007. On the genus "*Potamon*" and allies in Indochina (Crustacea: Decapoda: Brachyura: Potamidae). In: S. H. TAN & P. K. L. NG (eds.), Crustacean Supplement I. Raffles Bulletin of Zoology, (Supplement) **16**: 273-308.

First received 12 March 2009.
Final version accepted 27 May 2009.

GUINOTHUSA, A NEW GENUS OF INDOCHINESE FRESHWATER CRAB (DECAPODA, GECARCINUCIDAE)

BY

DARREN C. J. YEO[1,2]) and PETER K. L. NG[1,3])

[1]) Department of Biological Sciences, National University of Singapore, 14 Science Drive 4, Singapore 117543, Republic of Singapore

ABSTRACT

A new genus, *Guinothusa*, is established for three gecarcinucid freshwater crab species previously placed in *Heterothelphusa* Ng & Lim, 1986: *Guinothusa beauvoisi* (Rathbun, 1902), *G. harmandi* (Rathbun, 1902), and *G. phimaiensis* sp. nov. These species, from eastern Thailand and southern Vietnam, share a distinctive suite of diagnostic carapace, male abdomen, and G1 characters that distinguish them from the remaining species of *Heterothelphusa*, which are known from the area along the Thai-Malaysian border. *Guinothusa phimaiensis* was previously a nomen nudum and a junior synonym of *G. beauvoisi*, and the name was never validated. A key to the species of *Guinothusa* is included.

RÉSUMÉ

Le nouveau genre *Guinothusa* est établi pour trois espèces de crabes d'eau douce gecarcinucidés classés auparavent dans *Heterothelphusa* Ng & Lim, 1986 : *Guinothusa beauvoisi* (Rathbun, 1902), *G. harmandi* (Rathbun, 1902), and *G. phimaiensis* sp. nov. Ces espèces, de l'est Thailand et du sud Vietnam, partagent un ensemble de caractères sur la carapace, l'abdomen male et le G1, qui les séparent des espèces restantes d'Heterothelphusa connues des abords de la frontière Thailande-Malaisie. *Guinothusa phimaiensis* était un nomen nudum mais, comme il était considéré comme un junior synonyme de *G. beauvoisi*, ce nom ne fut jamais validé. Une clef des espèces de *Guinothusa* est présentée.

INTRODUCTION

The gecarcinucid freshwater crab genus *Heterothelphusa* Ng & Lim, 1986, was originally established for one species, *H. insolita* Ng & Lim, 1986, from northern Peninsular Malaysia. Ng & Naiyanetr (1993) subsequently trans-

[2]) Corresponding author; e-mail: darrenyeo@nus.edu.sg

[3]) e-mail: peterng@nus.edu.sg

© Koninklijke Brill NV, Leiden, 2010

ferred *Potamon* (*Parathelphusa*) *beauvoisi* Rathbun, 1902 ["Cochinchine" (= southern Vietnam)], to the genus, with Ng (1997) describing a third species, *H. fatum* Ng, 1997, from southern Thailand. Two other taxa, *Potamon* (*Parathelphusa*) *harmandi* Rathbun, 1902 ["Cochinchine" (= southern Vietnam)], and "*Siamthelphusa phimaiensis* Naiyanetr, 1978" nomen nudum [eastern Thailand], were regarded as junior subjective synonyms of *H. beauvoisi* (see Ng, 1988; Naiyanetr & Ng, 1990; Ng & Naiyanetr, 1993, 1997). Ng et al. (2008) recognized *H. harmandi* as a valid species of *Heterothelphusa* but without discussion.

In recent years, we have had the chance to re-examine and compare all the extant material of species that have been referred to *Heterothelphusa*. The present study confirms that *P.* (*P.*) *harmandi* and "*Siamthelphusa phimaiensis*" are in fact distinct species. These two taxa, together with *H. beauvoisi*, form a distinct morphological and geographical grouping, separate from the remaining *Heterothelphusa* species. We believe that these differences are of generic significance, warranting the establishment of a separate genus for these three species.

A new genus, *Guinothusa*, is consequently established herein for *H. beauvoisi*, *H. harmandi*, and the revalidated "*Siamthelphusa phimaiensis*", now formally named as *Guinothusa phimaiensis* sp. nov. *Heterothelphusa* is restricted to *H. fatum* and *H. insolita*.

The following abbreviations are used: G1 for male first pleopod, G2 for male second pleopod. Measurements are of carapace width and length, respectively. The terminology used essentially follows Ng (1988). All measurements are in millimetres. The classification follows that proposed in Ng et al. (2008), with the difference being that Gecarcinucidae Rathbun, 1904, is now regarded as a senior subjective synonym of Parathelphusidae Alcock, 1910 (cf. Klaus et al., 2009). Specimens are deposited in the Muséum national d'Histoire naturelle, Paris (MNHN); National Museum of Natural History, Smithsonian Institution, Washington, D.C. (USNM); and the Zoological Reference Collection (ZRC) of the Raffles Museum of Biodiversity Research, National University of Singapore.

TAXONOMY

GECARCINUCIDAE Rathbun, 1904

Guinothusa gen. nov.

Type species. — *Guinothusa phimaiensis* sp. nov., by present designation.

Diagnosis. — Carapace dorsal surfaces relatively flat, appearing sunken or gently concave, especially in branchial regions, smooth; epigastric, postorbital cristae sharp, epigastric cristae distinctly anterior to postorbital cristae, with outer edges greatly overlapping inner edges of postorbital cristae; postorbital cristae reaching to but not confluent with base of third epibranchial tooth; anterolateral margins with 3 epibranchial teeth (excluding external orbital angle), with first tooth much reduced or poorly developed compared to second and third teeth. Fingers of major chela relatively stout, shorter than palm, with small gap when closed. Merus of ambulatory legs with distinct dorsal subdistal spine. Suture between thoracic sternal somites 2, 3 distinct, complete. Male abdomen T-shaped, somite 6 without prominent constriction, lateral margins almost straight, parallel. G1 curved inwards; distal part not clearly demarcated from proximal part, weakly tapered towards tip, almost cylindrical in appearance, strongly curved inwards at tip; proximal part weakly dilated. G2 with elongate basal segment, distal segment less than half length of basal segment.

Etymology. — The genus is named after Danièle Guinot, in arbitrary combination with the genus name *Thelphusa*. This is in recognition not only of her many substantial contributions to brachyuran systematics but also for being a wonderful and most helpful colleague over the years. Gender feminine.

Remarks. — *Guinothusa* gen. nov. is most similar to *Heterothelphusa* in general carapace physiognomy and overall G1 shape (curved inwards distally), but can nevertheless be easily distinguished by the following combination of external carapace and G1 characters: sunken or gently concave branchial regions (figs. 1, 2A, B, 4A, B; flat to gently but distinctly convex branchial regions in *Heterothelphusa*; Ng & Lim, 1986, pl. 1 fig. 1A; Ng, 1997, fig. 2a, e); low, weakly developed first epibranchial tooth, much smaller than second and third teeth (figs. 1, 2A, B, 4A, B; relatively well-developed first epibranchial tooth, only slightly smaller than second and third teeth in *Heterothelphusa*; Ng & Lim, 1986, pl. 1 fig. 1A; Ng, 1997, fig. 2a, e); fingers of the adult male major chela shorter than the palm (figs. 2E, 4F; subequal to or longer than the palm in *Heterothelphusa*; Ng & Lim, 1986, fig. 1I; Ng, 1997, fig. 2g); lack of a prominent constriction in the sixth male abdominal somite due to straight and parallel lateral margins (figs. 2E, F, 4B; gentle but distinct constriction resulting from concave lateral margins in *Heterothelphusa*; Ng & Lim, 1986, fig. 1J; Ng, 1997, fig. 2d); G1 distal part more weakly tapered, appearing almost cylindrical for the most part, and more strongly curved at the tip (figs. 3, 5B, C; relatively more distinctly tapered, appearing more conical,

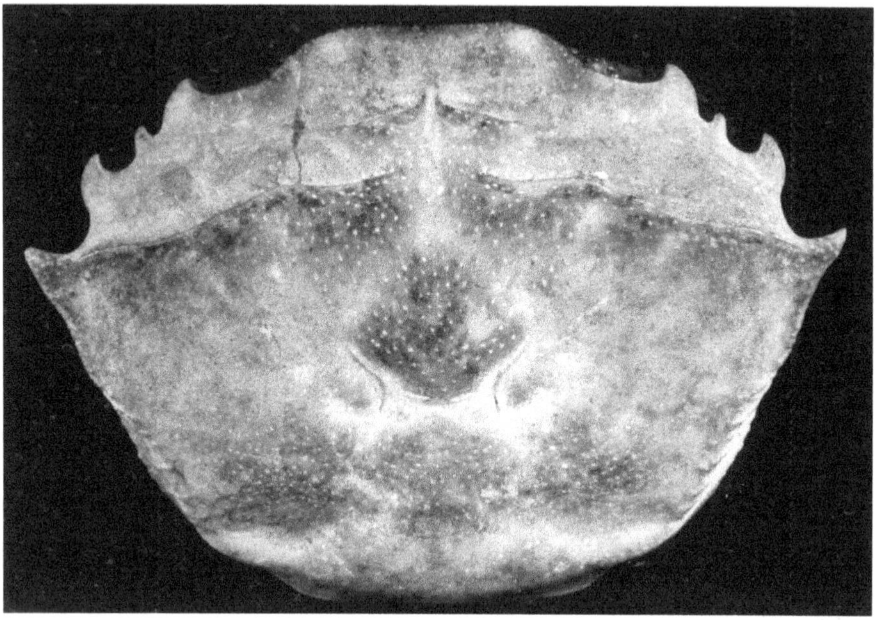

Fig. 1. *Guinothusa beauvoisi* (Rathbun, 1902), lectotype male, 27.1 by 20.0 mm, MNHN-B 5066. Dorsal view of carapace.

and only slightly curved at the tip in *Heterothelphusa*; Ng & Lim, 1986, fig. 2A-C; Ng, 1997, fig. 3a-d), and G1 proximal part weakly dilated (figs. 3, 5B, C; more dilated in appearance in *Heterothelphusa*; Ng & Lim, 1986, fig. 2A; Ng, 1997, fig. 3a, b). These morphological differences are discrete and there are no intermediate conditions. In addition, the geographical distributions of the species of *Guinothusa* and *Heterothelphusa* are considerably disjunct, with the latter occurring in northern Peninsular Malaysia to southernmost Thailand, compared to *Guinothusa*, which ranges from eastern Thailand to southern Vietnam. It is noteworthy that no known species of *Heterothelphusa* or *Guinothusa* are known in the areas in-between.

Guinothusa also superficially resembles *Siamthelphusa* Bott, 1968 (from Thailand and northern Peninsular Malaysia) in external morphology, but is immediately separated by the G1 being distinctly curved inwards (figs. 3, 5B, C; G1 straight or slightly bent outwards in *Siamthelphusa*; Ng & Naiyanetr, 1997, e.g., figs. 2A-D, 4F-I, 12A-D), with a relatively acuminate tip (figs. 3, 5B, C; tip of G1 rounded or truncate *Siamthelphusa*; Ng & Naiyanetr, 1997, e.g., figs. 2A-D, 4F-I).

Guinothusa species are similar to *Heterothelphusa* and *Siamthelphusa* in their habitat, being primarily aquatic and generally occurring in fast- to medium-flowing, lowland creeks and rivers (see also Ng & Lim, 1986; Ng, 1988, 1997; Ng & Naiyanetr, 1997). Many such waterways are connected to rice fields in this region, and consequently these crabs may occasionally be found in rice fields as well.

Geographical distribution. — Eastern Thailand and southern Vietnam.

Guinothusa beauvoisi (Rathbun, 1902) (figs. 1, 3A, D)

Potamon (*Parathelphusa*) *beauvoisi* Rathbun, 1902: 185.
Potamon (*Parathelphusa*) *beauvaisi* — Rathbun, 1905: 253, fig. 65, pl. 12 fig. 14.
Siamthelphusa paviei — Bott, 1968: 413, figs. 23-24, 36 (part), 1970: 116, pl. 22 figs. 60-62, pl. 30 fig. 87 (part); Chuensri, 1974: 17 (part) (not *Parathelphusa paviei* De Man, 1898).
Siamthelphusa beauvaisi — Naiyanetr, 1988: 6, pl. 3 fig. 6 (part).
Siamthelphusa beauvoisi — Ng, 1988: 102; Naiyanetr & Ng, 1990: 235 (part); Naiyanetr, 1992: 48 (part), 1998: 107 (part).
Heterothelphusa beauvoisi — Ng & Naiyanetr, 1993: 43, figs. 33, 68 (part); Ng, 1997: 718; Yeo & Ng, 1999: 642 (part); Ng et al., 2008: 70.

Material examined. — Lectotype (here designated): male, 27.1 by 20.0 mm, MNHN-B 5066, "Cochinchine" [= southern Vietnam], Saigon, coll. Harmand, no date. Paralectotypes: 1 male, 1 female, MNHN-B 5066, same data as lectotype.

Diagnosis. — Carapace distinctly transverse, about 1.36 times as broad as long; dorsal surface relatively flat, appearing gently sunken or concave, especially in branchial regions; epibranchial cristae distinctly anterior to and separated from postorbital cristae, with outer edge distinctly overlapping inner edge of postorbital cristae; postorbital cristae sinuous, gently sloping posterolaterally; external orbital angle with outer margin distinctly convex; anterolateral margin strongly converging anteriorly, with 3 epibranchial teeth, first tooth very small, low, second tooth relatively broad, with distinctly convex outer margin, third tooth directed obliquely outwards, relatively broad; posterolateral margins strongly converging posteriorly, with short but distinct oblique striae. Fingers of major chela shorter than palm, forming small gap when closed. Ambulatory legs long, slender; merus with distinct subdistal spine. Sixth male abdominal somite with lateral margins straight. G1 with distal part slightly shorter than proximal part, tip relatively blunt, strongly curved inwards; proximal part with straight inner margin, demarcated from distal part of G1 by distinct angle or shelf.

Remarks. — Bott (1970) synonymized *Potamon* (*Parathelphusa*) *beauvoisi* Rathbun, 1902, under *Siamthelphusa paviei* (De Man, 1898), without explanation or examination of the types. Ng (1988) and Naiyanetr & Ng (1990) questioned this action, preferring instead to tentatively recognize *beauvoisi* as a valid species of *Siamthelphusa* Bott, 1968.

Ng & Naiyanetr (1993) later transferred *Siamthelphusa beauvoisi* to *Heterothelphusa* as a senior synonym of a nomen nudum, *Siamthelphusa phimaiensis* Naiyanetr, 1978. Naiyanetr & Ng (1990) and Ng & Naiyanetr (1997) also transferred *Potamon* (*Parathelphusa*) *harmandi* Rathbun, 1902 to the synonymy of *H. beauvoisi*, although this was not reflected in Ng & Naiyanetr (1993). In any case, the present study shows that both species (*G. harmandi* and *G. phimaiensis*) are distinct from *Guinothusa beauvoisi*, differing from one another in carapace and G1 characters (see below).

Rathbun (1902, 1905) did not designate a holotype for the species. Therefore, in order to avoid future confusion, the larger male type specimen (27.1 by 20.0 mm, MNHN-B 5066) of *Potamon* (*Parathelphusa*) *beauvoisi* is hereby designated as the lectotype.

Geographical distribution. — "Cochinchine" (= southern Vietnam).

Guinothusa harmandi (Rathbun, 1902) (figs. 2, 3B, C, E, F)

Potamon (*Parathelphusa*) *harmandi* Rathbun, 1902: 185.
Potamon (*Parathelphusa*) *harmandi* — Rathbun, 1905: 255, fig. 66, pl. 12 fig. 13.
Siamthelphusa paviei — Bott, 1968: 413, figs. 23-24, 36 (part), 1970: 116, pl. 22 figs. 60-62, pl. 30 fig. 87 (part); Chuensri, 1974: 17 (part) (not *Parathelphusa paviei* De Man, 1898).
Siamthelphusa beauvaisi — Naiyanetr, 1988: 6 (part), pl. 3 fig. 6 (not *Potamon* (*Parathelphusa*) *beauvoisi* Rathbun, 1902).
Siamthelphusa beauvoisi — Ng, 1988: 102 (part); Naiyanetr & Ng, 1990: 235 (part) (not *Potamon* (*Parathelphusa*) *beauvoisi* Rathbun, 1902).
Heterothelphusa beauvoisi — Ng & Naiyanetr, 1993: 43, figs. 33, 68 (part); Yeo & Ng, 1999: 642 (part) (not *Potamon* (*Parathelphusa*) *beauvoisi* Rathbun, 1902).
Heterothelphusa harmandi — Ng et al., 2008: 70.

Material examined. — Lectotype (here designated): male, 13.2 by 12.0 mm, MNHN-B 5171, "Cochinchine" (= southern Vietnam), coll. Harmand, 1876. Paralectotypes: 4 males, 3 females, largest male 12.8 by 11.1 mm, MNHN-B 5171, 2 males, 1 female, larger male 12.3 by 11.5 mm, USNM 30017, same data as lectotype.

Diagnosis. — Carapace quadrate, about 1.2 times as broad as long; dorsal surface generally flat, appearing gently sunken or concave, especially in branchial regions, but with very gently convex epigastric region; epibranchial cristae distinctly anterior to and separated from postorbital cristae, with outer edge distinctly overlapping inner edge of postorbital cristae; postorbital cristae sinuous, gently sloping posterolaterally; external orbital angle with outer margin gently convex; anterolateral margin gently converging anteriorly, with 3 epibranchial teeth; first tooth very small, low, sometimes absent, second tooth spiniform, with gently convex outer margin, third tooth directed obliquely outwards, relatively narrow; posterolateral margins hardly converging posteriorly,

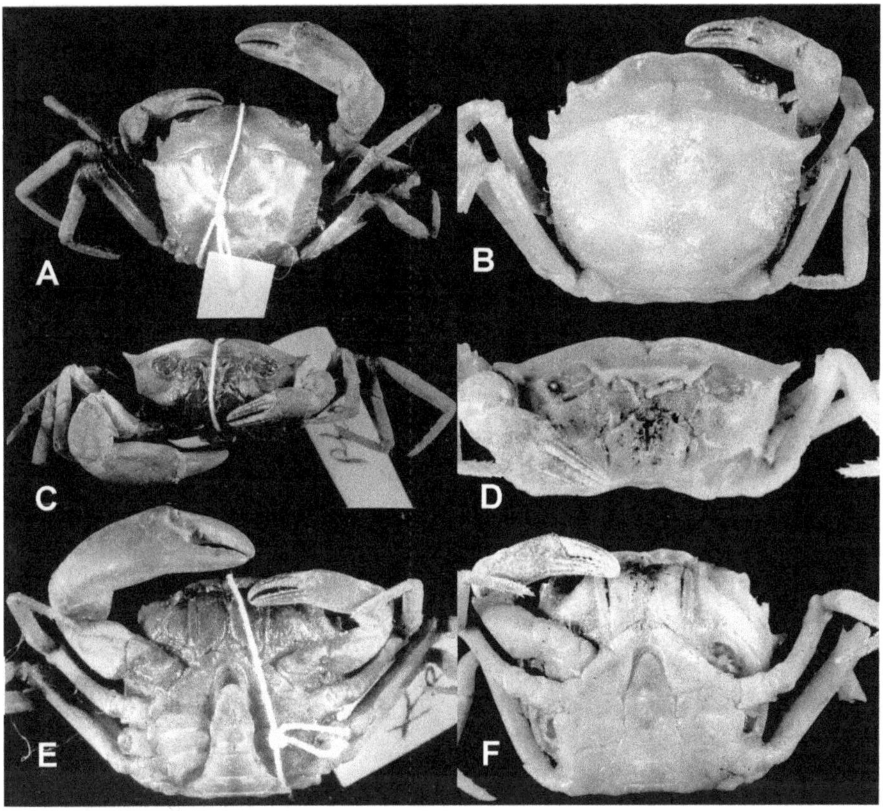

Fig. 2. *Guinothusa harmandi* (Rathbun, 1902). A, C, E, lectotype male, 13.2 by 12.0 mm, MNHN-B 5171; B, D, F, paralectotype male, 12.3 by 11.5 mm, USNM 30017. A, B, dorsal view; C, D, frontal view; E, F, ventral view.

almost parallel, with short but distinct oblique striae. Fingers of major chela shorter than palm, forming small gap when closed. Ambulatory legs long, slender; merus with distinct subdistal spine. Sixth male abdominal somite with lateral margins straight. G1 with distal part subequal to or slightly longer than proximal part, tip relatively strongly curved inwards, relatively acuminate proximal part with straight inner margin, demarcated by distinct angle or shelf.

Remarks. — Like the previously listed species, *Potamon* (*Parathelphusa*) *harmandi* Rathbun, 1902, had also been synonymized under *Siamthelphusa paviei* (De Man, 1898) by Bott (1970). It was subsequently regarded as a junior synonym of *Heterothelphusa beauvoisi* (Rathbun, 1902), which had also been collected in Cochinchina (= southern Vietnam) (Naiyanetr & Ng, 1990; Ng & Naiyanetr, 1997; Yeo & Ng, 1999). Most recently, however, it was

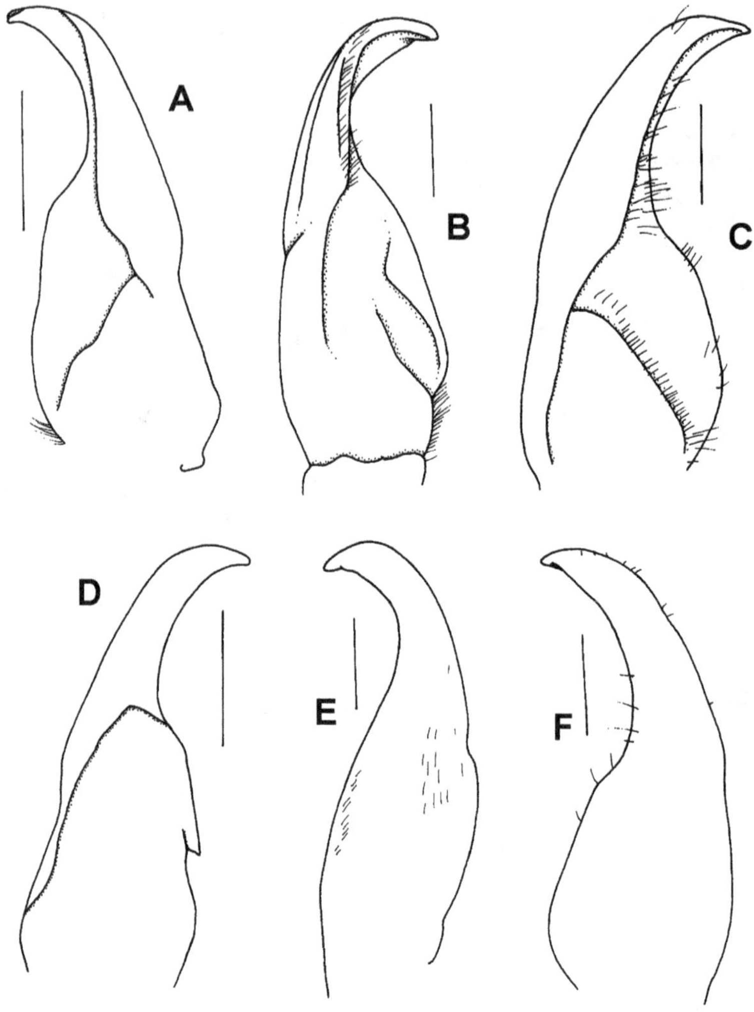

Fig. 3. A, D, *Guinothusa beauvoisi* (Rathbun, 1902), left G1, lectotype male, 27.1 by 20.0 mm, MNHN-B 5066; B, C, E, F, *Guinothusa harmandi* (Rathbun, 1902), right G1: B, E, lectotype male, 13.2 by 12.0 mm, MNHN-B 5171; C, F, paralectotype male, 12.3 by 11.5 mm, USNM 30017. A-C, ventral view of G1; D-F, dorsal view of G1. Scale bars: 1.0 mm in A, D; 0.5 mm in B, C, E, F.

recognized (but not discussed) as a valid species of *Heterothelphusa* by Ng et al. (2008). The differences between *G. harmandi* and *G. beauvoisi* listed below now confirm that both are separate species.

Guinothusa harmandi can be immediately distinguished from *G. beauvoisi* by its distinctly more quadrate carapace, about 1.1-1.2 times as broad as long (carapace distinctly more transverse, about 1.36 times as broad as long in *G. beauvoisi*), with posterolateral margins hardly converging posteriorly, appearing almost parallel (strongly converging posteriorly in *G. beauvoisi*); and distinctly spiniform and narrower second and third epibranchial teeth (broader, triangular in *G. beauvoisi*), with the outer margin of the second epibranchial tooth being gently convex (distinctly convex in *G. beauvoisi*) (figs. 1, 2A, B). These carapace differences are not likely to be size-related as gecarcinucid crabs, including closely related genera such as *Heterothelphusa* and *Siamthelphusa*, do not exhibit this degree of carapace differentiation from allometric growth (see Ng & Lim, 1986; Ng, 1997; Ng & Naiyanetr, 1997). The G1 of *G. harmandi*, is very similar to that of *G. beauvoisi* and the character is more difficult to use to differentiate the two species. The distal part of *G. harmandi*, however, appears to be proportionately slightly longer (subequal to or slightly longer than proximal part), with a relatively more acuminate tip than that of *G. beauvoisi*, in which the distal part appears proportionately slightly shorter than the proximal part, with a relatively blunter tip (fig. 3).

Guinothusa harmandi also closely resembles *G. phimaiensis* sp. nov. (= "*Siamthelphusa phimaiensis* Naiyanetr, 1978"), in the quadrate appearance of the carapace. However, they can be separated by other carapace characters as well as by the G1 (see Remarks for *G. phimaiensis* sp. nov.).

Rathbun (1902, 1905) did not designate a holotype for this species. To avoid future confusion, the largest male specimen among the MNHN types (13.2 by 12.0 mm, MNHN-B 5171) is, therefore, designated as the lectotype of *Potamon* (*Parathelphusa*) *harmandi*.

Geographical distribution. — "Cochinchine" (= southern Vietnam).

Guinothusa phimaiensis sp. nov. (figs. 4, 5)

Siamthelphusa phimaiensis — Naiyanetr, 1978a: 84 (nomen nudum), 1978b: 29 (nomen nudum), 1980: 50 (nomen nudum); Ng, 1988: 102 (nomen nudum).

Siamthelphusa beauvaisi — Naiyanetr, 1988: 6, pl. 3 fig. 6 (part) (not *Potamon* (*Parathelphusa*) *beauvoisi* Rathbun, 1902).

Siamthelphusa beauvoisi — Naiyanetr & Ng, 1990: 235 (part); Naiyanetr, 1992: 48 (part), 1998: 107 (part) (not *Potamon* (*Parathelphusa*) *beauvoisi* Rathbun, 1902).

Heterothelphusa beauvoisi — Ng & Naiyanetr, 1993: 43, 46, figs. 33, 68 (part); Yeo & Ng, 1999: 642 (part); Ng et al., 2008: 70 (part) (not *Potamon* (*Parathelphusa*) *beauvoisi* Rathbun, 1902).

Material examined. — Holotype: male, 17.2 by 14.3 mm, ZRC 1991.1888, Saigham, Phimai District, Nakhon Ratchasima Province, Thailand, coll. P. Naiyanetr, 29 Nov. 1974. Paratype: 1 female, 17.1 by 13.9 mm, ZRC 1992.8347, same data as holotype (= male lectotype and female paralectotype of *Siamthelphusa phimaiensis* Naiyanetr, 1978, nomen nudum).

Diagnosis. — Carapace quadrate, about 1.2 times as broad as long; dorsal surface generally flat, appearing gently sunken or concave, especially in

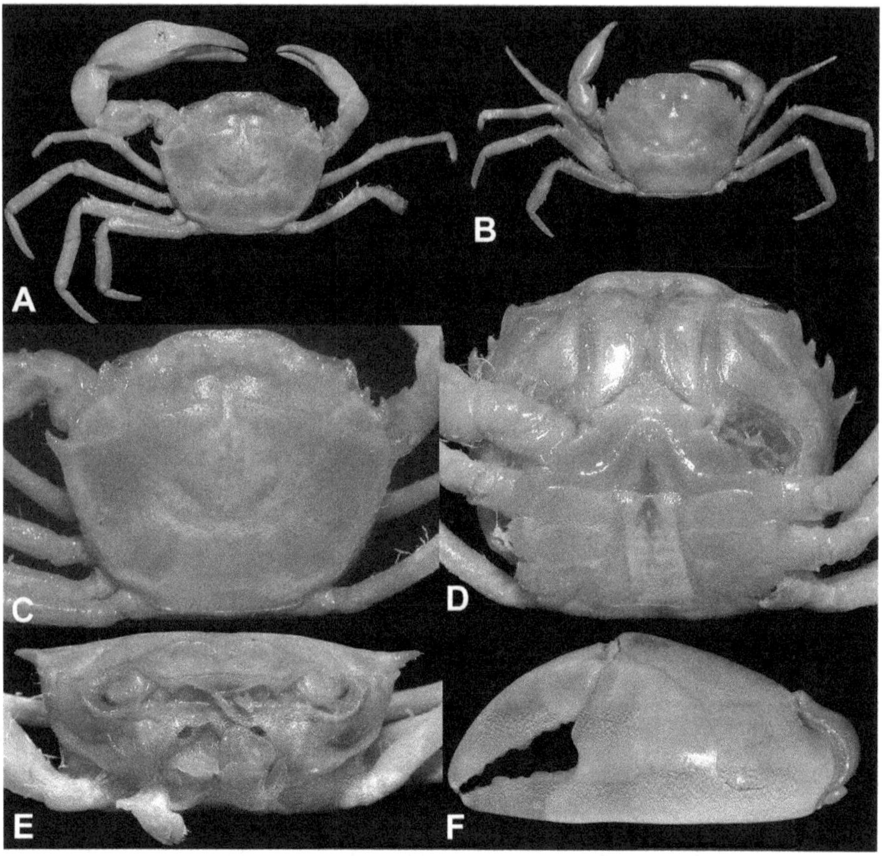

Fig. 4. *Guinothusa phimaiensis* sp. nov. A, C-F, holotype male, 17.2 by 14.3 mm, ZRC 1991.1888; B, paratype female, 17.1 by 13.9 mm, ZRC 1992.8347. A, B, dorsal view; C, dorsal view of carapace; D, ventral view; E, frontal view; F, outer view of left chela.

branchial regions, epigastric region flat; epibranchial cristae distinctly anterior to and separated from postorbital cristae, with outer edge distinctly overlapping inner edge of postorbital cristae; postorbital cristae sinuous, gently sloping posterolaterally; external orbital angle with outer margin gently convex; anterolateral margin gently converging anteriorly, with 3 epibranchial teeth; first tooth very small and low, sometimes absent, second tooth spiniform, with gently convex outer margin, third tooth directed obliquely outwards, relatively narrow; posterolateral margins hardly converging posteriorly, almost parallel, with short but distinct oblique striae. Fingers of major chela shorter than palm, forming small gap when closed. Ambulatory legs long, slender; merus with distinct subdistal spine. Sixth male abdominal somite with lateral margins straight. G1 with distal part subequal to or slightly longer than proximal part,

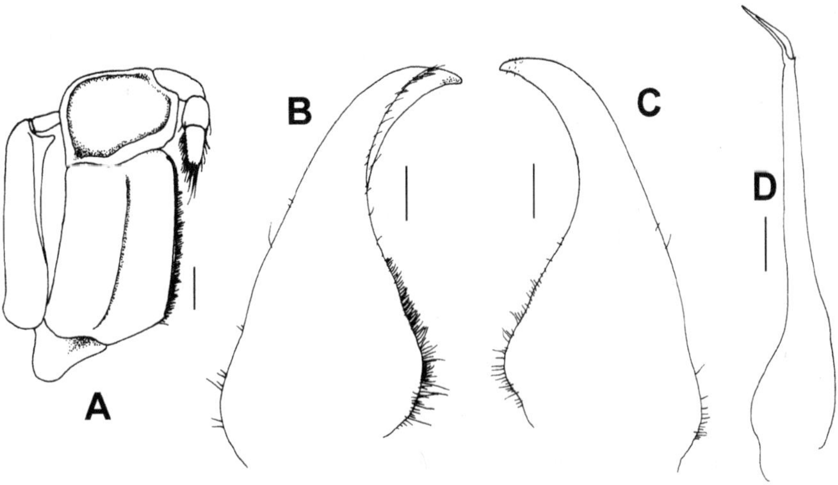

Fig. 5. *Guinothusa phimaiensis* sp. nov., holotype male, 17.2 by 14.3 mm, ZRC 1991.1888. A, right third maxilliped; B, ventral view of right G1; C, dorsal view of right G1; D, right G2. Scale bars: 2.0 mm. [After Ng & Naiyanetr, 1993, fig. 68.]

tip relatively strongly curved inwards, relatively acuminate proximal part with convex inner margin, lacking a distinct angle or shelf. See also diagnosis and illustrations for "*Heterothelphusa beauvoisi*" in Ng & Naiyanetr (1993: 43, figs. 33, 68).

Etymology. — The species is named after its type locality, Phimai District.

Remarks. — Ng & Naiyanetr (1993: 43) pointed out that the name "*Siamthelphusa phimaiensis*" was a nomen nudum as its original citation by Naiyanetr (1978a) did not constitute a valid publication, as was the case for subsequent citations of the name (Naiyanetr, 1978b, 1980; Ng, 1988). Ng & Naiyanetr (1993), however, made no attempt to validate the name as they had regarded it a junior subjective synonym of *Heterothelphusa beauvoisi* (Rathbun, 1902) (see Ng & Naiyanetr, 1993).

Re-examination of the types of *Guinothusa beauvoisi* and "*S. phimaiensis*", however, shows the latter does indeed represent a separate species; the differences, though subtle, are distinct. *Guinothusa phimaiensis* sp. nov. is, thus, validated as the name of the species. The account of *H. beauvoisi* in Ng & Naiyanetr (1993) was based on the type material of "*S. phimaiensis*", and hence their diagnosis and figures for "*Heterothelphusa beauvoisi*" now apply to *G. phimaiensis* (see figs. 4, 5; Ng & Naiyanetr, 1993: 43, figs. 33, 68).

Guinothusa phimaiensis can be readily separated from *G. beauvoisi* by its distinctly more quadrate carapace, about 1.2 times as broad as long (fig. 4A-C; distinctly more transverse, about 1.36 times as broad as long in *G. beauvoisi*;

fig. 1), with posterolateral margins hardly converging posteriorly, appearing almost parallel (fig. 4A-C; margins strongly converging posteriorly in *G. beauvoisi*; fig. 1); and inner margin of the G1 proximal part being convex, without an angle or shelf (fig. 5B, C; inner margin straight, demarcated by a distinct shelf in *G. beauvoisi*; fig. 3A, D). The above-mentioned G1 character also differentiates *G. phimaiensis* from *G. harmandi* (Rathbun, 1902), which has a similarly quadrate carapace. The dorsal surface of the carapace of *G. phimaiensis* is also distinctly flatter, while that of *G. harmandi* is very gently convex in the epigastric region (figs. 2C, D, 4A-C; Ng & Naiyanetr, 1993, fig. 33).

KEY TO THE SPECIES OF *GUINOTHUSA*

1 – Carapace distinctly more transverse, about 1.36 times as broad as long; epibranchial teeth relatively broader, triangular; posterolateral margins strongly converging posteriorly. Distal part of G1 proportionately shorter, tip relatively blunter; proximal part with straight inner margin, demarcated by distinct shelf (figs. 1, 3A, D); occurring in southern Vietnam . . . . *G. beauvoisi*

– Carapace more quadrate, about 1.2 times as broad as long; epibranchial teeth relatively narrower, spiniform; posterolateral margins hardly converging posteriorly, almost parallel. Distal part of G1 proportionately longer, tip relatively more acuminate; proximal part with straight or convex inner margin, with or without distinct shelf 3

2 – Epigastric region distinctly flatter. G1 proximal part with convex inner margin, without shelf (figs. 4A-C, 5B, C; Ng & Naiyanetr, 1993, figs. 33A, 68B, C); occurring in eastern Thailand . . . . *G. phimaiensis*

– Epigastric region very gently convex. G1 proximal part with straight inner margin, demarcated by distinct shelf (figs. 2C, D, 3B, C, E, F); occurring in southern Vietnam . . . . *G. harmandi*

ACKNOWLEDGEMENTS

The authors thank Danièle Guinot (MNHN), Rafael Lemaitre (USNM), Phaibul Naiyanetr (Chulalongkorn University), and Kelvin Lim (ZRC) for access to the important material in their collections. We also thank H. K. Yip for his help processing many of the photographs. The present study was supported by a research grant to the authors from the National University of Singapore.

LITERATURE CITED

ALCOCK, A., 1910. On the classification of the Potamonidae (Telphusidae). Records of the Indian Museum, **5**: 252-261.

BOTT, R., 1968. Parathelphusiden aus Hinterindien (Crustacea, Decapoda, Parathelphusidae). Senckenbergiana Biologica, Frankfurt, **49** (5): 403-422.

— —, 1970. Die Süsswasserkrabben von Europa, Asien, Australien und ihre Stammesgeschichte. Eine Revision der Potamoidea und Parathelphusoidea (Crustacea, Decapoda). Abhandlungen der Senckenbergischen Naturforschenden Gesellschaft, Frankfurt, **526**: 1-338, pls. 1-58.

CHUENSRI, C., 1974. Key to freshwater crabs (excluded Pseudothelphusidae and Potamocarcinidae): 1-52. (College of Fisheries, Kasetsart University, Bangkok).

KLAUS, S., D. BRANDIS, P. K. L. NG, D. C. J. YEO & C. D. SCHUBART, 2009. Phylogeny and biogeography of Asian freshwater crabs of the family Gecarcinucidae (Brachyura: Potamoidea). In: J. W. MARTIN, K. A. CRANDALL & D. FELDER (eds.), Decapod crustacean phylogenetics. Crustacean Issues, **18**: 509-531.

MAN, J. G. DE, 1898. Notes sur quelques *Thelphusides* recueillis par M. Pavie dans l'Indo-Chine. Bulletin de la Société philomathique, Paris, (8) **10**: 36-52.

NAIYANETR, P., 1978a. Six new species of freshwater crabs of Thailand. Abstracts of the National Conference on Agricultural and Biological Science: 84. (Kasetsart University, Bangkok).

— —, 1978b. The geographic distribution of freshwater crabs in Thailand. Geographical Journal of the Geographical Association of Thailand, Bangkok, **3** (3): 24-43.

— —, 1980. Crustacean Fauna of Thailand (Decapoda and Stomatopoda): 1-73. (Department of Biology, Faculty of Science, Chulalongkorn University, Bangkok).

— —, 1988. Freshwater crabs in Thailand: 1-15, pls. 1-8. In: Book published in memory of the Royal Cremation of Associate Professor Dr. Praphun Chitachumnong, Mahidol University. (Phaisalsilpa Press, Bangkok).

— —, 1992. Crustacean Fauna of Thailand (Decapoda and Stomatopoda): 1-71. (Department of Biology, Faculty of Science, Chulalongkorn University, Bangkok).

— —, 1998. Checklist of the Crustacean Fauna in Thailand (Decapoda and Stomatopoda). Office of Environmental Policy and Planning Biodiversity Series, **5**: 1-161.

NAIYANETR, P & P. K. L. NG, 1990. *Siamthelphusa holthuisi* spec. nov., a new species of gecarcinucoid freshwater crab (Crustacea: Brachyura: Decapoda) from Thailand. Zoologische Mededelingen, Leiden, **63** (17): 235-239.

NG, P. K. L., 1988. The freshwater crabs of Peninsular Malaysia and Singapore: i-viii, 1-156, figs. 1-63, 4 col. pls. (Department of Zoology, National University of Singapore, Shinglee Press, Singapore).

— —, 1997. On two new species of freshwater crabs of the genera *Sayamia* and *Heterothelphusa* (Brachyura: Parathelphusidae) from southern Thailand. Crustaceana, **70** (6): 710-719.

NG, P. K. L., D. GUINOT & P. J. F. DAVIE, 2008. Systema Brachyurorum: Part I. An annotated checklist of extant brachyuran crabs of the world. Raffles Bulletin of Zoology, (Supplement) **17**: 1-286.

NG, P. K. L. & R. P. LIM, 1986. Description of a new genus and species of lowland freshwater crab from Peninsular Malaysia (Crustacea, Decapoda, Brachyura, Parathelphusidae). Indo-Malayan Zoology, **3** (1): 97-103.

NG, P. K. L. & P. NAIYANETR, 1993. New and recently described freshwater crabs (Crustacea: Decapoda: Brachyura: Potamidae, Gecarcinucidae and Parathelphusidae) from Thailand. Zoologische Verhandelingen, Leiden, **284**: 1-117, figs. 1-68.

— — & — —, 1997. Revision of the genus *Siamthelphusa* Bott, 1968 (Crustacea: Decapoda: Brachyura: Parathelphusidae), with descriptions of five new species from Thailand. Journal of Natural History, London, **31**: 1751-1784.

RATHBUN, M. J., 1902. Description des nouvelles espèces de *Parathelphusa* appartenant au Muséum de Paris. Bulletin du Muséum national d'Histoire naturelle, Paris, **1902** (3): 184-187.

— —, 1904. Les crabes d'eau douce. Nouvelles Archives du Muséum d'Histoire Naturelle, Paris, (4) **6**: 225-312, pls. 9-18.

— —, 1905. Les crabes d'eau douce. Nouvelles Archives du Muséum d'Histoire Naturelle, Paris, (4) **7**: 159-323, pls. 13-22.

YEO, D. C. J. & P. K. L. NG, 1999. The state of freshwater crab taxonomy in Indochina (Decapoda, Brachyura). In: F. R. SCHRAM & J. C. VON VAUPEL KLEIN (eds.), Crustaceans and the biodiversity crisis, Proceedings of the Fourth International Crustacean Congress, 1998, **1**: 637-646.

First received 11 March 2009.
Final version accepted 22 June 2009.

www.ingramcontent.com/pod-product-compliance
Lightning Source LLC
LaVergne TN
LVHW010624300926
849162LV00001B/49